Ferroelectric Phenomena in Crystals

Springer
Berlin
Heidelberg
New York
Barcelona
Budapest
Hong Kong
London
Milan
Paris
Santa Clara
Singapore
Tokyo

B. A. Strukov A. P. Levanyuk

Ferroelectric Phenomena in Crystals

Physical Foundations

With 153 Figures and 14 Tables

Springer

Professor Boris A. Strukov
Department of Physics
Lomonosov Moscow State University
Moscow 119899, Russia
E-Mail: strukov@ofefs.phys.msu.su

Professor Arkadi P. Levanyuk
Department of Physics of Condensed Matter
Madrid Autonomous University
28049 Madrid, Spain
E-Mail: Levanyul@uam.es

ISBN-13: 978-3-642-64340-8 e-ISBN-13: 978-3-642-60293-1
DOI: 10.1007/978-3-642-60293-1

Library of Congress Cataloging-in-Publication Data
Strukov, Boris Anatol'evich.
Ferroelectric phenomena in crystals: physical foundations / B. A. Strukov; A. P. Levanyuk.
p. cm. Includes bibliographical references and index.
ISBN-13: 978-3-642-64340-8 (alk. paper)
1. Ferroelectricity. 2. Ferroelectric crystals. I. Levaníuk, A. P., 1933- . II. Title.
QC596.5.S74 1998 548'.85-dc21 97-35793 CIP

Softcover reprint of the hardcover 1st edition 1998

Typesetting: Data conversion by Satztechnik Katharina Steingraeber, Heidelberg
Cover design: *design & production* GmbH, Heidelberg
SPIN 10574433 57/3144 – 5 4 3 2 1 0 – Printed on acid-free paper

Preface

The expansion of the application of ferroelectric crystals in engineering as well as of a number of fundamental problems of solid-state physics, which have not yet been solved and which bear a direct relation to ferroelectricity, has lately stimulated much interest in the problem of ferroelectricity. In courses of solid-state physics ferroelectricity is studied today along with traditional disciplines, such as magnetism, superconductivity, and semiconducting phenomena. Moreover, new specialities have been born concerned directly with the development and utilization of ferroelectric materials in optics, acoustics, computer technology, and capacitor engineering. Special courses in the physics of ferroelectrics are read in a number of colleges and universities.

The study of the nature of ferroelectricity has currently reached such a level of development that we may speak of having gained a rather deep insight into the physical essence of a number of phenomena, which contribute to the generation of a spontaneous electric polarization in crystals. It is exactly at this level that it has become possible to single out that part of the problem, the physical picture of which can be depicted in a rather unsophisticated manner and which is the foundation for the construction of a building of "complete understanding". The specialists engaged in the field of solid-state physics are well aware that this building would hardly be covered with a roof in the nearest future for a number of reasons, the main one being the unusual complexity of describing and predicting the properties of a real solid by proceeding from "first principles". Those who wish to delve into the problem as quickly as possible are mainly interested in the part that has already been built — the firmly established conceptions and ideas underlying the modern science of ferroelectricity.

The purpose of the present book is just to expound the foundations of the physics of ferroelectric phenomena in crystals on the basis of a sufficiently general and as simple an approach as possible. This possibility is provided, first of all, by a description of the properties of ferroelectrics based on the Landau phenomenological theory of second-order phase transitions, which was applied for the first time to ferroelectric crystals as far back as 1949 by V.L. Ginzburg. Althouqh this approach is traditional in the literature on ferroelectricity, we devoted much more attention than is usually done to the elucidation of the physical content and the range of applicability of this approach and to the possibilities of its application to the description of various types of structural phase transitions.

As regards the microscopic theory, we are mainly concerned with sufficiently simple models which make it possible, on the one hand, to illustrate the major interactions leading to the generation of a spontaneous polarization and, on the other, to trace out the relationship to phenomenological theory.

Since we tried to render our treatment understandable and consistent, we had to limit the range of problems to be discussed and to omit a number of topics that are considered to be traditional in the literature on ferroelectricity — light and neutron scattering, sound absorption, photogalvanic phenomena, polarization switching, etc. Otherwise, the discussion of these subjects would be unjustifiably brief and fragmentary because of the limited space. We hope that we have carried the content of the book far enough to provide the reader with a good foundation for the acquaintance with the topics indicated. Moreover, there are a number of monographs which cover these subjects in sufficient detail. We have not touched at all upon the history of the development of the science of ferroelectricity.

Since phase transitions are basically classical phenomena, we have used the elements of quantum mechanics only for the solution of certain special problems in connection with the strictly quantum-mechanical effects.

This book is designed to serve as a textbook and reflects, to a considerable extent, the content of lectures read by the authors to fourth-year students at the Department of the Physics of Crystals (Moscow University). The reader is supposed to have a knowledge of the foundations of crystallography and crystal physics to the extent covered, for example, in J. Nye, *Physical Properties of Crystals. Their Representation by Tensors and Matrices*, Oxford, 1964. Since this is a textbook, we have ruled out references to the the original literature. We restricted ourselves to giving only a list of the works that will help the reader to continue a detailed study of the problems considered in the book and also of publications from which some photographs and diagrams were borrowed.

We are very grateful to our colleagues for their helpful comments and critical remarks during the preparation of the manuscript.

The authors wish to thank Mr. A.A. Beknazarov for his fine translation of the manuscript into English and Mr. I.V. Shnaidshtein for the preparation of its computer version.

Moscow, Madrid
November 1997

B.A. Strukov
A.P. Levanyuk

Contents

1. General Characteristics of Structural Phase Transitions in Crystals

1.1 First- and Second-Order Structural Phase Transitions

The great majority of crystals can exist in several crystalline phases. As a rule, such phases appear to be stable in a certain range of temperatures and pressures and the transition from one phase to another is accompanied by significant discontinuous changes in the volume and entropy of the crystal. This involves the displacement of various structural elements to distances of the order of the unit cell dimensions and no limitations are imposed on a change in the crystal symmetry. The temperature hysteresis of a phase transition, i.e., the difference between the transition temperatures upon cooling and heating of the crystal, may reach hundreds of degrees and is an indication of the possible formation of relatively stable metastable states: one of the crystalline phases can exist in the region of temperatures and pressures in which another phase is more stable.

Figure 1.1 shows the variation of the thermodynamic potentials φ_1 and φ_2 of two phases of the crystal as a function of temperature and pressure in the region of such a phase transition. The two functions of the two variables are given in the form of intersecting surfaces. The line of intersection AB corresponds to the equality of the specific thermodynamic potentials of the phases $\varphi_1 = \varphi_2$, and its projection on to the pT plane is the line of phase equilibrium. When the crystal temperature decreases at constant pressures (for example, at $p = 0$), the phase transition can occur at temperature T_1 (line 1–2–3–4–5), whereas upon heating (line 5–4–6–2–1) it can take place at temperature T_2.

The temperature hysteresis $\Delta T = T_2 - T_1$ depends on the rate of temperature change. Under the conditions of complete thermodynamic equilibrium in the system, which is practically never realized under real experimental conditions (with an infinitely slow change of temperature), a phase transition would have occurred at temperature T_0 corresponding to the condition of equality of the specific potentials. Structural phase transitions of this type are called *first-order transitions.* A more-or-less random intersection of the thermodynamic functions in such phase transitions is responsible for the insensitivity of the properties of the crystal to how close it is to the phase-

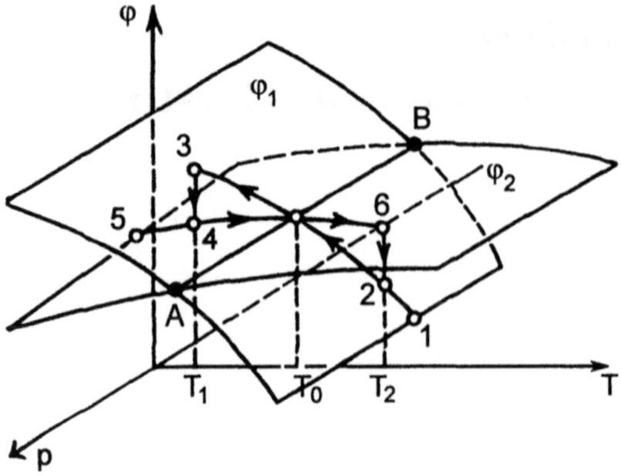

Fig. 1.1. Surfaces of the specific thermodynamic potentials of the phases of a crystal with a first-order phase transition.

transition point. Basic information about phase transitions of this kind is contained in the Clausius–Clapeyron equation, which relates the slope of the phase equilibrium curve to jumps in the volume Δv and entropy ΔS upon transition from one phase to another:

$$\frac{\mathrm{d}T}{\mathrm{d}p} = \frac{\Delta v}{\Delta S}. \tag{1.1}$$

In what follows we will be mainly concerned with phase transitions associated with relatively slight distortions of the crystal lattice, which nevertheless cause a change in the symmetry of the crystal. In those cases when these distortions occur continuously, we speak of *second-order phase transitions*; it is essential that in this case the state of the crystal is changed in a continuous, smooth way. Although the positions of atoms at the point of first-order phase transition are changed in a jumpwise manner, near the second-order transition the distortion of the structure in each unit cell may be as small as desired. Second-order phase transitions do not display a temperature hysteresis and are not accompanied by a discontinuous change in the volume and entropy, and the symmetry groups of the phases exhibit "continuity": as the temperature falls off and the crystal structure is distorted at the phase-transition point the crystal loses some of the symmetry elements of the symmetric (usually high-temperature) phase. Hence, the crystal symmetry changes from a group into a subgroup.

Second-order phase transitions are accompanied by pretransition anomalies of the various physical properties. By anomalies we mean, as a rule, an unusual temperature dependence or unusual values of the various characteristics of the crystal, heat capacity, elastic moduli, dielectric constant, thermal expansion coefficient, etc., near the transition point as compared with those observed far from the transition point. Various anomalies, which occur not only in structural phase transitions in crystals but also in a number of other physical systems, in which phase transitions are continuous (liquid helium, superconductors, ferromagnetics, at the critical liquid–gas point, etc.) constitute the basic subject matter of the physics of phase transitions. It is especially important that second-order phase transitions are often accompanied by the appearance of a new physical quality in the system (superconductivity, ferromagnetism, ferroelectricity, superfluidity, etc.), a fact that makes the object of investigation attractive from not only a fundamental scientific but also a practical standpoint.

If the most general viewpoint is adopted, it would seem that there are no formidable difficulties in obtaining the temperature dependence of the heat capacity, for example. The recipe for calculating the thermodynamic functions of any system composed of N particles, for which the law of interaction is known, was given by J.W. Gibbs. According to Gibbs, the thermodynamic potential of a system is given by[1]

$$\Phi = -k_{\mathrm{B}} T \ln Z, \tag{1.2}$$

$$Z = \int_{-\infty}^{\infty} \exp\left[-\frac{H(\boldsymbol{p}_n, \boldsymbol{r}_n)}{k_{\mathrm{B}} T}\right] \prod_n \mathrm{d}\boldsymbol{p}_n \mathrm{d}\boldsymbol{r}_n . \tag{1.3}$$

Here $H(\boldsymbol{p}_n, \boldsymbol{r}_n) = \sum_{n=1}^{N} (p_n^2/2m) + U(\boldsymbol{r}_1, \ldots, \boldsymbol{r}_N)$ is the Hamiltonian of the system; $\boldsymbol{p}_n$ is the momentum; $\boldsymbol{r}_n$ is the radius vector of the nth particle; and k_{B} is Boltzmann's constant.

Thus, in order to obtain an answer, it is necessary to take the integral (1.3); it can be evaluated only if the multiple integral can be split into the product of singular integrals, which is possible only for a limited number of systems, say for an ideal gas. But if we try to take into account the potential energy of interaction between the particles, we will no longer be able to break up (1.3) into the product of singular integrals and we will have to use perturbation theory with respect to the quantity U, i.e., the representation of the answer in the form of a series in the interaction energy. This is possible in cases where this energy is low, i.e., if the theory is suitable for only a slightly nonideal gas. The problem of evaluating the statistical integral (1.3) becomes practically insoluble when, for example, the system is about to make a transition from the gaseous to the liquid state and there are no grounds for considering the interaction energy to be low.

There exists a case where, in spite of the strong interaction between the particles, we succeed in breaking up the integral (1.3) into the product of simple integrals. This is the case of a so-called harmonic crystal, where the potential energy is a quadratic function of atomic displacements. For the harmonic crystal it is possible, by using a simple linear transformation, to

[1] Like other thermodynamic functions, the thermodynamic potential is proportional to the volume of the system, $\Phi = \varphi v$. For convenience, we will consider, as a rule, a unit volume, emphasizing the difference between Φ and φ only if necessary. Formula (1.2) determines both the thermodynamic potential and the free energy of the crystal, depending on how the Hamiltonian of the system is determined. For more detail, see M.A. Leontovich, *Introduction to Thermodynamics and Statistical Physics* (in Russian), Nauka, Moscow, 1983.

pass over from the variables $\boldsymbol{r}_i$ to other variables ξ_i, which are devoid of the distinct sense of the coordinates of individual particles, in the representation of which the function $U(\boldsymbol{r}_1, \ldots, \boldsymbol{r}_N)$ assumes the following form:

$$U = \frac{1}{2} \sum_{i=1}^{3N} k_i \xi_i^2 , \qquad (1.4)$$

where k_i are certain coefficients and ξ_i are new variables – the normal coordinates. However, assuming the harmonic approximation for a crystal, we cannot describe a number of important effects, such as the thermal expansion of a crystal. It is necessary to go beyond the framework of the harmonic approximation, i.e., we will have to take into account, in the potential energy, terms of higher order in atomic displacements. In this case, the transition to the normal coordinates does not allow one to split the variables and it will be necessary to develop perturbation theory, this time in small anharmonic corrections. Here, it turns out, for example, that the heat capacity is not constant at high temperatures, as follows from the Debye theory, and increases with temperature by the law $C = C_0 + aT$. This result is valid provided that $aT \ll C_0$, i.e., if the temperature dependence is weak. It is clear that in order to account for the strong temperature dependence of the heat capacity in the anomalous region near the second-order phase transition (Fig. 1.2), we can no longer limit ourselves to small anharmonic corrections, i.e., the anharmonic terms cannot be considered to be small. Therefore, the problem of calculating the anomalies in the region of the phase transition within the framework of statistical physics is extremely complicated.

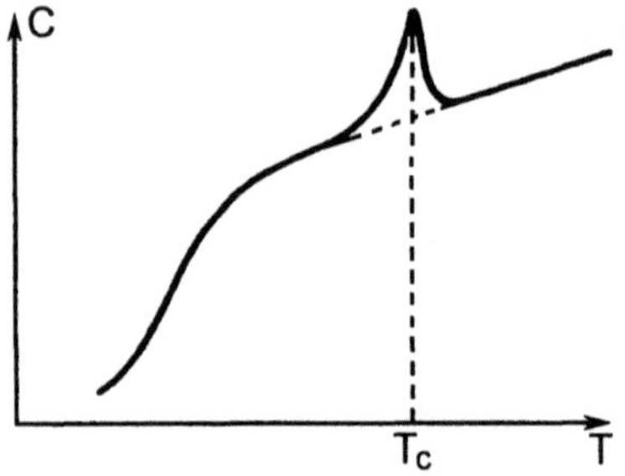

Fig. 1.2. Temperature dependence of the heat capacity of a crystal in a wide temperature range, including the Curie point of a second-order phase transition T_c.

We did not intend to discuss this problem in detail because it goes far beyond the scope of this book. As a matter of fact, this book is intended to be the first stage in the study of the problem.

Thus, the study of phase transitions in general, and structural phase transitions in particular, provides, for both theoreticians and experimentalists, the unique possibility of dealing with systems that become complex gradually, upon approach to the phase transition point. In such a case, these systems themselves become unusually mobile and their properties are strongly changed under the action of external factors – electric and magnetic fields, homogeneous and inhomogeneous mechanical stresses, etc. This opens up wide possibilities for practical applications, which are realized especially successfully in the case of ferroelectric crystals.

1.2 Structural Phase Transitions of Displacive and Order–Disorder Types

Let us consider structural changes that occur in second-order phase transitions, limiting ourselves for the present to the simplest models. The planar model of a diatomic rhombic crystal is shown in Fig. 1.3a. The point group of this structure includes the following symmetry operations: 1, $\bar{1}$, 2, m_1, m_2 (1 is an identity operation and $\bar{1}$ is the center of inversion). We will not consider the symmetry elements that "invert" the plane of the figure, assuming that the "lower side" of atoms is tinted in a different color. Suppose that at a certain temperature one of the sublattices is shifted relative to another in the direction $\boldsymbol{a}_1$ (Fig. 1.3b). After such an infinitesimal displacement the symmetry group of our model, apart from translations, will contain only two elements, 1 and m. In what follows we will call the original phase (Fig. 1.3a) symmetric and the new, distorted one (Fig. 1.3b) nonsymmetric.

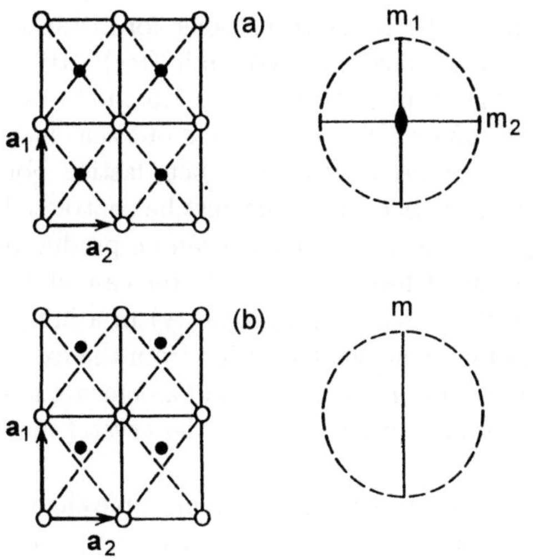

Fig. 1.3. (**a**) A planar model of a diatomic crystal and the elements of its point symmetry. (**b**) A lowering of the symmetry as a result of a relative displacement of sublattices and elements of the point symmetry of the distorted phase.

The coordinates of the lattice points in the crystallographic system of coordinates will change as a result of the displacement as follows:

(1) the coordinates of atoms before the displacement:

$n_1\boldsymbol{a}_1, n_2\boldsymbol{a}_2$ for atoms ∘,

$$\left(n_1 + \frac{1}{2}\right) \boldsymbol{a}_1, \left(n_2 + \frac{1}{2}\right) \boldsymbol{a}_2 \qquad \text{for atoms } \bullet;$$

(2) the coordinates of atoms after the displacement:

$$n_1 \boldsymbol{a}_1, n_2 \boldsymbol{a}_2 \qquad \text{for atoms } \circ,$$

$$\left(n_1 + \frac{1}{2} + \eta\right) \boldsymbol{a}_1, \left(n_2 + \frac{1}{2}\right) \boldsymbol{a}_2 \qquad \text{for atoms } \bullet$$

Here η is the fraction of translation $\boldsymbol{a}_1$ to which the atoms of the "internal" sublattice were displaced. The quantity η will be called the *order parameter* or the *Characteristic phase transition parameter.* The value $\eta = 0$ corresponds to the symmetrical phase and $\eta \neq 0$ to the nonsymmetrical phase.

Figure 1.4 shows variants of the temperature dependence of η in the region of a structural phase transition. It is clear that the dependence given in Fig. 1.4a corresponds to a second-order phase transition. The phase transition may also occur in a jumpwise manner, in which case the temperature dependence of η will have the form shown in Fig. 1.4b and c. In the case depicted in Fig. 1.4b one speaks of a first-order phase transition close to a second-order transition. Here η arises jumpwise but has an appreciable temperature dependence in the nonsymmetrical phase; transitions of this kind could have been, according to the character of the structural rearrangement, second-order transitions. They can become second-order transitions with an increase in the pressure, for example. It should be noted that such a possibility does not exist for all structural first-order phase transitions; for instance, the transition between γ-iron, which has a face-centered cubic (fcc) lattice, and α-iron, with a body-centered cubic (bcc) lattice. Figure 1.5 shows two neighboring unit cells of the fcc lattice and a volume is singled out that may be regarded as a unit cell of the body-centered tetragonal (bct) lattice. For this volume to move into the unit cell of the bcc lattice, it must be contracted along the z axis by $\sqrt{2}$ times. Thus, though the bcc lattice can be produced from the fcc lattice by way of a certain deformation, the latter cannot be continuous; it must occur (by virtue of the structural geometry) in a jumpwise manner. At the same time, as can be seen, the transition from the fcc to the bct lattice can be a second-order transition – here as small a deformation as possible will be sufficient, though under real conditions the crystal may "prefer" a jump to a continuous change.

A second-order phase transition may be accompanied by a multiple change of the unit cell volume. If the displacements of the central atoms in neighboring rows of the unit cell of the model shown in Fig. 1.3 occur in opposite directions, the translational symmetry will be changed (Fig. 1.6). It can be seen that the new basis translations (in a nonsymmetrical phase) are $\boldsymbol{a}_1' = 2\boldsymbol{a}_1$, $\boldsymbol{a}_2' = \boldsymbol{a}_2$, i.e., the "volume" of the unit cell is doubled.

The above models of structural second-order phase transitions are characterized by the fact that at the phase transition point the atoms begin to

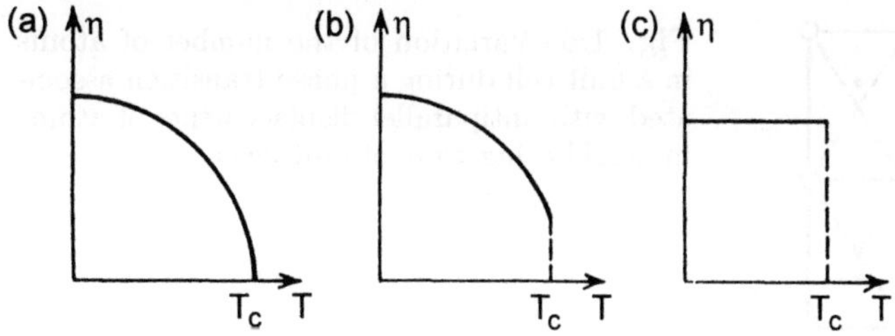

Fig. 1.4. (**a**) Temperature dependences of the order parameter for second-order phase transitions and (**b**) first-order phase transitions close to second-order transitions and (**c**) first-order transitions.

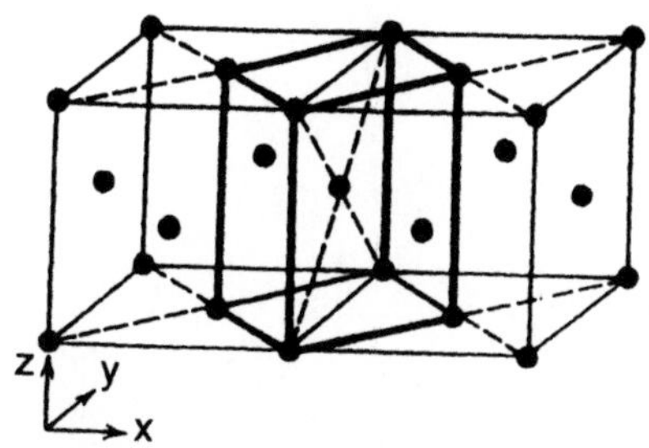

Fig. 1.5. Change of a crystal unit cell during a phase transition of the bct–bcc type and the fcc–bcc lattice.

be displaced, this leading to a change in crystal symmetry. Therefore, such transitions are called *phase transitions of the displacive type.*

The crystal symmetry can change as a result of the redistribution of the probability of finding atoms in different equilibrium positions in the unit cell. Suppose that the inner atoms in our model (see Fig. 1.3) are moving in a potential relief pictured in Fig. 1.7.

At rather high temperatures both positions 1 and 2 are occupied uniformly and the probability of an atom being in one of them is equal to 1/2. Since, on average, an inner atom is in the center of the unit cell, the structural symmetry here is the same as that shown in Fig. 1.3a. Of course, for transitions of the displacive type we in fact also mean the mean positions averaged over the thermal motion. With a decrease in temperature the phase transition manifests itself in the appearance of the difference of the probabilities of finding a particle in positions 1 and 2 or, what is the same thing, the difference of the number of particles in these positions. At the phase transition point the following quantity becomes different from zero:

$$\eta = \frac{N_1 - N_2}{N_1 + N_2}, \tag{1.5}$$

where N_1 and N_2 are the numbers of atoms in positions 1 and 2, respectively. The quantity η may be called the order parameter more naturally than in the case of phase transitions of the displacive type. Structural phase transitions, in which a change of symmetry occurs as a result of the redistribution of particles over equiprobable positions, are termed *phase transitions of the*

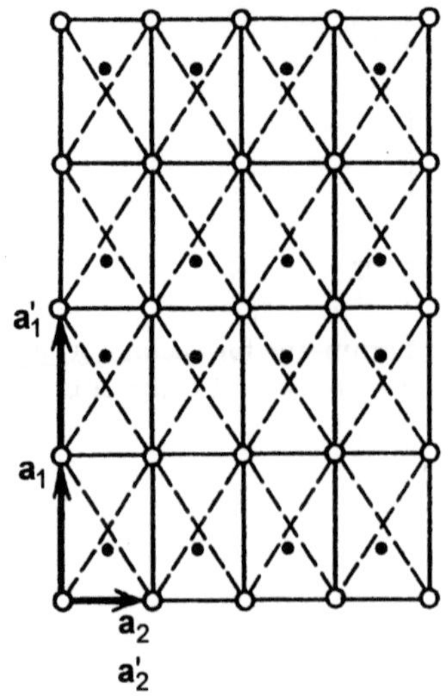

Fig. 1.6. Variation of the number of atoms in a unit cell during a phase transition associated with antiparallel displacements of atoms in neighboring rows of unit cells.

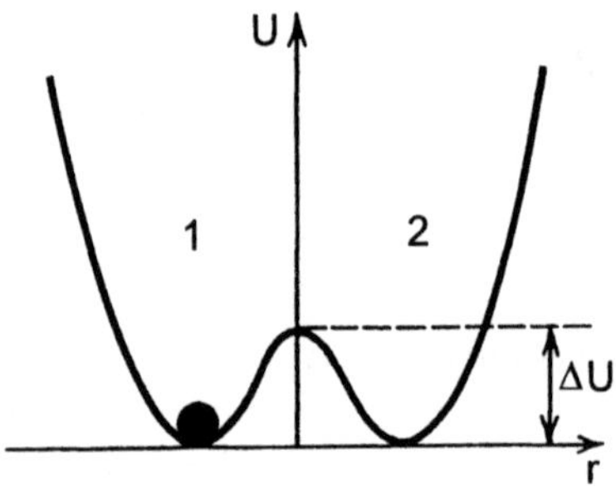

Fig. 1.7. The potential relief in which the atoms of a disordered sublattice are moving.

order–disorder type.[2] It must be stressed here that no sharp boundary line can be drawn between displacive and order–disorder phase transitions. From a symmetry standpoint there is no difference at all between them: while considering the symmetry of a structure, we always mean, as has already been pointed out, the mean positions of atoms; it is not essential whether the averaging is performed over a discrete or continuous set of positions. As regards some other properties, especially dynamic properties, systems with a phase transition of the displacive or order–disorder type may differ appreciably. However, in any case there may exist "intermediate" systems. As a matter of fact, the height of the potential barrier ΔU in Fig. 1.7 may vary within wide limits and when it becomes comparable with $k_B T$, the presence of the barrier is no longer important. Note, for example, that for phase transitions of the displacive type we can determine the order parameter using basically the same method that was employed for transitions of the order–disorder type.

[2] Both displacive and order–disorder phase transitions of the second order and of the first order close to the second order will be referred to in the text as the Curie points.

Namely, we may assume that for the model shown in Fig. 1.3a the quantity η is given by

$$\eta = \frac{1}{a_1 N} \sum_i x_i, \tag{1.6}$$

where x_i is the displacement of the atom along the x axis in the ith unit cell (at a given moment of time) and N is the number of unit cells in the crystal. This definition is quite equivalent to that used earlier, but it is more convenient from the standpoint of further treatment.

1.3 The Domain Structure

Returning to Fig. 1.3, it should be pointed out that the energy equivalence of the states of a crystal with "upward" and "downward" displacements of atoms of the internal sublattice is obvious. This means that its thermodynamic potential must be invariant with respect to the replacement of η by $-\eta$. Upon cooling of the crystal through the phase transition point the displacements of atoms in crystalline regions sufficiently remote from one another occur independently; as a result the crystal appears to be divided into regions which differ by the sign of the order parameter η (Fig. 1.8). Such regions are called *domains* and the surfaces that separate them are *domain walls*. The thickness and orientation of domain walls are determined by many factors. In the simplified diagram of Fig. 1.8 the domain walls are devoid of thickness, i.e., the order parameter changes from η to $-\eta$ in neighboring unit cells which belong to different domains. As a matter of fact, the interdomain boundary may appear to be a spatially extended region, in which there occurs a gradual change of the order parameter; moreover, the arrangement of domain walls along certain crystallographic planes is evidently energetically favoarable. In

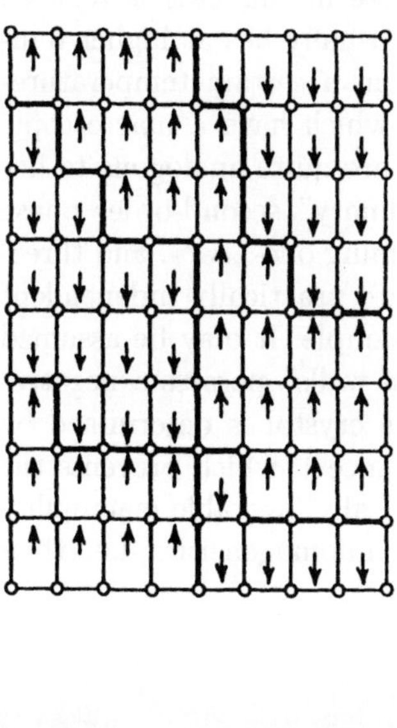

Fig. 1.8. The domain structure of a crystal in a nonsymmetrical phase.

most cases, as a result of a phase transition, a certain configuration of domains and interdomain boundaries, which corresponds to a relative minimum of its thermodynamic potential, is formed in the crystal. Domain walls make a positive contribution to this potential since they possess a finite and positive surface energy, which is temperature dependent; as a rule, the spatial inhomogeneity of the order parameter increases the energy of the system as compared with its homogeneous state. Therefore, to the natural, though very slow as a rule, process of evolution of the crystal to the equilibrium state (and of the thermodynamic potential to the absolute minimum) there must correspond a gradual decrease of the domain-wall density to zero. In ferroelectrics, as will be shown below, the equilibrium domain configuration is determined by the electrical boundary conditions in which the crystal exists.

Introducing the concepts of first- and second-order phase transitions in crystals, we became convinced of the usefulness and importance of the concept of the phase-transition parameter, which may be associated with particular atomic displacements – the distortion of the crystal lattice, which are homogeneous throughout the entire volume of the crystal (or within a single domain). The continuous change of the quantity η upon cooling or heating of the crystal through the phase-transition point is responsible for a continuous change of the crystal state at $T = T_c$. Hence, we may assert that a second-order phase transition is a continuous transition, meaning that at the phase transition point the characteristics of the crystal, such as entropy, density and volume, undergo a continuous change.

It is appropriate here to point out that the concept of the "continuous phase transition" is broader than that of the second-order transition. At present there is convincing evidence that there exist structural phase transitions in crystals in which the state is changed continuously and the phases on both sides of the transition differ substantially in their physical properties. At the same time, we may only conditionally speak of the characteristic parameter of such a phase transition in the sense indicated above. The continuity in the temperature dependence of the state in this case is realized due to the fact that in one of the phases (which usually lies at higher temperatures) there appears and exists at equilibrium a certain temperature-dependent number of identical inhomogeneities, which have a macroscopic extent. By their structure these inhomogeneities are quite analogous to domain walls in the nonsymmetrical phase in "ordinary" second-order phase transitions and are arranged in a regular way, forming one-, two-, and three-dimensional superstructures, whose parameters are practically independent of either boundary conditions or the size of the sample. It may be assumed that in the case under consideration the "domain walls" possess a negative energy and that their equilibrium number in the crystal is determined by the energy of interaction (repulsive at small distances), which prevents the "oversaturation" of the crystal with these energetically favorable macroinhomogeneities. Actually, the density of macroscopic inhomogeneities, i.e., their number n per unit length of the crystal (Fig. 1.9) may be considered to be a temperature-dependent quantity (and analogous to the order parameter in this respect). It is evident that $n = L^{-1}$, where L is the period of the corresponding suprastructure. In Fig. 1.9, d is the linear size of the inhomogeneity (domain-wall width) and A is a certain physical quantity which characterizes the distribution of atoms in the crystal structure of a phase lying at lower temperatures. It is essential that the possibility of a continuous transition of one phase (homogeneous) to another (with a regular system of domain walls) can be realized when the quantity n approaches to zero at the phase-transition point (and, hence, with the period L of the suprastructure tending to infinity) (Fig. 1.10).

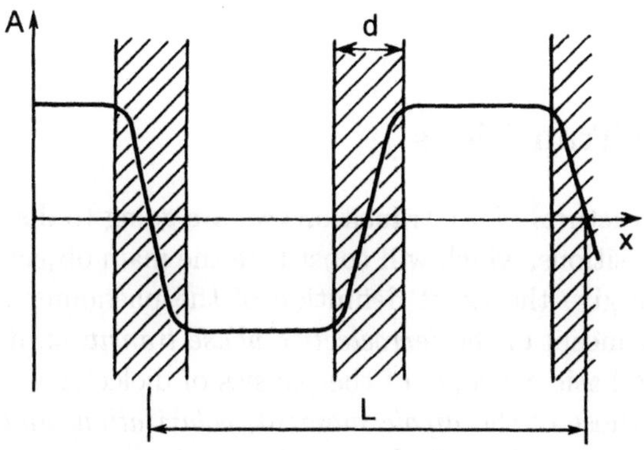

Fig. 1.9. The distribution of a physical quantity A (for example, the displacements of atoms of a single species) along the crystallographic direction x during the nucleation of an equilibrium system of macroscopic inhomogeneities (domain walls) in a crystal.

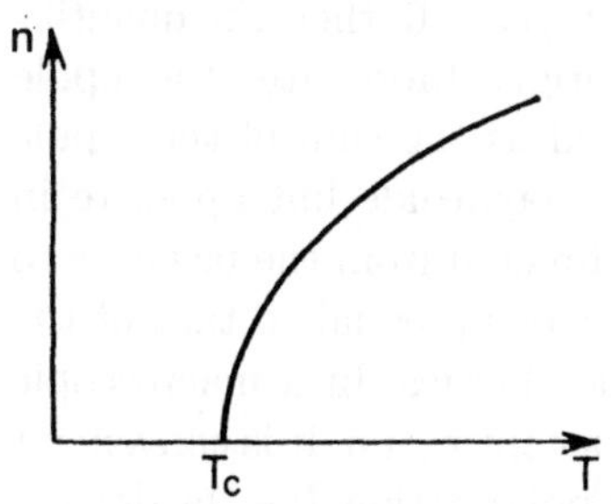

Fig. 1.10. Temperature dependence of the density of domain walls.

As regards the crystal symmetry and its change at the phase-transition point T_c, from Fig. 1.10 it follows that the macrosymmetry of the crystal with $T > T_c$ ($n \neq 0$) corresponds to the condition $\langle A \rangle = 0$ (the brackets $\langle \ldots \rangle$ are used here and henceforth to denote the value of a quantity averaged over the crystal volume). With $T < T_c$, when the macroinhomogeneities disappear, $\langle A \rangle \neq 0$ and the macrosymmetry of the crystal is lowered. Hence, in this case too, just as in ordinary second-order phase transitions, at the phase transition point the symmetry is lowered from a group to a subgroup. However, this is not valid for translational symmetry.

The presence of regularly (with period L) arranged macroinhomogeneities in an ideal crystal implies that the presence or absence in the crystal of translations in a direction perpendicular to the plane of "domain walls" is determined by whether the period of the "basis" lattice is a rational part of the suprastructural period ($L/a = m/l$, where m and l are integers) or not. In the latter case, one speaks of the *incommensurability* of the basis translation and of the period of repetition of regular macroinhomogeneities; the change of translational symmetry upon transition from a phase with a superstructure to a homogeneous phase has nothing in common with the group–subgroup relationships.

1.4 Ferroelectric Phase Transitions

Among the wide diversity of structural phase transitions we are going to distinguish ferroelectric phase transitions, which will constitute the main object of our investigation. Before we give the exact definition of the phenomena that may be embraced by the concept of the *ferroelectric phase transition*, it is appropriate to consider some basic concepts of the physics of dielectrics.

The key concepts here are those of the *dipole moment, polarization, and bound and free charges*. For a system of point charges the dipole moment is defined as

$$\boldsymbol{p} = \sum_i e_i \boldsymbol{r}_i, \tag{1.7}$$

where e_i is the charge of the ith particle and $\boldsymbol{r}_i$ is the radius-vector of this charge. If the system is electrically neutral, i.e., $\sum_i e_i = 0$, then the quantity $\boldsymbol{p}$ is independent of the choice of the coordinate origin. More often the dipole moment of the system is conveniently represented as the sum of the dipole moments of dipoles – pairs of charges of equal magnitude but opposite in sign. The dipole moment of a dipole is a vector directed from the negative to the positive charge, the magnitude of this vector being equal to that of the charge multiplied by the distance separating the charges. In a macroscopic description of a dielectric, when it is regarded as a continuum, it is convenient to describe its electrical state by introducing the polarization $\boldsymbol{P}$ – the density of the dipole moment, i.e., the dipole moment of a unit volume. Since the electrical state of the crystal depends on the magnitude and direction of the electric field in it, an important characteristic is the dielectric susceptibility of the crystal $\chi_{ij} = \partial P_i/\partial E_i$.

In order to elucidate the notion of the bound charge, suppose we have a homogeneous dielectric consisting of identical dipoles which occupy the entire space (Fig. 1.11). The dipoles are assumed to be aligned in one direction.

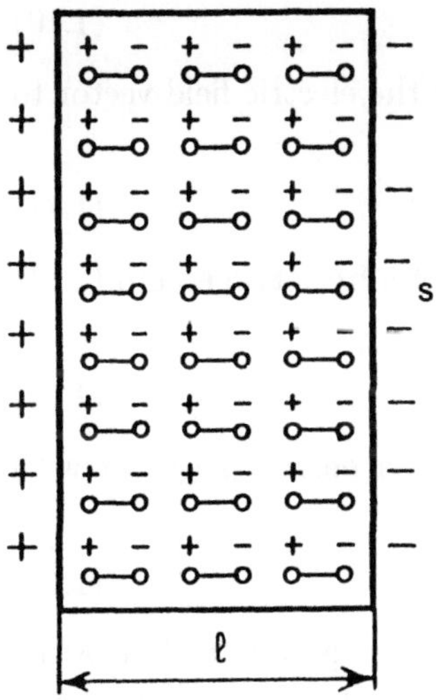

Fig. 1.11. A model of a polar dielectric consisting of dipolar molecules.

Since the dipoles are arranged in a tail-to-head manner and since the head and tail carry charges of equal magnitude but opposite sign, the net charge at the head-to-tail junction is equal to zero. Such a situation will, however, occur for the internal points of the sample only since the charges of "heads" and "tails" emerging at the surface are not compensated. The sample surfaces are thus found to be charged. For the surface perpendicular to the direction of the vector $\boldsymbol{P}$ the magnitude of the surface charge density is equal to the magnitude of the vector $\boldsymbol{P}$. This can easily be seen if the sample is considered to be a plate of thickness l and area s. Denoting the surface charge density as σ_s, we find that the dipole moment of the plate is $\sigma_s sl$. On the other hand, it can be found if we multiply the density of the dipole moment by the volume sl. Thus, we have

$$\sigma_s sl = Psl \tag{1.8}$$

that is, $\sigma_s = P$. The quantity σ_s determines the surface density of a bound charge which appears only where the polarization is changed in space – in the example considered it took place on the crystal surface. But if the crystal is polarized inhomogeneously, a bound charge also appears inside the crystal; in this case one is speaking of the volume density of the bound charge ρ_v. When the vector $\boldsymbol{P}$ is aligned for all points of the medium along a single axis (let it be the x axis), then

$$\rho_v = -\frac{\partial P}{\partial x}. \tag{1.9}$$

It can easily be seen that (1.8) is a special case of (1.9). To this end, it is convenient to assume that the polarization on the crystal surface is changed not in a jumpwise manner but within a layer of thickness l, i.e., in such a way that $\rho_v = -\partial P/\partial x \approx -P/l$, and to take into account that $\sigma_s = \rho_v l$. In the general case we have

$$\rho_v = -\operatorname{div} \boldsymbol{P}. \tag{1.10}$$

Let us now write down the equation which relates the electric field vector to the charge density:

$$\operatorname{div} \boldsymbol{E} = 4\pi\rho. \tag{1.11}$$

Substituting $\rho = \rho_f + \rho_v$, where ρ_f is the density of a free charge, into (1.11) and using (1.10), we get

$$\operatorname{div} \boldsymbol{D} = 4\pi\rho_f, \tag{1.12}$$

where $\boldsymbol{D} = \boldsymbol{E} + 4\pi\boldsymbol{P}$ is the vector of electric displacement. Together with the equation

$$\operatorname{rot} \boldsymbol{E} = 0 \tag{1.13}$$

(1.12) constitutes the Maxwell system of equations as applied to electrostatics. In order to solve it, it is necessary, of course, to know also the dependence $\boldsymbol{P}(\boldsymbol{E})$ or $\boldsymbol{D}(\boldsymbol{E})$, i.e., the electrical equation of state.

Let us now find the field of the homogeneously polarized plate considered in the above example (see Fig. 1.11). From (1.12) it follows that the vector $\boldsymbol{D}$ is constant throughout the entire space (since $\rho_f = 0$), i.e., $\boldsymbol{D} = \boldsymbol{C}$, where $\boldsymbol{C}$ is a certain constant vector. Thus, $\boldsymbol{E} = \boldsymbol{C} - 4\pi\boldsymbol{P}$ inside and $\boldsymbol{E} = \boldsymbol{C}$ outside the plate. It is clear that we should put $\boldsymbol{C} = 0$ since otherwise we will find that a field different from zero also exists at $\boldsymbol{P} = 0$, i.e., in the absence of any charges. Thus, $\boldsymbol{E} = -4\pi\boldsymbol{P}$ within the plate and $\boldsymbol{E} = 0$ outside of it. The same answer can be obtained if we calculate the field of the plate as the field of a capacitor with the charge density on the plates being $\sigma = P$. The capacitor must be regarded as a vacuum since the presence of a dielectric has already been taken into account by introducing bound charges.

The macroscopic polarization $\boldsymbol{P}$ appearing in the Maxwell equations is correlated in a natural way with the electrical charge distribution in the unit cell of the crystal. In the approximation of the point charges of ions the dipole moment of a unit cell of volume v_{uc} is equal to $\boldsymbol{p} = \sum_i e_i \boldsymbol{r}_i$, where the summation is over the charges and radius-vectors of ions belonging to the unit cell chosen, and

$$\boldsymbol{P} = \frac{1}{v_{\text{uc}}} \sum_i e_i \boldsymbol{r}_i. \tag{1.14}$$

At first sight this formula causes no objections; it is well known, however, that the choice of the unit cell is somewhat arbitrary. As a matter of fact, it is necessary and sufficient that the unit cell chosen be electroneutral in the first place ($\sum_i e_i = 0$) and, in the second, have the size and form of an elementary parallelepiped built on the vector basis of the lattice $\boldsymbol{a}_1$, $\boldsymbol{a}_2$, $\boldsymbol{a}_3$, which repeats itself by way of three-dimensional translations. The basis to be chosen should be translations that are the shortest and reflect the lattice symmetry in the best way.

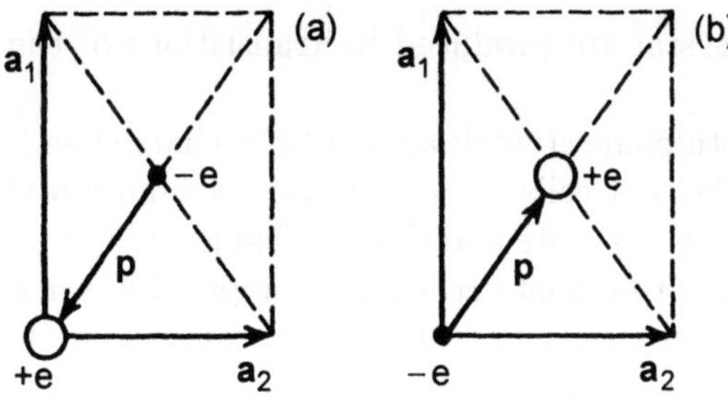

Fig. 1.12a,b. Various ways of choosing the unit cell of the model rhombic lattice of a planar crystal shown in Fig. 1.3a.

Let us consider in more detail the situation for a plane rhombic crystal depicted schematically in Fig. 1.3a. It is evident that all neccessary conditions for the choice of the unit cell will be fulfilled, irrespective of whether the origin is represented by a light (positive) or dark (negative) atom; two atoms, a light and a dark atom, must belong to a single unit cell.

The two possibilities are presented in Fig. 1.12. They are equivalent in all respects, except that the conditionally chosen diatomic "molecule" of a plane crystal has in these two cases a dipole moment of equal magnitude but opposite direction. This means that, in accordance with (1.14) and (1.8), on the crystal surface there must appear a bound charge of the corresponding magnitude, which, generally speaking, contradicts the macroscopic symmetry of the crystal: the point is that the presence of the planes m_1 and m_2 and a center of inversion renders the "top" and "bottom" of the crystal in Fig. 1.3a equivalent in all respects. This also follows from Fig. 1.3a, which, however, requires some detailing since the structure shown in this figure, which has all the symmetry elements listed, is represented by a charged ionic configuration – nine light and four black atoms have a net charge of $+5e$. A more accurate representation of the structure is given in Fig. 1.13, where the electrically

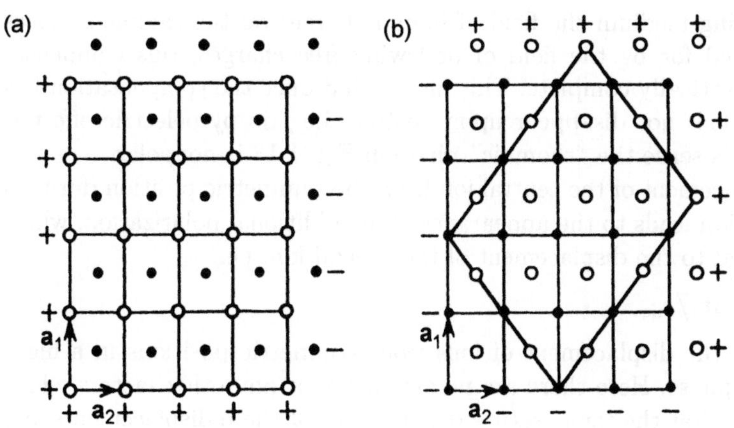

Fig. 1.13a,b. Electroneutral elements of crystal lattices produced by way of translations of unit cells chosen from Fig. 1.12.

neutral structures of the plane crystal are produced by translations of the unit cells depicted in Fig. 1.3.

It is seen that the "faces" of both plane crystals appear to be charged and, hence, to the dipole moments of the unit cells there corresponds a quite real bound charge on the external faces and a quite real electric field generated by these bound charges. The problem is thus that the samples prepared in some definite way from crystals of practically any symmetry may possess an electric polarization in the sense of (1.8) and (1.14). Of course, it is also possible to prepare an electrically neutral sample devoid of electric polarization (the region delineated by a boldface line in Fig. 1.13b).

It is essential, however, that such a bound charge on the surface of a sample of a nonpolar dielectric cannot be the cause of a pyroelectric effect. Let us consider a three-dimensional analog of the rhombic crystal sample shown in Fig. 1.13a, which is built on the vector basis $\boldsymbol{a}_1$, $\boldsymbol{a}_2$, $\boldsymbol{a}_3$ (the projection of the dipole moment of the unit cell on to a direction perpendicular to a charged plane of area $N_2a_2 \times N_3a_3$ will be equal to $(a_1/2)e$). The net charge on the surface a_2a_3 is given by

$$Q = P_n N_2 a_2 N_3 a_3 = \frac{p_n}{v_{\text{uc}}} N_2 a_2 N_3 a_3,$$

where p_n is the projection of the vector of the dipole moment of the unit cell on to the direction of the vector $\boldsymbol{a}_1$. Thus, the charge magnitude is

$$Q = \frac{(a_1/2)eN_2a_2N_3a_3}{a_1a_2a_3} = \frac{eN_2N_3}{2},$$

that is, it is independent of the lattice constant and, hence, of temperature. A somewhat more sophisticated treatment shows that even if the effective charges of ions depend on temperature, no pyroelectric effect will occur in the situation described. Moreover, this polarization cannot be determined experimentally. Besides, because of the finite conductivity of real crystals and the surrounding medium the field of bound charges on the sample surfaces is compensated for by the field of onflowing free charges, this compensation being practically complete. This means that once the compensation has occurred, it does not disappear upon heating, i.e., no pyroelectric effect is present; in this sense the "rhombic" phase in Fig. 1.13 is nonpolar.

The displacement of the central ion from the symmetric position during a phase transition leads to the appearance of an additional polarization, which is proportional to the displacement of the central ion, i.e.,

$$\Delta \boldsymbol{P} = e_i \Delta \boldsymbol{r}_i / v_{\text{uc}},$$

where $\Delta \boldsymbol{r}_i$ is the displacement of ions from symmetric positions in a high-temperature phase. Here there are no symmetry or any other factors which would require that the temperature dependences of these displacements and the lattice constant be the same. Evidently, the compensation of the electric fields is now upset with a change in temperature, i.e., a pyroelectric effect arises in the crystal. Usually, the polarization that already exists in a nonpolar phase but does not manifest itself in any way is simply ignored, and the *spontaneous polarization* is defined as a quantity numerically equal to the surface density of a bound charge on the sample surface, which is due to the relative displacement of atoms in a phase transition.

Thus, with what has been said above being taken into account, we think that the most consistent definition of the ferroelectric phase transition is as follows.

The ferroelectric phase transition is a structural phase transition, as a result of which a spontaneous polarization occurs in the crystal, i.e., a component of the electric polarization vector, which is caused by relative displacements of atoms in each of the unit cells of the crystal and which results in the appearance of a pyroelectric effect in the crystal.

We have pointed out above that in a low-symmetric phase there are equiprobably realized states corresponding to order-parameter values of different sign, as a result of which the crystal can be split into domains. In ferroelectric crystals, to ionic displacements of different sign there evidently correspond different directions of the vector of spontaneous polarization in neighboring domains. It is for this reason that the macroscopic polarization of a crystal divided into domains will be equal to zero. Let us see how the macroscopic polarization will change when such a crystal is placed in an electric field. For the sake of clarity, suppose we have a sample of a ferroelectric crystal of rectangular shape; the metal electrodes applied to the faces of the sample perpendicular to the direction of the spontaneous polarization vector $\boldsymbol{P}_{\mathrm{s}}$ form a plane-parallel capacitor filled completely with the crystal. Let us trace out the value of macroscopic polarization of the crystal using a simple diagram shown in Fig. 1.14. As can be seen from this figure, an electric voltage is applied to the crystal, which varies quasistatically by a linear law from 0 up to $+U_0$ and then from $+U_0$ to $-U_0$; the cycle is terminated with a linear increase of the voltage from $-U_0$ to 0. Then, the electric voltage changes periodically by the same law. It can be seen from the diagram that this voltage is fed to the vertical plates of an oscillograph; the horizontal plates are supplied with a voltage from the capacitor C_0 which is series-connected with the crystal. This voltage is proportional to the charge on capacitors C_{c} and C_0 and, hence, to the polarization of the crystal. Using this circuit (which is known as the Sawyer–Tower circuit), we can observe the character of the change of macropolarization during the variation of the electric field in the crystal.

It is quite obvious that in an electric field the domains in which the spontaneous polarization vector is directed oppositely to the vector $\boldsymbol{E}$ possess a certain excess energy as compared to domains with a different direction of $\boldsymbol{P}_s$. Therefore, with a linear increase of the electric field at a certain value of the latter the domain walls are brought into motion and there occurs a redistribution of the volumes of energetically favorable and unfavorable regions in favor of the former (portion OB of the curve in Fig. 1.15). The point

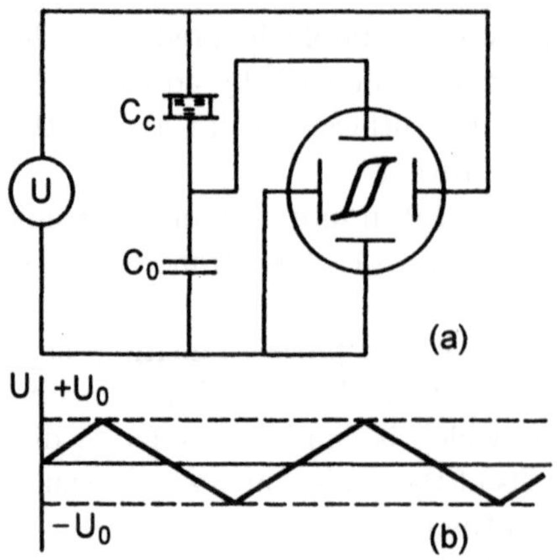

Fig. 1.14. (a) The simplest electric circuit for observation of the dependence of electric polarization on electric field (Sawyer–Tower circuit) and (b) the shape of voltage applied to the crystal.

B corresponds to a monodomain crystal and further increase of polarization on portion BC occurs only due to the induced polarization which is linear with respect to the field. The magnitude of the field E_m corresponds to the maximum voltage U_0 on the crystal; the value of E begins to decrease, but at point D, where $E = 0$, the crystal remains monodomain. The change of sign does not lead to an immediate change of the direction of the vector $\boldsymbol{P}_s$; this process (switching of polarization) starts when the field reaches a certain critical value E_c (*coercive field*) at which regions with a "favorable" direction of the spontaneous polarization vector (portion FL) appear and rapidly expand. At point L the crystal again consists of a single domain, but the vector $\boldsymbol{P}_s$ is found to be aligned in the same direction as the vector $\boldsymbol{E}$. The processes then occur in an analogous manner. It is characteristic that we never get to point O from which the polarization of the "fresh" crystal started (the "fresh" crystal is one to which an electric field has not yet been applied). The *hysteresis loop* shown in Fig. 1.15 clearly illustrates the energy equivalence of two possible directions of the vector $\boldsymbol{P}_s$ in the crystal and the possibility of "switching" the spontaneous polarization by an external electric field. It is easy to see that the crystal's properties associated with switching processes are directly related to the dynamics of domain walls, their motion under the influence of an electric field.

At the same time it should be kept in mind that the transition of the crystal from the single-domain state with a certain direction of spontaneous polarization to an analogous state with a different direction of the vector $\boldsymbol{P}_s$ under the action of an electric field can also take place without formation of domain walls in the crystal, i.e., simultaneously throughout the entire volume of the crystal. The electric field in which the energetically unfavorable configuration of atoms becomes absolutely unstable (a theoretical coercive field) usually exceeds real experimental coercive fields by more than an order of magnitude. Just as in the case of a metastable oversaturated solution being cooled, a multiple crystallization begins long before the limiting oversatura-

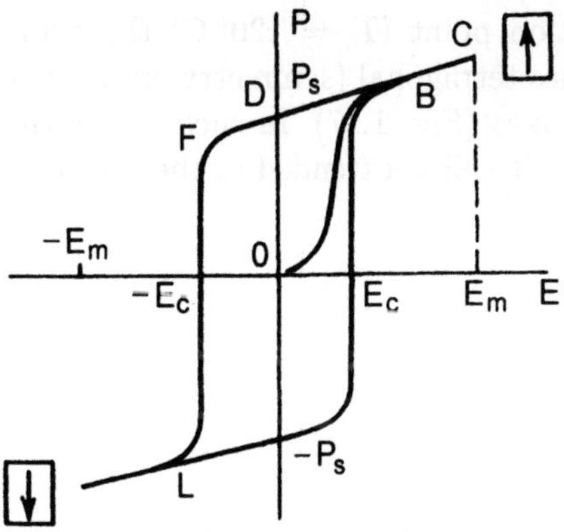

Fig. 1.15. Hysteresis dependence of the polarization of a ferroelectric crystal on electric field.

tion is attained – when the electric field in the single-domain crystal increases in a direction opposite to the direction of the vector $\boldsymbol{P}_s$ there begins a multiple nucleation and growth of regions with a "favorable" direction of spontaneous polarization before the original state becomes absolutely unstable.

1.5 Basic Types of Ferroelectric Crystals

Let us consider some concrete examples of ferroelectric and nonferroelectric transitions and certain attendant phenomena. We begin with crystals of barium titanate, $BaTiO_3$, which have a cubic perovskite-type structure and a symmetry $m3m$ at a temperature above 120° C in the symmetrical phase (Fig. 1.16).

There is one formula-unit of $BaTiO_3$ per unit cell; the atoms are in the following positions:

Ba^{2+} : (0, 0, 0),
Ti^{4+} : (1/2, 1/2, 1/2),
O^{2-} : (1/2, 1/2, 0); (1/2, 0, 1/2); (0, 1/2, 1/2).

The cubic unit cell parameter is of the order of 4 Å.

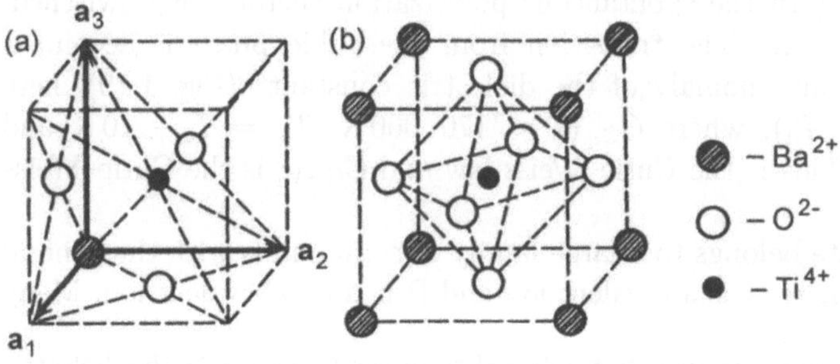

Fig. 1.16. (a) The unit cell and (b) structural elements of the $BaTiO_3$ crystal.

Below the ferroelectric phase transition point ($T_c = 120°$ C) the cubic lattice is distorted and the crystal becomes tetragonal (symmetry $4mm$); the atoms undergo shifts along the fourfold axes (Fig. 1.17). In such a case the oxygen octahedron is distorted and the unit cell is extended in the direction of atomic displacement.

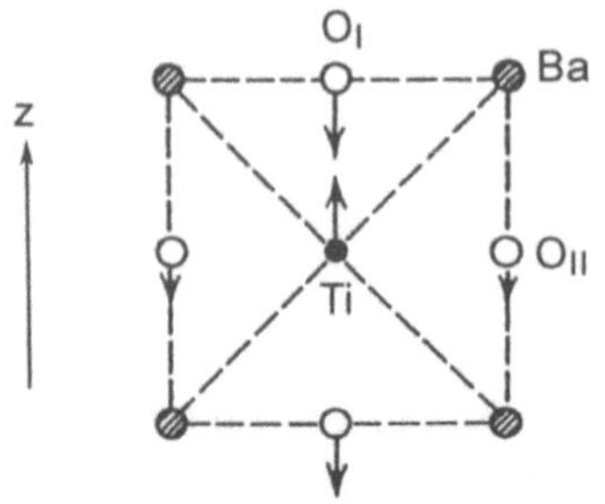

Fig. 1.17. Projection of the $BaTiO_3$ structure on to the (010) plane. The arrows indicate the directions of displacement of Ti and O atoms relative to the sublattice of Ba atoms assumed to be immobile.

It is evident that the transition is a ferroelectric phase transition of the displacive type; the spontaneous polarization can be determined by the following expression:

$$P_s = \frac{1}{v_{uc}} \sum_i e_i \Delta z_i \approx 25\,\mu\text{C} \cdot \text{cm}^{-2}.$$

A spontaneous polarization occurs at 120° C; upon further cooling the crystal experiences another two phase transitions at 0° C and −70° C into phases with symmetries $mm2$ and $3m$, respectively. In these phase transitions the spontaneous polarization changes its direction and becomes successively oriented along the fourfold, twofold and threefold axes (Fig. 1.18a). From Fig. 1.18b it can be seen that all transitions in the $BaTiO_3$ crystal are first-order transitions, the transition from the cubic to the tetragonal phase being a first-order transition close to a second-order phase transition.

The symmetry change $m3m \to 4mm$ could be accomplished by way of a second-order phase transition, whereas it is impossible to visualize a continuous change of the crystal state with symmetry changes $4mm \to mm2$ and $mm2 \to 3m$ with the spontaneous polarization vector being switched, as shown in Fig. 1.18. The transition from the cubic phase is accompanied by a significant anomaly of the dielectric constant[3] (Fig. 1.19); here $\varepsilon = C_{C-W}/(T - T_0)$, where $C_{C-W} \approx 170{,}000\,\text{K}$, $T_0 = T_c - 10\,\text{K}$ and $T > T_c = 393\,\text{K}$. This is the Curie–Weiss law and C_{C-W} is the Curie–Weiss constant.

Barium titanate belongs to a large family of compounds with the general formula ABO_3, where A is a divalent ion and B is a quadrivalent ion. Many

[3] The dielectric constant is also often referred to in the literature as the dielectric permittivity or simply permittivity.

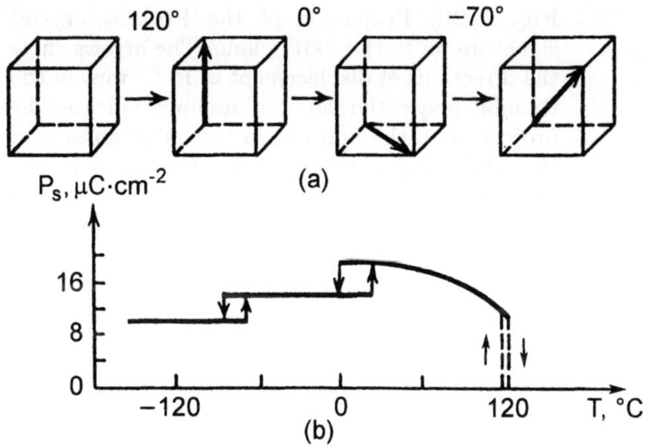

Fig. 1.18. (**a**) Change of the direction of the spontaneous polarization vector in phase transitions and (**b**) the temperature dependence of spontaneous polarization along the [100] direction in $BaTiO_3$ crystals.

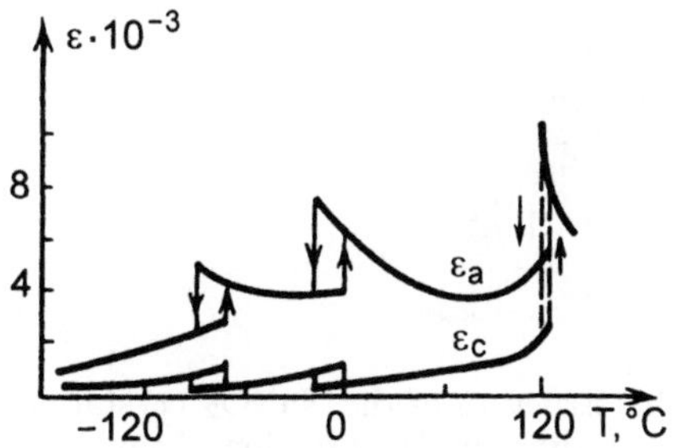

Fig. 1.19. Temperature dependence of the dielectric constant of $BaTiO_3$ crystals along the polar axis (ε_c) and in the direction perpendicular to the axis (ε_a).

such compounds undergo structural phase transitions of the ferroelectric or nonferroelectric type (lead titanate, $PbTiO_3$; lead zirconate, $PbZrO_3$; strontium titanate, $SrTiO_3$, etc.). In particular, nonferroelectric ordering occurs at a temperature of 230° C in crystals of lead zirconate ($PbZrO_3$). Below the phase transition point the crystal undergoes a rhombic distortion and a symmetry change to the group $mm2$ with antiparallel displacements of lead and zirconium atoms in the plane (001) along the [110] axis leading to an eightfold increase of the unit-cell volume (Fig. 1.20).

We will now consider examples of phase transitions of the order–disorder type. Note that, as a rule, order–disorder ferroelectrics possess a rather complex crystalline structure. The ordering mentioned above takes place in a certain sublattice but it brings about the displacement of atoms in other sublattices. In this case the spontaneous polarization can be governed by displacements of ions that do not belong to the sublattice being ordered.

Crystals of sodium nitrite ($NaNO_2$), potassium dihydrogen phosphate (KDP, KH_2PO_4) and triglycine sulphate (TGS, $(CH_2NH_2COOH)_3 \cdot H_2SO_4$)

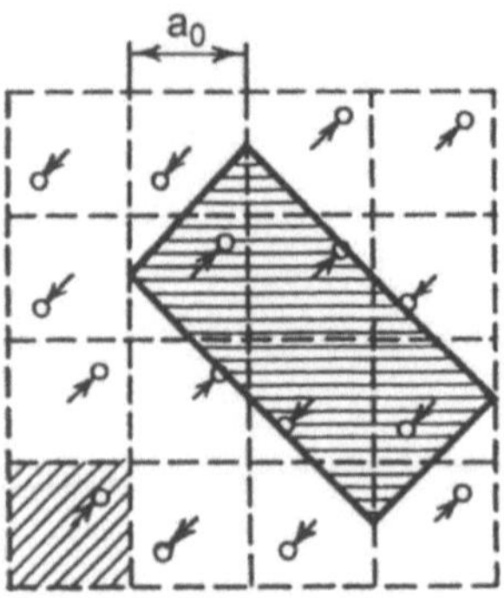

Fig. 1.20. Projection of the $PbZrO_3$ crystal structure on to the (001) plane. The arrows show the directions of displacement of Pb^{2+} ions in the rhombic phase; the slanting hatching indicates the projection of the unit cell in the cubic phase and the horizontal hatching shows the same in the rhombic phase.

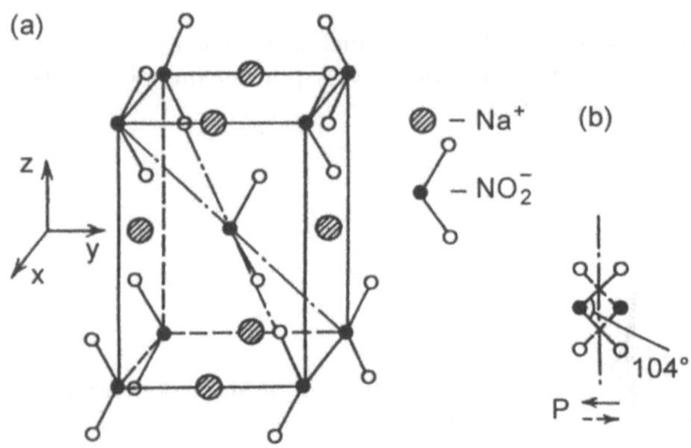

Fig. 1.21. (**a**) Arrangement of atoms in the structure of the $NaNO_2$ crystal and (**b**) possible positions of the NO_2 group, which differ in the direction of the dipole moment.

are typical representatives of ferroelectrics with a phase transition of the order–disorder type.

Figure 1.21 shows the structure of sodium nitrite ($NaNO_2$) in a paraelectric phase. At $T > 164°$ C the crystal has an orthorhombic symmetry mmm; the NO_2 groups form a body-centered lattice. It is essential that the atoms of the NO_2 group are not arranged on a single line along the z axis.

Such an arrangement is realized in the symmetrical phase only "on an average" since the angle between the NO bonds in the group O–N–O is not equal to 180° and the nitrogen atom performs jumps in the direction of the y axis in the yz plane (Fig. 1.21b). Such motion corresponds to the reorientation of the dipole moment of the polar group NO_2.

At a temperature of 164° C there takes place a ferroelectric phase transition associated with an increase of the probability of the nitrogen atom being on the right or on the left from the symmetry plane xz. This plane ceases to be a plane of symmetry and the crystal symmetry is lowered to the polar group $mm2$ and a spontaneous polarization occurs along the y axis. A partial

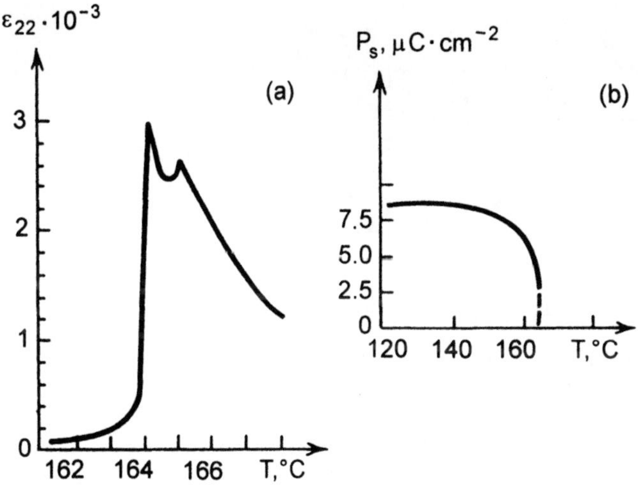

Fig. 1.22. (**a**) Temperature dependence of the dielectric constant and (**b**) spontaneous polarization of the $NaNO_2$ crystal along the polar [010] axis.

ordering of NO_2 groups is accompanied by the displacement of sodium atoms along the y axis, which makes a significant contribution to the spontaneous polarization. Note that two phase transitions take place in $NaNO_2$ crystals in the temperature range 164–165° C, this observation being supported by the appearance of two maxima in the temperature dependence $\varepsilon(T)$ (Fig. 1.22). In the transient phase, the temperature range for which is only 1.5 K, there occurs a nonuniform distribution of the polarization, which is characteristic of the so-called incommensurate phase (Chap. 11).

A large number of ferroelectric crystals, which contain *hydrogen bonds* are known. These bonds appear when the hydrogen atom is found to be a neighbor of a strongly electronegative atom – oxygen, nitrogen, fluorine, chlorine, etc.

The most thoroughly studied crystals are those of potassium dihydrogen phosphate, KH_2PO_4(KDP) and triglycine sulphate, $(CH_2NH_2COOH)_3{\cdot}H_2SO_4$ (TGS). In most cases, the potential energy of the proton of the hydrogen bond is described by a curve with two maxima separated by a potential barrier (see Fig. 1.7). If the hydrogen bond links atoms of a single species, the curve is symmetrical and the reasoning that has been given above for phase transitions of the order–disorder type are equally applicable to the proton sublattice. As a rule, hydrogen bonds do not introduce a direct contribution to spontaneous polarization if they are oriented at an angle close to 90° in the direction of spontaneous polarization induced by displacements of other ions. But the ordering of protons on hydrogen bonds serves, as it were, as a "triggering mechanism" of a phase transition. It is exactly this situation that takes place in the KH_2PO_4 crystal, whose structure is depicted in Fig. 1.23.

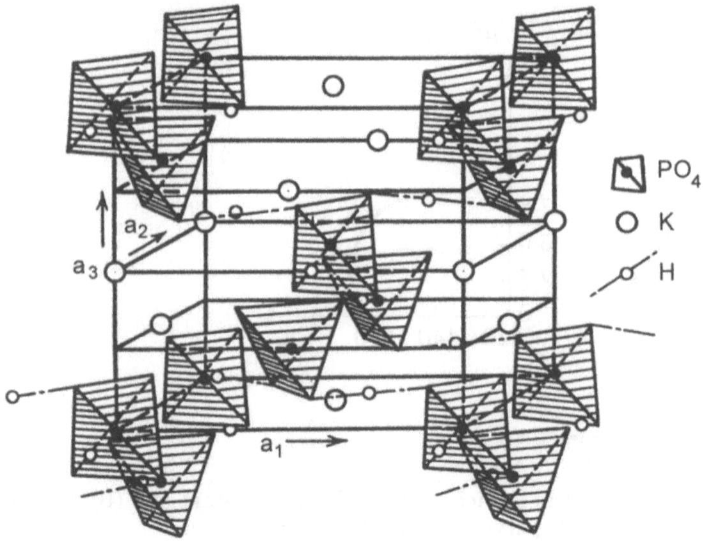

Fig. 1.23. A general view of the structure of the KH_2PO_4 crystal.

The symmetry of the crystal above $T_c = 123\,K$ belongs to the group $\bar{4}2m$; as can be seen from Fig. 1.23, the structure consists of K^+ ions and almost regular tetrahedral PO_4^{3-} groups. The potassium and phosphorus atoms lie one under the other on fourfold axes at a distance equal to half the unit-cell parameter in the direction of the z axis. The "lower" oxygen atoms of the tetrahedrons of one column of PO_4 groups lie at the same level with the "upper" oxygen atoms of the tetrahedrons of the adjacent column. This gives rise to a net of PO_4 tetrahedra linked together by symmetrical hydrogen bonds O–H ... O (Fig. 1.24). Each of the PO_4 groups is linked to the other four tetrahedra by hydrogen bonds. The entire net of hydrogen bonds is almost parallel to the (001) plane. In the symmetrical phase the hydrogen atoms are statistically uniformly distributed over hydrogen bonds in such a way that the occupancy of the "right-hand" and "left-hand" minima is the same.

At a temperature of 123 K the KDP crystals undergo a phase transition, as a result of which the symmetry group is lowered to the rhombic symmetry $mm2$ and the crystal lattice becomes polarized along the z axis. Below the phase-transition temperature the protons are partly ordered, being located near the upper (or lower) oxygen atoms of PO_4 groups. Since the reorientation of the dipoles attached to hydrogen bonds occurs in a plane perpendicular to the z axis, they introduce no direct contribution to the spontaneous polarization. The major contribution to the polarization is made by the displacements of heavy ions caused by their interaction with protons. X-Ray diffraction analysis shows that in the ferroelectric phase the atoms K and P are displaced along the z axis in opposite directions by 0.04 Å and

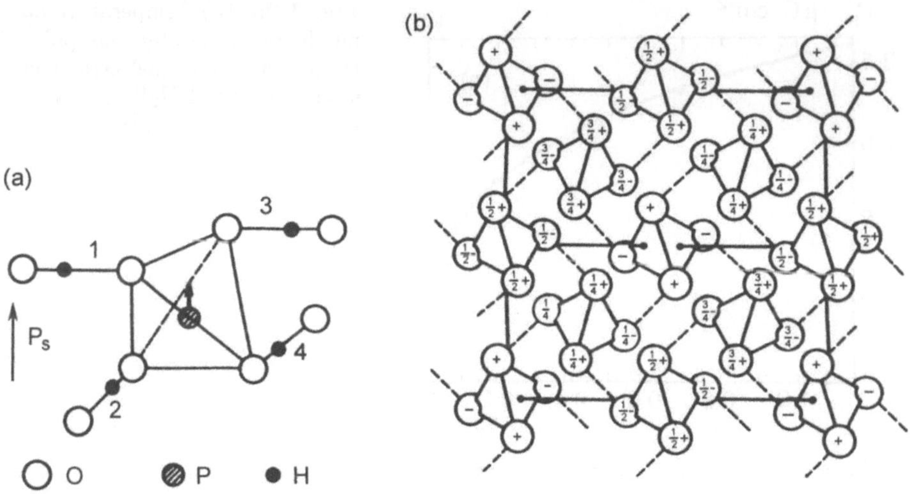

Fig. 1.24. (**a**) The system of hydrogen bonds in KH_2PO_4 crystals: the PO_4 tetrahedron with hydrogen bonds to nearest neighbors and (**b**) the projection of a net of tetrahedra on to the (001) plane.

0.08 Å, respectively, from those oxygen atoms near which the protons are predominantly located. Figure 1.25 shows the temperature dependences of the spontaneous polarization and low-frequency dielectric constant of the KDP crystal along the z axis. Note that, just as in the case of the $BaTiO_3$ crystal, the polarization occurs discontinuously at the phase-transition point, i.e., a first-order transition close to a second-order one occurs, and the dielectric constant above T_c obeys the Curie–Weiss law: $\varepsilon_{33} = C_{C-W}/(T - T_0)$, where $C_{C-W} = 3500\,K$ and the temperature T_0 is lower than the temperature T_c of the maximum ε_{33} by 0.2 K.

The group of hydrogen-containing ferroelectrics with a phase transition of the order–disorder type includes crystals of triglycine sulphate (TGS). Above 49° C in the paraelectric phase single crystals of TGS belong to the symmetry group $2/m$. There are two formula units per unit cell. The TGS structure is complex: it is a net of glycine molecules CH_2NH_2COOH and SO_4 tetrahedra linked together by hydrogen bonds of the types O–H...O and N–H...O (Fig. 1.26). The glycine groups are asymmetric and possess a significant dipole moment, but in the paraelectric phase the glycine ions lying in the planes $y = 1/4$ and $y = 3/4$ can evidently occupy two equivalent symmetric positions at an angle of 12.5° to the plane of symmetry. Thus, above the Curie point the TGS structure consists of an equal number of unit cells with dipole moments of opposite polarity and there are planes of mirror reflection ($y = 1/4$ and $y = 3/4$) in the structure. There is a viewpoint according to which the "triggering mechanism" of the phase transition in TGS crystals is the ordering of protons on hydrogen bonds that link the glycine groups III and II in Fig. 1.26. To the phase transition at $T_c = 49°$ C

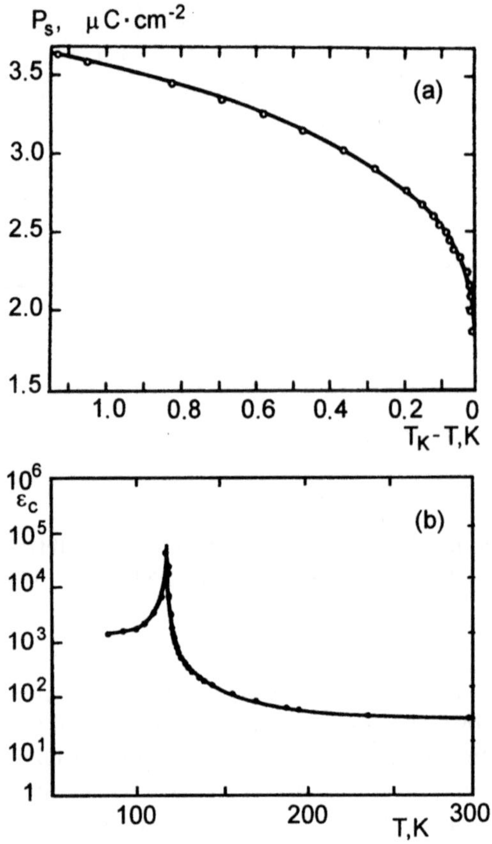

Fig. 1.25. (**a**) Temperature dependence of spontaneous polarization and (**b**) dielectric constant of the KH_2PO_4 crystal along the polar axis [1, 2].

in TGS there corresponds a partial ordering of protons on hydrogen bonds and of the mobile glycine groups I. In this case, naturally, the planes of mirror reflection disappear and the crystal symmetry is lowered to the polar group 2. The phase transition is accompanied by an anomaly of only the component ε_{22} of the dielectric constant tensor (the y axis is directed along the polar axis). The spontaneous polarization is continuously temperature-dependent and the phase transition exhibits all the features of a second-order phase transition (Fig. 1.27).

The dielectric constant obeys the Curie–Weiss law with the constant $C_{\mathrm{C-W}} = 3200\,\mathrm{K}$.

The three ferroelectrics considered above differ completely in composition, structure and symmetry. At the same time, we can clearly see that there is a similarity in temperature dependences of the spontaneous polarization and dielectric constants of all three crystals in the region of the phase transition. Another feature common to these crystals is the property of ferroelectrics indicated above – the spontaneous polarization vector may be changed under the action of an external electric field.

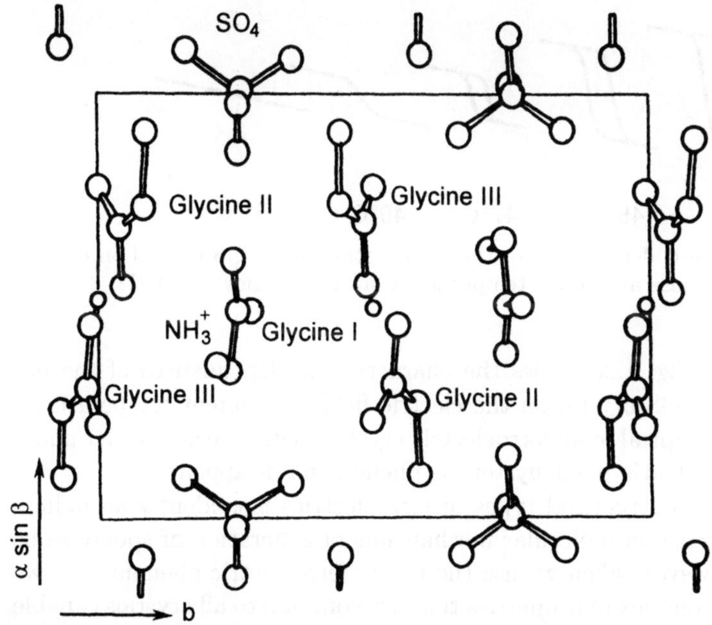

Fig. 1.26. Schematic representation of the structure of a TGS crystal: a and b are the unit-cell parameters, β is the monoclinic angle.

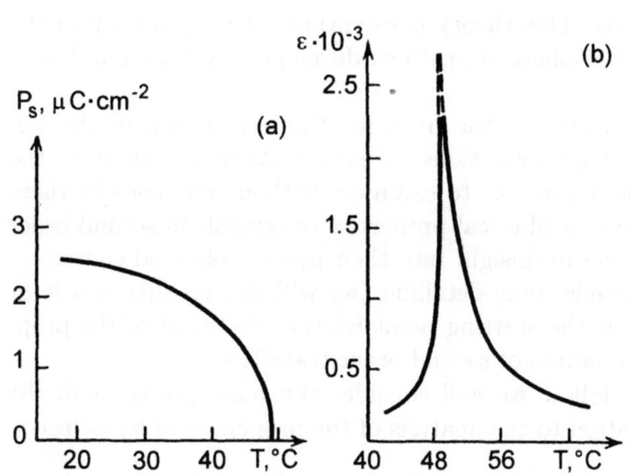

Fig. 1.27. (**a**) Temperature dependence of spontaneous polarization and (**b**) dielectric constant along the polar axis 2 of a TGS crystal.

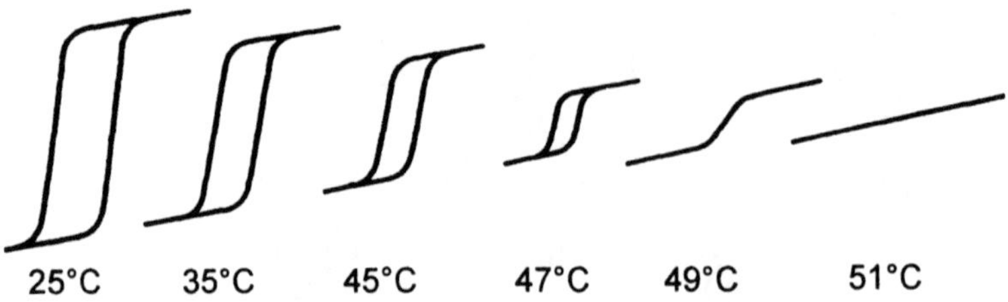

Fig. 1.28. Hysteresis dependence of the polarization of a TGS crystal on electric field in the polar phase at various temperatures at a frequency of 0.1 Hz.

To illustrate, Fig. 1.28 shows the characteristic dependences of the polarization of the TGS crystal on the electric field at different temperatures, these plots being typical of all ferroelectric crystals with a second-order phase transition. At the Curie point hysteresis phenomena disappear.

The diversity of structural types of ferroelectrics is evidently an indication of the difference in molecular mechanisms of generation of spontaneous polarization. However, when we use the term "ferroelectric phenomena", we mean the entire complex of properties that are common to all crystals capable of undergoing ferroelectric phase transitions.

One can clearly see that the macroscopic behavior of all substances shows a similar pattern in the region of phase transformations. This accounts for the fact that the phenomenological theory of ferroelectric phenomena has been developed extensively. This theory is essentially the application of the phenomenological theory of phase transitions developed by L.D. Landau to ferroelectrics.

While not limiting, of course, the need for the elucidation of the micromechanisms of phase transformations in each particular substance, the Landau theory has made it possible to examine, without reference to these mechanisms, the anomalies of physical properties of crystals in second-order phase transitions and to get an insight into their general physical causes.

Though this theory needs some detailing (we will discuss this at a later time), it should be used as the starting point for the treatment of the properties of crystals near the points of second-order transitions.

In the chapters that follow we will consider the basic positions of the Landau theory in application to the analysis of the properties of ferroelectric crystals.

Concluding this introductory chapter, we give a table of ferroelectric crystals, which lists the most representative types and provides a notion of the structural and compositional diversity of ferroelectrics (Table 1.1).

Table 1.1. Chemical formulas and Curie points of some ferroelectric crystals.

Name of substance	Chemical formula	Curie point
Barium titanate	$BaTiO_3$	120° C
Lithium niobate	$LiNbO_3$	1210° C
Antimony sulfoiodide	SbSI	20° C
Potassium dihydrogen phosphate (KDP)	KH_2PO_4	– 150° C
Ammonium fluoberyllate	$(NH_4)_2BeF_4$	– 98° C
Methylammonium aluminum alum (or sulphate) (MASD)	$CH_3NH_3Al(SO_4)_3 \cdot 12H_2O$	– 96° C
Colemanite	$Ca_2B_5O_{11} \cdot 5H_2O$	– 7° C
Strontium dicalcium propionate	$Ca_2Sr(CH_3CH_2COO)_6$	8.5° C
Rochelle salt (RS)	$KNaC_4H_4O_6 \cdot 4H_2O$	– 18° C; 24° C
Triglycine sulphate (TGS)	$(NH_2CH_2COOH)_3 \cdot H_2SO_4$	49° C
Trisarcosine calcium chloride	$(CH_3NHCH_2COOH)_3 \cdot CaCl_2$	– 146° C
Sodium nitrite	$NaNO_2$	164° C
Lead germanate	$Pb_5Ge_3O_{11}$	180° C

2. Phenomenological Theory of Second-Order Structural Transitions in Crystals

2.1 The Incomplete Thermodynamic Potential

Having established how the distortion of the crystal lattice, which occurs at the Curie point of a second-order phase transition (or a first-order transition close to it), can be described by the order parameter, we come to the following important conclusion: in phase transitions of this type the non-symmetric phase may be represented as a distorted symmetrical phase. This statement is the basis of the phenomenological theory of second-order phase transitions. Below we will try to explain how the character of variation of the physical properties of a crystal in the temperature range, including the Curie point, can be predicted if we know the change of its symmetry at the Curie point. Especially important is the result of the phenomenological theory, which shows that the information about a "symmetry jump", i.e., about the symmetry elements lost by the crystal at the Curie point, is sufficient for a description of the anomalies of practically all thermodynamic properties of the crystal.

It is known that, depending on the convenience and the particular conditions of the problem, one can analyze the thermodynamic properties of an equilibrium system by considering its thermodynamic functions expressed in terms of various variables. An important condition is that the change of the corresponding thermodynamic function must determine the work performed over the system by external forces.

The elementary work of polarization of a dielectric

$$\mathrm{d}A = \frac{1}{4\pi} E_i \mathrm{d}D_i = E_i \mathrm{d}P_i + \mathrm{d}\left(\frac{E_i^2}{8\pi}\right) \tag{2.1}$$

(E_i are the components of the vector of the electric field strength; D_i are those of the displacement vector; P_i are those of the polarization vector) can be treated as the sum of the work of polarization proper $E_i \mathrm{d}P_i$ and the work associated with the excitation of an electric field between the capacitor plates $\mathrm{d}(E_i^2/8\pi)$. Therefore, the differential of the internal energy of a dielectric will have the form

$$\mathrm{d}U(S, v, P_i) = T\mathrm{d}S - p\mathrm{d}v + E_i \mathrm{d}P_i \tag{2.2}$$

and the differential of the Gibbs thermodynamic potential is given by

$$\mathrm{d}\Phi(T, p, E_i) = -S\mathrm{d}T + v\mathrm{d}p - P_i\mathrm{d}E_i. \tag{2.3}$$

The function $\Phi = \Phi(T, p, E_i)$ determines the properties of a dielectric crystal placed between the capacitor plates which are kept at a constant potential difference; S is the entropy; T is the temperature; v is the volume; and p is the pressure.

The electrical equation of state of the unit volume of the dielectric can be obtained in the form

$$P_i = -\left(\frac{\partial \Phi}{\partial E_i}\right)_{p,T}. \tag{2.4}$$

This equation enables one to determine its dielectric susceptibility:

$$\chi_{ij}^{p,T} = \left(\frac{\partial P_i}{\partial E_j}\right)_{p,T} = -\left(\frac{\partial^2 \Phi}{\partial E_i \partial E_j}\right)_{p,T}. \tag{2.5}$$

The second derivatives of the thermodynamic potential determine the material constants of the substance:
(a) the heat capacity at constant pressure and electric field:

$$C_{p,E} = -T\left(\frac{\partial^2 \Phi}{\partial T^2}\right)_{p,E}; \tag{2.6}$$

(b) the volume compressibility at constant temperature and electric field:

$$\zeta_{T,E} = \left(\frac{\partial^2 \Phi}{\partial p^2}\right)_{T,E}; \tag{2.7}$$

(c) the volume thermal expansion coefficient (or volume expansivity) at constant electric field:

$$K_E = -\left(\frac{\partial^2 \Phi}{\partial p \partial T}\right)_E; \tag{2.8}$$

(d) the pyroelectric coefficients at constant pressure:

$$\gamma_i^p = -\left(\frac{\partial^2 \Phi}{\partial E_i \partial T}\right)_p; \tag{2.9}$$

and others.

For what follows it will be useful to recall how the equilibrium thermodynamic functions can be found within the framework of statistical physics. Special attention should be drawn to the possibility of introducing new variables in addition to those that govern the thermodynamic potential (in addition to the thermodynamic forces and coordinates p, v, T, S, E_i, P_i).

Let us return to the system of N interacting particles, the potential energy of which is specified in the function of the radius-vectors of particles $U = U(\boldsymbol{r}_1, \ldots, \boldsymbol{r}_N)$. At $T \neq 0$ to the equilibrium configuration of the system at fixed pressure there will correspond the minimal value of the thermodynamic potential Φ. The probability that the value of the radius-vector of the first particle lies between $\boldsymbol{r}_1$ and $\boldsymbol{r}_1 + \mathrm{d}\boldsymbol{r}_1$, that of the second between $\boldsymbol{r}_2$ and $\boldsymbol{r}_2 + \mathrm{d}\boldsymbol{r}_2$, and so on, can be determined by

$$\mathrm{d}w = C \exp\left[-U(\boldsymbol{r}_1, \dots, \boldsymbol{r}_N)/(k_B T)\right] \prod_i \mathrm{d}\boldsymbol{r}_i. \tag{2.10}$$

Integration over all the variables yields unity on the left, so that the constant $C = Z^{-1}$, and

$$Z = \int_{-\infty}^{\infty} \exp\left[-U(\boldsymbol{r}_1, \dots, \boldsymbol{r}_N)/(k_B T)\right] \prod_i \mathrm{d}\boldsymbol{r}_i. \tag{2.11}$$

According to Gibbs, the thermodynamic potential is given by

$$\Phi(p, T) = -k_\mathrm{B} T \ln Z \tag{2.12}$$

and the probability distribution has the following form:

$$\mathrm{d}w = \exp\left\{[\Phi - U(\boldsymbol{r}_1, \dots, \boldsymbol{r}_N)]/(k_\mathrm{B} T)\right\} \prod_i \mathrm{d}\boldsymbol{r}_i. \tag{2.13}$$

Here Φ is the equilibrium thermodynamic potential of the system or, more exactly, the part of the thermodynamic potential associated with the potential energy of interaction of the system's particles.

The term thermodynamic potential is also used to describe the nonequilibrium states. Naturally, for nonequilibrium states to be described, a larger number of variables (not only p and T) will be required. We will explain how the thermodynamic potential, which is dependent on such variables, is constructed. The introduction of new variables will be done with the aid of a linear transformation:

$$\boldsymbol{r}_1, \dots, \boldsymbol{r}_N \to \xi_1, \dots, \xi_{3N}. \tag{2.14}$$

These variables are chosen so that the variable of interest, η, which describes the nonequilibrium state, is one of them, say $\xi_1 = \eta$.

For the new variables the probability distribution of their values being within the range from η to $\eta + \mathrm{d}\eta$, from ξ_2 to $\xi_2 + \mathrm{d}\xi_2$, etc., will have the form

$$\mathrm{d}w = \exp\left\{[\Phi - U(\eta, \xi_2, \dots, \xi_{3N})]/(k_\mathrm{B} T)\right\} \mathrm{d}\eta \mathrm{d}\xi_2 \dots \mathrm{d}\xi_{3N}. \tag{2.15}$$

Suppose we are interested in the probability distribution of values of only one variable, say η . Then, evidently,

$$\mathrm{d}w(\eta) = \mathrm{d}\eta \int_{-\infty}^{\infty} \exp\left\{[\Phi - U(\eta, \xi_2, \dots, \xi_{3N})]/(k_\mathrm{B} T)\right\} \mathrm{d}\xi_2 \dots \mathrm{d}\xi_{3N} \tag{2.16}$$

is the probability of finding the system in a state in which the value of η lies in the interval between η and $\eta + \mathrm{d}\eta$. We introduce the notation

$$\Phi(\eta) = -k_\mathrm{B} T \ln \int_{-\infty}^{\infty} \exp\left\{[-U(\eta, \xi_2, \dots, \xi_{3N})]/(k_\mathrm{B} T)\right\} \mathrm{d}\xi_2 \dots \mathrm{d}\xi_{3N}. \tag{2.17}$$

Then

$$\mathrm{d}w(\eta) = \exp\{[\Phi - \Phi(\eta)]/(k_\mathrm{B}T)\}\mathrm{d}\eta. \tag{2.18}$$

It can be seen that the function $\Phi(\eta)$ yields the probability distribution of various values of η; to the equilibrium value of the variable η, i.e., the one corresponding to the maximum of the distribution function, there will correspond the minimal value of $\Phi(\eta)$. Knowing this function or having found its form on the basis of some physical considerations, we can directly pass over to the equilibrium function proper. From (2.18) we have

$$\Phi(p,T) = -k_\mathrm{B}T \ln \int_{-\infty}^{\infty} \exp\{[-\Phi(\eta)]/(k_\mathrm{B}T)\}\,\mathrm{d}\eta. \tag{2.19}$$

Let η_0 be the equilibrium value of the variable η, i.e., let it correspond to the minimum value of the function $\Phi(p,T,\eta)$ (Fig. 2.1). We expand the function $\Phi(\eta)$ into a series in η in the vicinity of the point $\eta = \eta_0$:

$$\Phi(\eta) = \Phi(\eta_0) + \Delta\Phi(\eta - \eta_0) = \Phi(\eta_0) + \frac{1}{2}A(\eta - \eta_0)^2 + \dots \tag{2.20}$$

and using (2.19) we obtain

$$\Phi(p,T) = \Phi(\eta_0) - k_\mathrm{B}T \ln \int_{-\infty}^{\infty} \exp\{[-\Delta\Phi(\eta - \eta_0)]/(k_\mathrm{B}T)\}\,\mathrm{d}\eta. \tag{2.21}$$

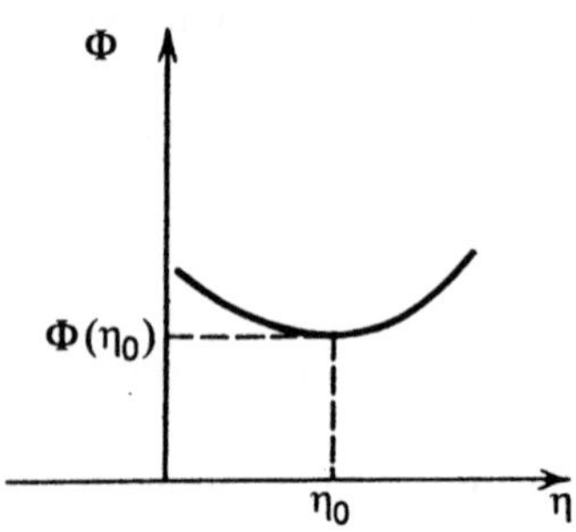

Fig. 2.1. Dependence of the function $\Phi(p,T,\eta)$ on η at given p and T.

The first term on the right-hand side of the equation is the minimum value of the thermodynamic potential of the system as a function of η (Fig. 2.1) and the second term represents the contribution to the thermodynamic potential associated with fluctuations of η, i.e., possible deviations of η from the most probable value of η_0. Evaluating the integral, we became convinced that this term is negligibly small as compared to the first term and corresponds in order of magnitude to the contribution of the thermal energy per one degree of freedom ($k_\mathrm{B}T/2$) to the thermodynamic potential. Therefore, putting

$$\Phi(p,T) = \Phi(p,T,\eta_0) \tag{2.22}$$

we will practically introduce no error. Note that the neglect of the fluctuations of one degree of freedom in a system consisting of a large ($\sim 10^{23}$) number of particles is quite natural. But if the number of degrees of freedom comparable to their total number in the system is left "not completely integrated" in the sense of (2.17), then the calculation of the equilibrium thermodynamic function according to the scheme of (2.22) and the neglect of the fluctuations become incorrect. In this case, it is necessary to use (2.19). It is precisely this situation which we will encounter in the analysis of the behavior of thermodynamic systems near the second-order phase transition point. Apart from the ordinary thermodynamic forces and coordinates, a few additional variables may appear in the thermodynamic potential. It simply means that we are speaking of the nonequilibrium thermodynamic functions that were left "not completely integrated" over those variables. Such thermodynamic functions are known as incomplete functions.

The *incomplete thermodynamic potential* of the system may be given in a general form as

$$\Phi(p,T,\eta_1,\ldots,\eta_n) = \Phi_0(p,T) + \Phi_1(p,T,\eta_1,\ldots,\eta_n) \tag{2.23}$$

where η_i are additional variables separated in an explicit form and $\Phi_0(p,T)$ is the part of the nonequilibrium thermodynamic function which is independent of these variables. The equilibrium values of η_{0i} are found by minimizing $\Phi(p,T,\eta_1,\ldots,\eta_n)$ with respect to the corresponding variables:

$$\frac{\partial \Phi}{\partial \eta_1} = 0, \ldots, \frac{\partial \Phi}{\partial \eta_n} = 0. \tag{2.24}$$

These equations determine a set of values $\eta_{0i}(p,T)$ which, after being substituted into (2.23), will determine the equilibrium thermodynamic function

$$\Phi(p,T) = \Phi_0(p,T) + \Phi_1(p,T,\eta_{01},\ldots,\eta_{0n}) \tag{2.25}$$

in which no account has been taken of the fluctuations of the η_{0i} values; the corresponding error will be insignificant if $n \ll 3N$ of the total number of degrees of freedom.

2.2 Structural Phase Transitions Described by a One-Component Order Parameter

As has been shown in Chap. 1, the structure of the nonsymmetric phase can be obtained from the structure of the symmetric phase by means of small distortions – displacements or distributions of atoms – described by the order parameter. In determining the order parameter within the scope of a particular model, we were dealing with the displacement of certain atoms or groups of atoms; it is natural, however, that one and the same lowering of symmetry in second-order phase transitions can occur with different types of displacement and ordering. In this sense, it would seem that the choice of the order parameter is ambiguous. As a matter of fact, it turns out that for the character of anomalies of the physical properties in phase transitions to be elucidated any particular relationship between the order parameter and the displacements of atoms is unimportant. It is only important that the appearance of the order parameter leads to a desirable change of the crystal's symmetry.

We will treat the order parameter η as the variable of the incomplete thermodynamic potential of the system. The very fact of the presence of a phase transition may be described as follows: the function $\Phi(\eta)$ has a minimum at $\eta = 0$ in the symmetric phase and at $\eta \neq 0$ in the nonsymmetric phase; here, as can be seen from the consideration of the models in Chap. 1, in the latter case the function $\Phi(\eta)$ exhibits two or more minima. Thus, the structural phase transition may be depicted by studying the change of the function

$$\Phi = \Phi(p, T, \eta). \tag{2.26}$$

It should be emphasized that this function is determined not only for the nonsymmetric but also for the symmetric phase. This imposes important limitations on the form of the dependence of Φ on η: the potential Φ as a scalar quantity characterizing the physical properties of crystals must be invariant with respect to any transformations of the symmetric phase:

$$\Phi(\eta) = \Phi(\eta^*) \tag{2.27}$$

where η^* is the order parameter transformed by symmetry operations.

In the vicinity of the second-order phase transition we may limit ourselves to the analysis of small lattice distortions (of small η) and it is therefore natural to consider the expansion of the thermodynamic potential into a series in η:

$$\Phi(\eta) = \Phi(0) + \Phi'\eta + \frac{1}{2}\Phi''\eta^2 + \dots \tag{2.28}$$

which must evidently satisfy the condition (2.27). It is clear that $\Phi(\eta)$ depends on temperature T and pressure p as parameters.

The law of transformation of the order parameter under the influence of the symmetry elements of the crystal predetermines, to a certain extent, the change of its physical properties. In particular, if the order parameter is transformed as a vector component, then, as will be shown below, the phase transition will be accompanied by a spontaneous polarization, i.e., will be ferroelectric. For a ferroelectric phase transition the order parameter is proportional to the component of the polarization vector. For this reason, a suitable component of the polarization vector is occasionally referred to as the order parameter.

For the sake of definiteness, we will further consider the concrete problem of the second-order phase transition into the ferroelectric state in a crystal that changes symmetry at $T = T_c$ as follows: $2/m \rightarrow 2$ (Fig. 2.2). At the transition temperature the crystal loses the inversion center and the plane of symmetry. In keeping with what has been said above, let us single out in the unit cell of the crystal lattice a displacement of atoms (or a group of atoms) corresponding to the order parameter η. This displacement really occurs in the nonsymmetric phase and is sufficient for the lowering of symmetry indicated. It should again be emphasized that only the transformation properties of this quantity will be essential for what follows, i.e., the law according to which it is transformed under the action of the symmetry elements of the crystal in the symmetrical phase.

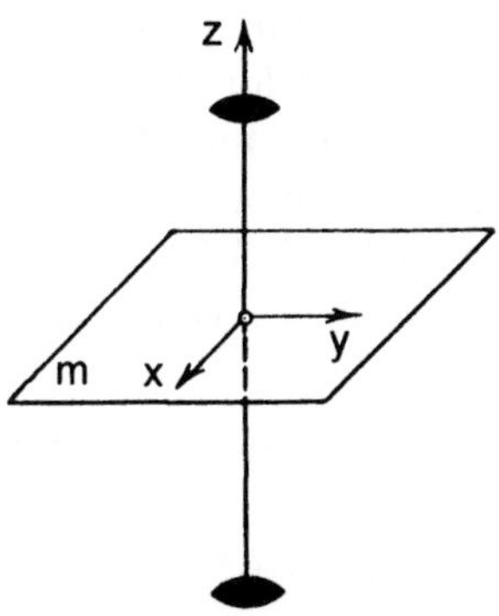

Fig. 2.2. The symmetry elements of the $2/m$ group: 1, $\bar{1}$, m, 2.

We take into account that the disappearance of the symmetry elements $\bar{1}$ and m is associated with the fact that it is exactly these symmetry operations that bring about a change of the sign of η and, hence, do not enter into the symmetry group of the low-temperature phase. Thus, the law of transformation of the order parameter may be characterized by Table 2.1. While analyzing the possible dependence of the thermodynamic potential on η, we must take into account that the quantity Φ cannot include terms linear in η: otherwise, the value of the potential would be changed in symmetry operations m and $\bar{1}$, which contradicts the condition (2.27).

Table 2.1. Transformation of the order parameter.

$2/m$	1	2	m	$\bar{1}$
η	η	η	$-\eta$	$-\eta$

So, the first nonzero and η-dependent term in the thermodynamic potential will be a term quadratic in η, of the type $A(p,T)\eta^2$, where $A(p,T)$ is a certain function of pressure and temperature. For the sake of simplicity, we will take into account only the temperature dependence, assuming that the pressure is a fixed parameter $p = p_c$. The thermodynamic potential of a crystal must "reflect" the following situation: there is the transition temperature T_c (the Curie point), on one side of which $\eta = 0$ corresponds to the minimum

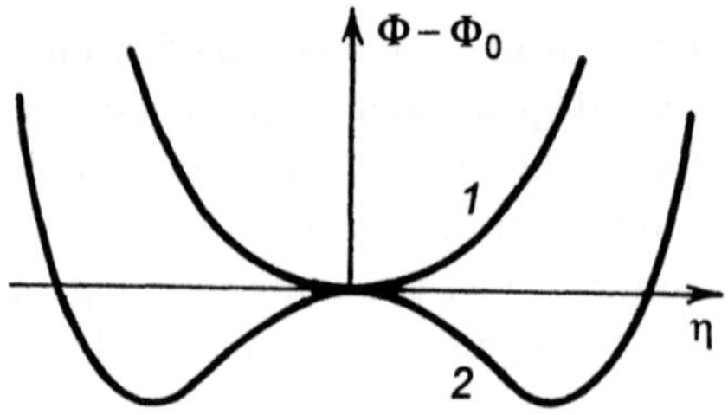

Fig. 2.3. Dependence of the nonequilibrium thermodynamic potential on the order parameter at $A(p,T) > 0$ (1) and at $Ap,T) < 0$ (2).

of Φ and on the other, low-temperature side, $\eta \neq 0$ is the equilibrium state. It means that with $T > T_c$ the coefficient $A(T) > 0$ and therefore Φ is at a minimum at $\eta = 0$ (Fig. 2.3, curve 1). But if $A(T) < 0$, then min Φ is attained for $\eta \neq 0$ (Fig. 2.3, curve 2) and to the equilibrium state of the crystal there will correspond a distorted crystal lattice of symmetry 2. We see that the necessary condition for the existence of the Curie point of a continuous phase transition must have the following form (Fig. 2.4):

$$A(p_c, T_c) = 0. \tag{2.29}$$

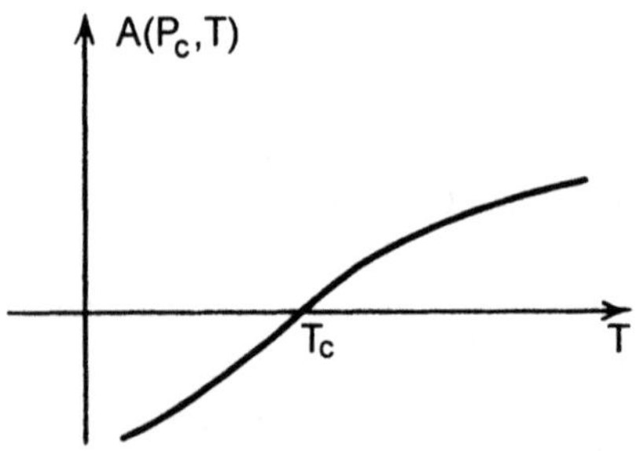

Fig. 2.4. Temperature dependence of the coefficient $A(p,T)$ in the vicinity of the second-order transition point; p_c is the pressure at which the transition temperature is T_c.

The system will be in a state of stable equilibrium in the symmetric phase if the function $\Phi(\eta)$ has a finite minimum for $\eta = 0$. It is therefore necessary to take into account terms of order higher than second in the thermodynamic potential. In the case under consideration no third-order term can exist since it will change sign in m and $\bar{1}$ operations (in accordance with Table 2.1). For the existence of the Curie point of a continuous phase transition it will be sufficient that, apart from the fulfilment of the condition (2.29), the thermodynamic potential contain the term $B(T)\eta^4$ with $B(T)$ being greater than zero. The condition (2.29) determines the line of the points of second-order phase transitions on the p–T diagram.

The thermodynamic potential assumes the following form:

$$\Phi(T, \eta) = \Phi_0(T) + A(T)\eta^2 + B(T)\eta^4 + \dots . \tag{2.30}$$

Now we can assert with certainty that for a second-order phase transition with a symmetry change $2/m \to 2$ to occur at $T = T_c$ and $p = p_c$, it is necessary that at this temperature the coefficient $A(p,T)$ vanish in a continuous way (Fig. 2.4). Though the exact form of the dependence of the coefficients on temperature and pressure is difficult to find, near the phase-transition temperature we can use their series expansion in powers of $T - T_c$ at $p = p_c$:

$$A(T) = A(T_c) + A'(T - T_c) + \frac{1}{2}A''(T - T_c)^2 + \dots \tag{2.31}$$

$$B(T) = B(T_c) + B'(T - T_c) + \frac{1}{2}B''(T - T_c)^2 + \dots . \tag{2.32}$$

Taking account of the above-indicated properties of the coefficients A and B ($A(T_c) = 0$ and $B(T_c) > 0$) and retaining the essential first expansion terms, we obtain at $p = p_c$

$$\Phi(T, \eta) = \Phi_0(T) + \frac{1}{2}\alpha(T - T_c)\eta^2 + \frac{1}{4}\beta\eta^4 + \dots \tag{2.33}$$

where $\alpha = 2A'$ and $\beta = 4B(T_c)$ (the numerical coefficients are introduced for the convenience of further calculations).

The potential $\Phi(T, \eta)$ is an incomplete thermodynamic function containing the variable η, which should be dropped out by using the equilibrium condition: the equilibrium function

$$\Phi(T) = \Phi(T, \eta_0(T))$$

where $\eta_0(T)$ can be found from the condition

$$\frac{\partial \Phi}{\partial \eta} = 0, \qquad \frac{\partial^2 \Phi}{\partial \eta^2} > 0.$$

Evidently, $\eta_0 = 0$ for $T > T_c$ and

$$\eta_0^2 = -\frac{\alpha(T - T_c)}{\beta} \qquad \text{for} \quad T < T_c.$$

The equilibrium thermodynamic function therefore has the form

$$\Phi(T) = \begin{cases} \Phi_0(T) & \text{for} \quad T > T_c \\ \Phi_0(T) - \frac{[\alpha(T - T_c)]^2}{4\beta} & \text{for} \quad T < T_c. \end{cases} \tag{2.34}$$

We will now take into account the dependence of the thermodynamic potential of a ferroelectric crystal of symmetry $2/m$, which undergoes a second-order phase transition into the group 2, on the electric field. This dependence may be represented as follows:

$$\Phi(T, \eta, E_2) = \Phi_0 + \frac{1}{2}\alpha(T - T_c)\eta^2 + \frac{1}{4}\beta\eta^4 - a\eta E_2 , \tag{2.35}$$

where E_2 is the component of the vector of electric field strength along the symmetry axis and a is a constant coefficient. It is obvious that the combination ηE_2 is an invariant of group $2/m$ since the law of transformation of the quantity E_2 as a vector component also corresponds to Table 2.1. In Chap. 3 we will analyze in detail the incomplete thermodynamic potential (2.35) and compare the thermodynamic results obtained with experimental data for the TGS crystal, which displays a change of symmetry $2/m \to 2$ at the Curie point of the second-order phase transition.

Note that in (2.35) we took into account only the lowest invariant in powers of E_2: account of invariants of higher powers of E_2 would have taken us beyond the framework of the approximations with respect to the coefficients A and B made in the derivation of (2.33). For instance, account of the invariant E_2^2 leads, as can be shown, to results that coincide, in the degree of approximation, with the refinements of the theory corresponding to terms proportional to $(T - T_c)^2$ in (2.31).

2.3 Structural Phase Transitions Described by Two- and Three-Component Order Parameters

We will now consider a more complicated case of ferroelectric phase transitions from the centrosymmetric group $4/mmm$ (Fig. 2.5). This group has 16 symmetry elements; two cases should be dealt with separately:

1. The change of symmetry $4/mmm \to 4mm$ is due to the displacement of atoms and the polarization of the crystal along the fourfold axis (Fig. 2.5a). Here the order parameter when acted upon by symmetry operations of group $4/mmm$ either remains unchanged or changes sign; the case under consideration is practically the same as the transition $2/m \to 2$ discussed above, and the nonequilibrium thermodynamic potential that describes the system has the form of (2.35).

2. The displacements of atoms occur in a plane perpendicular to the fourfold axis (Fig. 2.5b and c). Locating the origin as shown in Fig. 2.6a, we see that it is necessary to take into account that the displacements of atoms along the positive and negative directions of the axes x and y are crystallographically absolutely equal by virtue of the presence of the axis 4.

In order to take this into account, we introduce two components of the order parameter, η_1 and η_2, which correspond to displacements of atoms

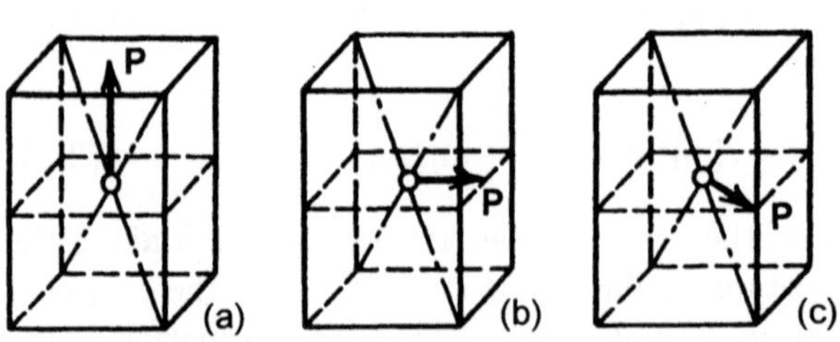

Fig. 2.5. (a) Directions of the displacements of atoms in the unit cell of a crystal of symmetry $4/mmm$ during the generation of a spontaneous polarization along the 4 axis; (b), (c) – in the basis plane of symmetry.

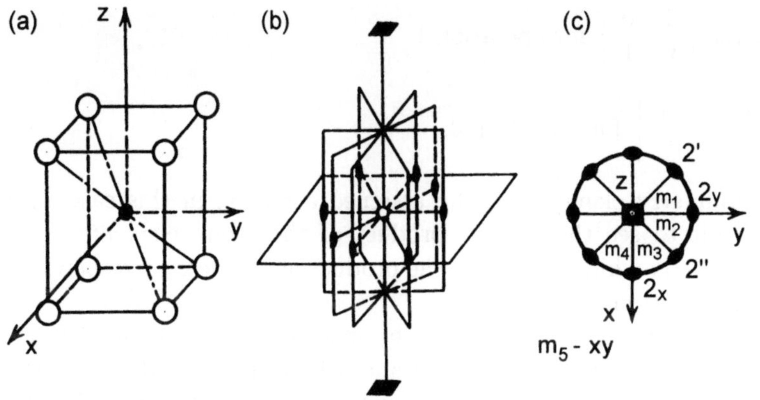

Fig. 2.6. (**a**) A structural element and (**b**), (**c**) operations of the group of point symmetry and of crystal of symmetry $4/mmm$ (16 symmetry operations: 1, 4, 4^{-1}, 2_x , 2_y, 2_z, $2'$, $2''$, m_1, m_2, m_3, m_4, m_5, $\bar{1}$, $\bar{4}$, $\bar{4}^{-1}$).

Table 2.2. Transformation properties of the components η_1 and η_2.

$4/mmm$	1	$\bar{1}$	4	4^{-1}	$\bar{4}$	$\bar{4}^{-1}$	$2'$	$2''$
η_1	η_1	$-\eta_1$	$-\eta_2$	η_2	$-\eta_2$	η_2	$-\eta_2$	η_2
η_2	η_2	$-\eta_2$	η_1	$-\eta_1$	η_1	$-\eta_1$	$-\eta_1$	η_1

2_x	2_y	2_z	m_1	m_2	m_3	m_4	m_5
η_1	$-\eta_1$	$-\eta_1$	$-\eta_2$	$-\eta_1$	η_2	η_1	η_1
$-\eta_2$	η_2	$-\eta_2$	$-\eta_1$	η_2	η_1	$-\eta_2$	η_2

along the axes x and y, respectively. The transformation properties of these components are presented in Table 2.2.

It should be pointed out that under symmetry transformations η_1 and η_2 not only change sign but also turn into each other. The "transformed" parameters may be expressed in terms of the old ones:

$$\eta_i' = a_{ij}\eta_j ,$$

where i, $j = 1$, 2 and the matrix of the coefficients

$$a_{ij} = \begin{vmatrix} a_{11} & a_{12} \\ a_{21} & a_{22} \end{vmatrix}$$

completely determines the transformation. For example, we have unit matrix $\begin{vmatrix} 1 & 0 \\ 0 & 1 \end{vmatrix}$ for operation 1,

matrix $\begin{vmatrix} 0 & -1 \\ 1 & 0 \end{vmatrix}$ for operation 4.

Thus, each symmetry element may be assigned a matrix in this case, and it is not difficult to see that to the product of the group elements there corresponds a matrix which is equal to the product of matrices corresponding to the elements multiplied.[1]

Since the quantities η_1 and η_2 must be incorporated into the thermodynamic potential in combinations invariant with respect to all symmetry operations of the group $4mmm$, the functional dependence $\Phi(\eta_1, \eta_2)$ must satisfy the condition

$$\Phi(\eta_1, \eta_2) = \Phi(\eta_1^*, \eta_2^*)$$

where η_1^* and η_2^* are the "transformed" components of the order parameter. From Table 2.2 it can be seen that the second-order invariant combination consisting of the quantities η_1 and η_2 will have the form

$$\eta_1^2 + \eta_2^2. \tag{2.36}$$

The third-order terms in the thermodynamic potential could have had the form η_1^3, $\eta_1\eta_2^2$, η_2^3, $\eta_1^2\eta_2$ and their linear combinations. It is easy to see that none of these combinations is an invariant for operations of the group $4/mmm$ and, hence, there will be no third-order terms in the potential. As regards fourth-order invariants, they can also be easily made up by taking account of Table 2.2:

$$(\eta_1^2 + \eta_2^2)^2; \quad \eta_1^4 + \eta_2^4; \quad \eta_1^2\eta_2^2. \tag{2.37}$$

We see that only two invariants are linearly independent, which is why the incomplete thermodynamic potential assumes the form

$$\begin{aligned} \Phi(T, \eta_1, \eta_2) &= \Phi_0(T) + A(T)(\eta_1^2 + \eta_2^2) \\ &\quad + B_1(T)(\eta_1^2 + \eta_2^2)^2 + B_2(T)\eta_1^2\eta_2^2 + \dots . \end{aligned} \tag{2.38}$$

For a second-order phase transition to exist it is necessary that the coefficient of an invariant of second degree vanish. As before,

$$\begin{aligned} \Phi(T, \eta_1, \eta_2) &= \Phi_0(T) + \frac{1}{2}\alpha(T - T_c)(\eta_1^2 + \eta_2^2) \\ &\quad + \frac{1}{4}\beta_1(\eta_1^2 + \eta_2^2)^2 + \frac{1}{2}\beta_2\eta_1^2\eta_2^2 + \dots \end{aligned} \tag{2.39}$$

[1] In group theory it is said that a given set of matrices realizes a two-dimensional representation of the group $4/mmm$ and the quantities η_1 and η_2 constitute the basis of this representation.

is the potential which is dependent on η_1 and η_2 and which satisfies all symmetry conditions. We have not yet found out, however, what symmetry will be exhibited by a crystal as a result of a phase transition at T_c. Assuming α, β_1 and β_2 to be constant and positive quantities, we can find values of the order parameter corresponding to the minimum of the thermodynamic potential. The relevant conditions have the following form:

$$\frac{\partial \Phi}{\partial \eta_1} = 0, \qquad \frac{\partial \Phi}{\partial \eta_2} = 0 \tag{2.40}$$

and

$$\begin{vmatrix} \partial^2\Phi/\partial\eta_1^2 & \partial^2\Phi/\partial\eta_1\partial\eta_2 \\ \partial^2\Phi/\partial\eta_2\partial\eta_1 & \partial^2\Phi/\partial\eta_2^2 \end{vmatrix} > 0$$

that is,

$$\frac{\partial^2\Phi}{\partial\eta_1^2} > 0, \qquad \frac{\partial^2\Phi}{\partial\eta_2^2} > 0, \qquad \frac{\partial^2\Phi}{\partial\eta_1^2}\frac{\partial^2\Phi}{\partial\eta_2^2} - \left(\frac{\partial^2\Phi}{\partial\eta_1\partial\eta_2}\right)^2 > 0. \tag{2.41}$$

From (2.40) we obtain

$$\eta_{01}[\alpha(T-T_c) + \beta_1(\eta_{01}^2 + \eta_{02}^2) + \beta_2\eta_{02}^2] = 0 \tag{2.42}$$

$$\eta_{02}[\alpha(T-T_c) + \beta_1(\eta_{01}^2 + \eta_{02}^2) + \beta_2\eta_{01}^2] = 0. \tag{2.43}$$

These equations have the following solutions:

$$\eta_{01} = 0, \qquad \eta_{02} = 0 \tag{2.44}$$

$$\begin{aligned} \eta_{01} &= 0, \quad \eta_{02}^2 = -\alpha(T-T_c)/\beta_1 \\ \eta_{02} &= 0, \quad \eta_{01}^2 = -\alpha(T-T_c)/\beta_1 \end{aligned} \tag{2.45}$$

$$\eta_{01}^2 = \eta_{02}^2 = -\alpha(T-T_c)/(2\beta_1+\beta_2). \tag{2.46}$$

It is easy to see that the solution (2.44) corresponds to the original symmetric phase $4/mmm$, the four solutions (2.45) to a phase with symmetry $mm2$ and the four solutions (2.46) also to a phase of symmetry $mm2$. In the general case two nonsymmetrical phases may have different symmetries. Though in both cases a lowering of symmetry occurs in this particular case into the group $mm2$, physically we obtain different low-symmetry phases (two-fold axes 2_x, 2_y and $2'$, $2''$, which are crystallographically nonequivalent, and the corresponding lattice distortions are different). By direct substitution into (2.41) we see that the solutions (2.44), (2.45) and (2.46) are stable at $T > T_c$, $T < T_c$ and $\beta_2 > 2\beta_1$, and at $T < T_c$ and $\beta_2 < 2\beta_1$, respectively. Thus, depending on the relative value of the coefficients β_1 and β_2, different structural rearrangements take place; hence, using the same expression for the thermodynamic potential, we can now describe two different second-order phase transitions. Moreover, proceeding from this expression, it is possible to describe a first-order phase transition from a configuration corresponding to the solution (2.45) into a configuration corresponding to the solution (2.46) or vice versa. To do this, we must assume that in the temperature range of interest $T < T_c$ the coefficients β_1 and β_2 are temperature dependent; to the first-order transition point there will correspond the condition $2\beta_1 = \beta_2$.

Thus, because of the higher symmetry of the original phase, we had to introduce a two-component order parameter in order to describe second-order transitions from the group $4/mmm$. There was no such need, as far as we were concerned with the low-symmetry monoclinic group $2/m$, whose symmetry elements changed only the sign of the order parameter. Ferroelectric phase transitions from the group $4/mmm$ may be depicted schematically (see Fig. 2.5): there exist only three nonsymmetric phases; transitions with the occurrence of a spontaneous polarization along the axis 4 and in a plane perpendicular to this axis can take place independently. However, a second-order phase transition from the group $4/mmm$ into m, for example (the occurrence of a spontaneous polarization in the plane of symmetry xz in a direction that does not coincide with the coordinate axes) is hardly probable. For this transition to take place, it is necessary that the coefficients $A_1(p,T)$ and $A_2(p,T)$ vanish at the same temperature and pressure at $\eta_1^2+\eta_2^2$ and η_3^2, respectively (η_3 is the order parameter corresponding to the polarization of the crystal along the fourfold axis). The two equations

$$A_1(p,T) = 0 \qquad \text{and} \qquad A_2(p,T) = 0 \tag{2.47}$$

determine not the line of phase transitions but an isolated point on the p, T diagram.

Finally, let us consider phase transitions with a spontaneous polarization taking place in a crystal of the cubic class $m3m$ (Fig. 2.7). This case differs from the one just considered in that here all the three components of the polarization vector are found to be completely equivalent, i.e., we are to introduce into our treatment three components of the order parameter, namely η_1, η_2, η_3. Acting on them by symmetry operations of the group $m3m$ (there are altogether 48 operations in this group), we detect that, for example, under rotations about fourfold axes the following transformation occurs:

$$\eta_1 \to -\eta_1, \pm\eta_2, \pm\eta_3; \quad \eta_2 \to -\eta_2, \pm\eta_1, \pm\eta_3; \quad \eta_3 \to -\eta_3, \pm\eta_1, \pm\eta_2.$$

Reasoning in a manner analogous to the example just considered, we arrive at the conclusion that the quantities η_1, η_2, η_3 must be included in the thermodynamic potential of the crystal in combinations that are not changed upon replacements, respectively:

$$\begin{aligned} &\eta_1 \quad \text{by} \quad -\eta_1, \pm\eta_2, \pm\eta_3 \\ &\eta_2 \quad \text{by} \quad -\eta_2, \pm\eta_3, \pm\eta_1 \\ &\eta_3 \quad \text{by} \quad -\eta_3, \pm\eta_1, \pm\eta_2. \end{aligned}$$

It is easy to check that such properties are displayed by the function

$$\begin{aligned} \Phi(T,\eta_1,\eta_2,\eta_3) = {} & \Phi_0(T) + \frac{1}{2}\alpha(T-T_c)(\eta_1^2+\eta_2^2+\eta_3^2) \\ & + \frac{1}{4}\beta_1(\eta_1^2+\eta_2^2+\eta_3^2)^2 + \frac{1}{2}\beta_2(\eta_1^2\eta_2^2+\eta_2^2\eta_3^2+\eta_3^2\eta_1^2) + \dots . \end{aligned} \tag{2.48}$$

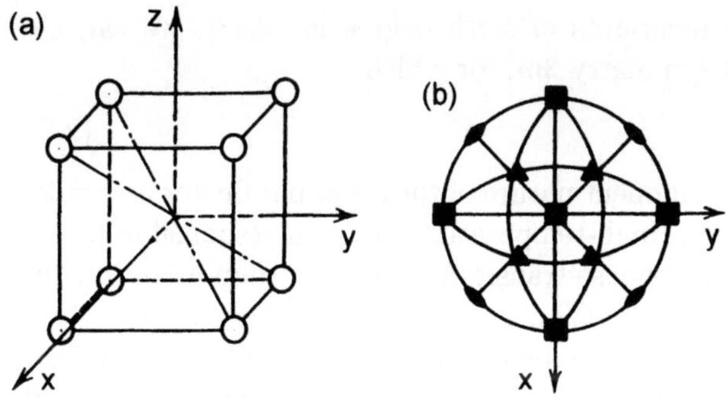

Fig. 2.7. (**a**) A structural element and (**b**) operations of the group of point symmetry of a cubic crystal of symmetry $m3m$. The group has 48 symmetry elements.

Here, as before, we will assume that α, β_1, $\beta_2 > 0$.

Minimizing the thermodynamic potential with respect to the variables η_1, η_2, η_3, we arrive at three equations that determine the equilibrium values of the components of the order parameter:

$$\frac{\partial \Phi}{\partial \eta_1} = 0, \qquad \frac{\partial \Phi}{\partial \eta_2} = 0, \qquad \frac{\partial \Phi}{\partial \eta_3} = 0. \tag{2.49}$$

The equations obtained have several different solutions, one of which, namely,

$$\eta_{01} = \eta_{02} = \eta_{03} = 0 \tag{2.50}$$

corresponds to the symmetrical phase $m3m$ and the others to different nonsymmetrical phases:

$$\eta_{01}^2 = \eta_{02}^2 = 0, \quad \eta_{03}^2 \neq 0$$

or

$$\eta_{01}^2 = \eta_{03}^2 = 0, \quad \eta_{02}^2 \neq 0 \tag{2.51}$$

or

$$\eta_{02}^2 = \eta_{03}^2 = 0, \quad \eta_{01}^2 \neq 0$$

to the tetragonal phase of symmetry $4mm$;

$$\eta_{01}^2 = \eta_{02}^2 \neq 0, \quad \eta_{03}^2 = 0$$

or

$$\eta_{01}^2 = \eta_{03}^2 \neq 0, \quad \eta_{02}^2 = 0 \tag{2.52}$$

or

$$\eta_{02}^2 = \eta_{03}^2 \neq 0, \quad \eta_{01}^2 = 0$$

to the rhombic phase of symmetry $mm2$.

If we take account of invariants of sixth degree in (2.48), we can also obtain a trigonal phase of symmetry $3m$, for which

$$\eta_{01}^2 = \eta_{02}^2 = \eta_{03}^2 \neq 0. \tag{2.53}$$

We see that the multicomponent nature of the order parameter associated with the high symmetry of a high-temperature phase is responsible for the possibility of the second-order phase transition from the group $m3m$ into one of the three nonsymmetric phases. Which one of these phases will actually be stable in a certain temperature range depends on the ratio of the coefficients of the thermodynamic potential. In a wide temperature range it is possible that each of the nonsymmetric phases becomes stable by turns. It is exactly this situation that we observe in $BaTiO_3$ crystals, in which the following sequence of crystalline phases is realized:

$$m3m \xrightarrow{120^\circ\,\mathrm{C}} 4mm \xrightarrow{0^\circ\,\mathrm{C}} mm2 \xrightarrow{-70^\circ\,\mathrm{C}} 3m.$$

Figures 1.18 and 1.19 showed the anomalies of physical properties associated with these transitions. Note that the phase transitions between low-symmetry phases are first-order transitions since for these phases the necessary principal condition of a second-order transition, which requires that the low-symmetry phase be represented as a minor distortion of the high-symmetry phase, is not fulfilled. The transitions $m3m \rightarrow 4mm$, $mm2$, $3m$ may be second-order transitions. As a matter of fact, the ferroelectric phase transition $m3m \rightarrow 4mm$ in $BaTiO_3$ crystals is also a first-order transition, though being close to a second-order transition. Concluding this chapter, it should be pointed out that a spontaneous polarization can exist only in 10 polar symmetry classes (Fig. 2.8), which are the subgroups of the limiting group ∞m. Therefore, the problem of searching for combinations of symmetry groups between which ferroelectric second-order transitions are possible, may seemingly be formulated in the following manner.

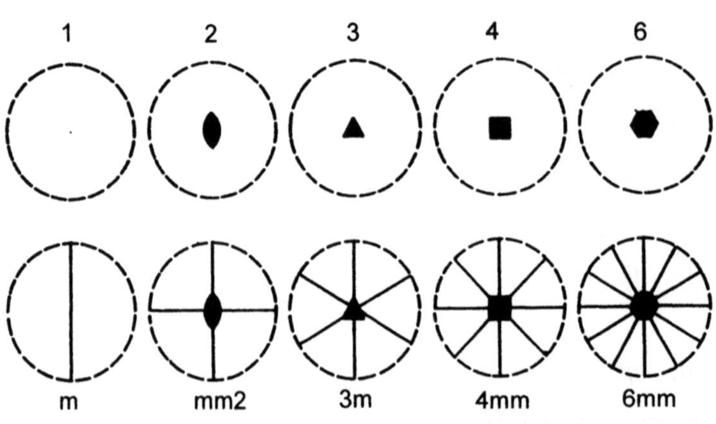

Fig. 2.8. Polar symmetry group of crystals allowing for a spontaneous polarization.

The symmetry group G_2 of an ordered low-symmetry phase contains all the symmetry elements that are common to the symmetric phase (group G_1) and the symmetry group of the polarization vector (symmetry group ∞m) at a given orientation of the latter relative to the symmetry elements of the group G_2. Thus, G_2 is the highest subgroup of the group G_1, which includes the common symmetry elements of G_1 and ∞m. Varying the direction of the ∞ axis relative to the symmetry elements of G_1, we can determine all possible symmetry changes in so-called proper ferroelectric phase transitions when the order parameter is a quantity proportional to the spontaneous polarization.

The results of such an approach, which makes it possible to determine the symmetry of polar phases in typical cases considered above are given in Table 2.3, the first row of which lists possible directions of the ∞ axis.

Table 2.3. Symmetry changes in ferroelectric phase transitions.

Original symmetry group	[001]	[111]	[110]	[$hk0$]	[hkk]	[hhl]	[hkl]
$2/m$, $2\|\|z$	2	1	m	m	1	1	1
$4/mmm$, $4\|\|z$	$4mm$	m	$mm2$	m	m	m	1
$m3m$, $4\|\|z$	$4mm$	$3m$	$mm2$	m	m	m	1

As can be seen, the results of this analysis are practically analogous to those obtained by means of the Landau theory discussed in this chapter; the only difference is that in Table 2.3 the equally probable variants are those in which a second-order phase transition can occur along the line of phase transitions and at isolated points on the p,T diagram. This method provides most of the observed ferroelectric phase transitions but not all of them as we shall see in Chap. 4.

3. Proper Ferroelectrics: Anomalies of Physical Properties in Phase Transitions

3.1 Anomalies of Thermal and Electrical Properties (One-Component Order Parameter)

Before we turn to the analysis of the main consequences of the theory of second-order phase transitions for ferroelectric crystals, we are going to discuss the conditions, the fulfilment of which leads us to believe that the inferences of the theory will be quantitative. The most important question is evidently the following: Is the representation of the nonequilibrium thermodynamic potential in the form of a power series of the order parameter at the point $T = T_c$ permissible? The answer to this question is known and it can be shown that this point is a singular point of the thermodynamic potential and the expansion coefficients vanish at this point or go to infinity. A special temperature dependence is also exhibited by the quantity $\Phi_0(p, T)$, which is the part of the thermodynamic potential containing the contributions of all degrees of freedom of a crystal, except η.

Thus, the theory described is not correct at the phase transition point proper and in its neighborhood. However, as a matter of fact, for ferroelectrics this neighborhood is very small and is very difficult to attain under the conditions of a real physical experiment. We will therefore use the expansion of the function $\Phi(\eta)$ into a series in η and near T_c; such an expansion of the coefficients $A(p, T)$ and $B(p, T)$ into a series in $T - T_c$ was made earlier, where only the first essential terms were used. It is clear that in moving away from the point of the phase transition to a temperature range of the order of a certain characteristic temperature it is necessary to take account of the other expansion terms as well. The most obvious characteristic temperature is T_c, which is why one may assume a priori that the adopted assumptions of the temperature dependence of the expansion coefficients are valid only in the region of $(T - T_c) \ll T_c$. In effect, this limitation arises in connection with the possibility of expanding the thermodynamic potential in a power series in η, which consists of a reasonably small number of terms.

We may thus believe that the inferences of the Landau theory will prove valid in a certain temperature range delimited both from below (by the singularity of the function $\Phi(p, T, \eta)$ at $T = T_c$) and from above (by possible limitations associated with the necessity of taking into account a larger number of terms in the expansion of the coefficients A and B in powers of $T - T_c$ and a larger number of terms in the expansion of the potential in η).

In order to find the equations of state, including the electrical one, i.e., the form of the function $P(E)$, we must substitute the equilibrium value of the order parameter into the thermodynamic potential (2.35). From (2.35) we have

$$P_2 = -\frac{\partial \Phi}{\partial E_2} = a\eta_0 \tag{3.1}$$

$$\chi_{22}^{p,T} = \frac{\partial P_2}{\partial E_2} = a\frac{\partial \eta_0}{\partial E_2}. \tag{3.2}$$

Note that, as follows from Chap. 1, the order parameter may be given the meaning of a certain displacement of an ion in the unit cell. The quantity a may therefore be treated as the density of the effective ionic charge. The equilibrium value of the order parameter can be determined by minimizing $\Phi(T, E_i, \eta)$ with respect to η:

$$\frac{\partial \Phi}{\partial \eta} = \alpha(T - T_c)\eta_0 + \beta\eta_0^3 - aE_2 = 0 \tag{3.3}$$

whence we find

$$\eta_0 = \eta_0(T - T_c, E_2). \tag{3.4}$$

At $E_2 = 0$ (the case of the absence of an external electric field)

$$\eta_0 = \begin{cases} 0 & \text{for} \quad T > T_c \\ \pm\left(-\frac{\alpha(T-T_c)}{\beta}\right)^{1/2} & \text{for} \quad T < T_c. \end{cases} \tag{3.5}$$

A stable solution corresponds to the condition $\partial^2\Phi/\partial\eta^2 > 0$; from (2.35) it follows that the first solution of (3.5) corresponds to the temperature region of $T > T_c$, i.e., to a symmetric phase, and the solution (3.5) to the condition $T < T_c$, i.e., to a nonsymmetric phase. Using (3.1), we then find that the equilibrium spontaneous polarization in the absence of the electric field is given by

$$P_s = \begin{cases} 0 & \text{for} \quad T > T_c \\ \pm a\left(-\frac{\alpha(T-T_c)}{\beta}\right)^{1/2} & \text{for} \quad T < T_c. \end{cases} \tag{3.6}$$

The derivative contained in (3.2) is easy to determine by using (3.3). Differentiating (3.3) with respect to E_2 , we obtain

$$\frac{\partial \eta_0}{\partial E_2} = \frac{a}{\alpha(T - T_c) + 3\beta\eta_0} = \begin{cases} a/[\alpha(T - T_c)] & \text{for} \quad T > T_c \\ -a/[2\alpha(T - T_c)] & \text{for} \quad T < T_c. \end{cases} \tag{3.7}$$

Substituting (3.7) into (3.2), we find the dielectric constant:

$$\varepsilon_{22}^{p,T} \approx 4\pi\chi_{22}^{p,T} = \begin{cases} 4\pi a^2/[\alpha(T - T_c)] & \text{for} \quad T > T_c \\ -2\pi a^2/[\alpha(T - T_c)] & \text{for} \quad T < T_c. \end{cases} \tag{3.8}$$

It can be seen that the dielectric constant of a crystal along the polar axis $\varepsilon_{22}^{p,T}$ depends on temperature according to the Curie–Weiss law; for $T > T_c$

$$\varepsilon_{22}^{p,T} = \frac{C_{C-W}}{T - T_c},$$

where $C_{C-W} = 4\pi a^2/\alpha$ is the Curie–Weiss constant.

The set of properties of a ferroelectric crystal near T_c described is characteristic of *proper ferroelectrics*.

The main feature of proper ferroelectrics is the proportionality between the order parameter and the component of the electric polarization vector, which follows from (3.1).

The corollaries of phenomenological theory pertaining to the temperature dependence of P_s and $\varepsilon_{22}^{p,T}$ have been thoroughly tested by experiment for a number of crystals; for TGS, for example, there have been obtained data indicating that the Curie–Weiss law is satisfied both above and below the phase transition point up to the immediate vicinity of T_c and that the relation $P_s \sim (T_c - T)^{1/2}$ holds for $T < T_c$ (Figs. 3.1 and 3.2).

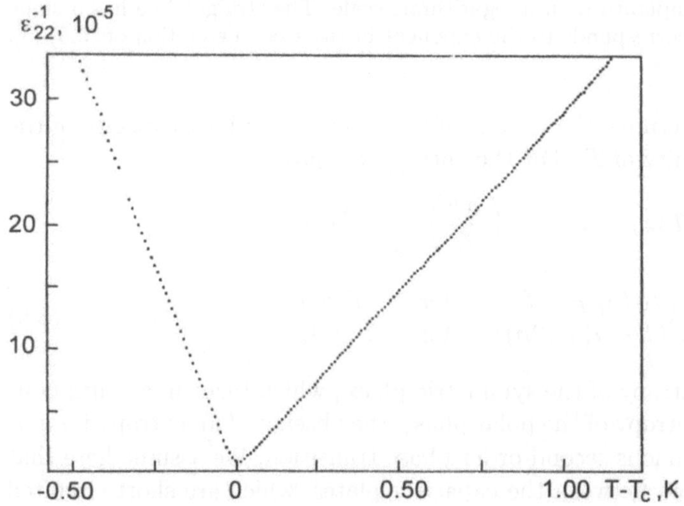

Fig. 3.1. Temperature dependence of the reciprocal dielectric constant along the polar axis of TGS [3].

We see that the slopes of the $\varepsilon_{22}^{-1}(T)$ plots are different on both sides of T_c (as follows from (3.8)) and the slope of $\ln P_s = f(\ln[T_c - T])$ near T_c is equal to 0.5 in keeping with the second equation of (3.6). The appreciable departures from this plot, which begin at $(T_c - T) > 1\,\mathrm{K}$, can easily be taken into account in considering the next term in the expansion of the thermodynamic potential in η.

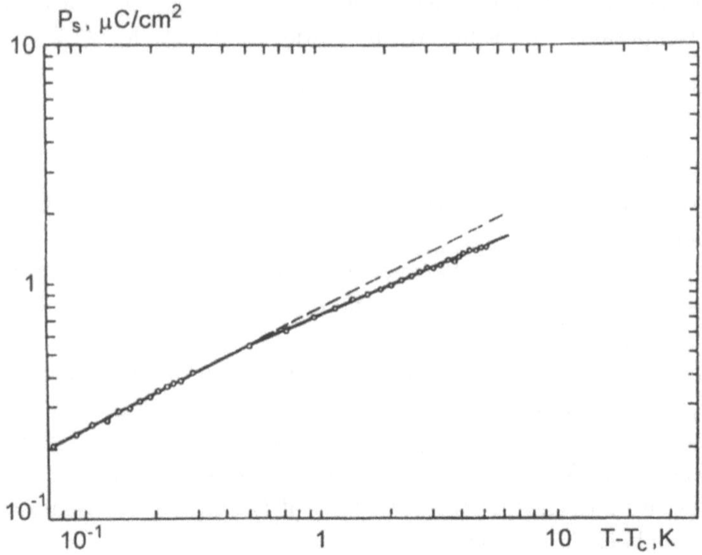

Fig. 3.2. Temperature dependence of the spontaneous polarization of TGS near the phase transition temperature on a logarithmic scale. The straight line has a slope close to 0.5, which corresponds to the exponent in the second equation of (3.5) [3].

Let us now determine the change of the entropy and heat capacity of the crystal in the vicinity of T_c. For the entropy we have

$$S = -(\partial\Phi_0/\partial T)_{p,E,\eta=\eta_0} = -\left(\frac{\partial\Phi}{\partial T}\right)_{p,E} - \frac{1}{2}\alpha\eta_0^2$$

$$= \begin{cases} -(\partial\Phi_0/\partial T)_{p,E} = S^0 & \text{for} \quad T > T_c \\ S^0 + \alpha^2(T - T_c)/(2\beta) & \text{for} \quad T < T_c\,, \end{cases} \tag{3.9}$$

where S^0 is the entropy of the symmetric phase, which turns in a continuous manner into the entropy of the polar phase; the absence of an entropy jump is the sign of a continuous second-order phase transition. We assume here that the crystal is placed between the capacitor plates, which are short-circuited (the electric field is equal to zero).

The heat capacity of the crystal is $C_{p,E} = T(\partial S/\partial T)_{p,E}$ and, hence, it is equal to

$$C_{p,E} = \begin{cases} T(\partial S^0/\partial T)_{p,E} = C^0_{p,E} & \text{for} \quad T > T_c \\ C^0_{p,E} + \alpha^2 T/2\beta & \text{for} \quad T < T_c. \end{cases} \tag{3.10}$$

At the phase transition point at $T = T_c$ the heat capacity at constant pressure in an electric field undergoes a jump by an amount $\alpha^2 T/(2\beta)$. Comparison of (3.10) with experiment shows (Fig. 3.3) that the heat capacity of the TGS crystal does experience a discontinuous change at the transition

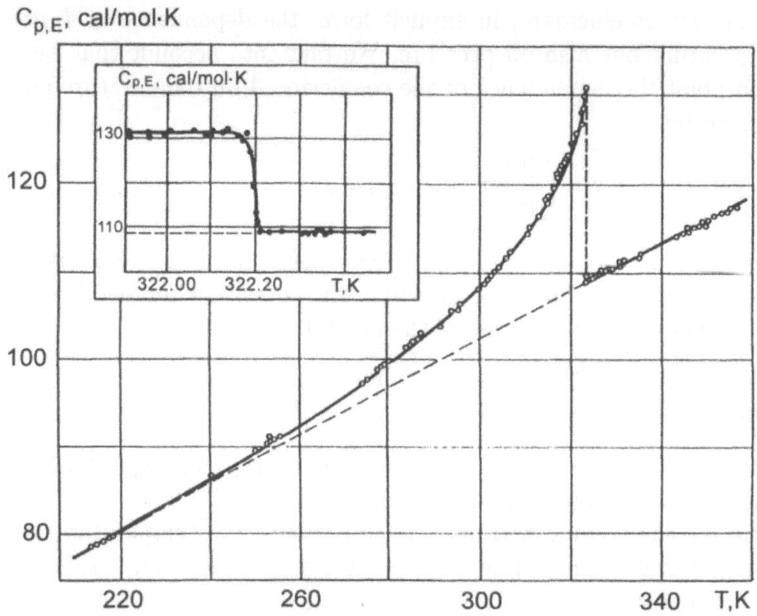

Fig. 3.3. Temperature dependence of the heat capacity of TGS in a wide temperature range and near T_c (inset) [4].

point, gradually decreasing to a "normal" value corresponding to the extrapolation of $C^0_{p,E}$ from the paraelectric phase. Here the λ-shaped character of the heat capacity anomaly in a wide temperature range and its rapid decline in moving away from T_c correlate with the experimental $P_s(T)$ plot (see Fig. 3.2).

It is easy to see that since all the three parameters, P_s, $\varepsilon_{22}^{p,T}$ and $C_{p,E}$, determined by us are expressed in terms of three constant coefficients in the expansion of the thermodynamic potential a, α and β, they are not independent, being interconnected near T_c by the relation

$$\frac{4\pi P_s^2}{(T_c - T)/T}(C_{p,T} - C^0_{p,T})^{-1} = 2C_{\text{C-W}}.$$

In the numerator of the left-hand side of this equation the value of P_s is taken at corresponding values of $(T_c - T)/T$ substituted into the denominator.

We will now consider the temperature dependence of another two important characteristics of the crystal – the volume compressibility

$$\zeta_{E,T} = 1/v(\partial^2\Phi/\partial p^2)_{E,T}$$

and the volume thermal expansion coefficient

$$K_E = -1/v(\partial^2\Phi/\partial p\partial T)_E.$$

Here v is the volume of the crystal.

To this end, let us elucidate, in explicit form, the dependence of Φ not only on temperature but also on pressure. We take into account that near the transition point the dependence of the coefficient A on temperature and pressure is given by

$$A = \left(\frac{\partial A}{\partial T}\right)_p (T - T_c) + \left(\frac{\partial A}{\partial p}\right)_T (p - p_c) + \dots .$$

Introducing the notation $(\partial A/\partial T)_p = \alpha_T$, $(\partial A/\partial p)_T = \alpha_p$ and taking into account that p_c is the pressure at which the transition temperature is equal to T_c, we find that the volume of the crystal is

$$v = \left(\frac{\partial \Phi}{\partial p}\right)_{E,T} = \begin{cases} (\partial \Phi_0/\partial p) = v_0 & \text{for} \quad T > T_c \\ v_0 + (2\alpha_p A/\beta) & \text{for} \quad T < T_c . \end{cases} \tag{3.11}$$

We see that the volume of the crystal (like its entropy and spontaneous polarization) is a continuous function of temperature at $T = T_c$.

Differentiating the volume with respect to pressure and temperature, we determine, respectively, the volume compressibility and the volume thermal expansion coefficients:

$$\zeta_{E,T} = \begin{cases} (1/v)(\partial v_0/\partial p) = \zeta^0_{E,T} & \text{for} \quad T > T_c \\ \zeta^0_{E,T} + 1/v(\alpha_p^2/\beta) & \text{for} \quad T < T_c \end{cases} \tag{3.12}$$

$$K_E = \begin{cases} -(1/v)(\partial v_0/\partial T) = K^0_E & \text{for} \quad T > T_c \\ K^0_E - 1/v(\alpha_p \alpha_T/\beta) & \text{for} \quad T < T_c . \end{cases} \tag{3.13}$$

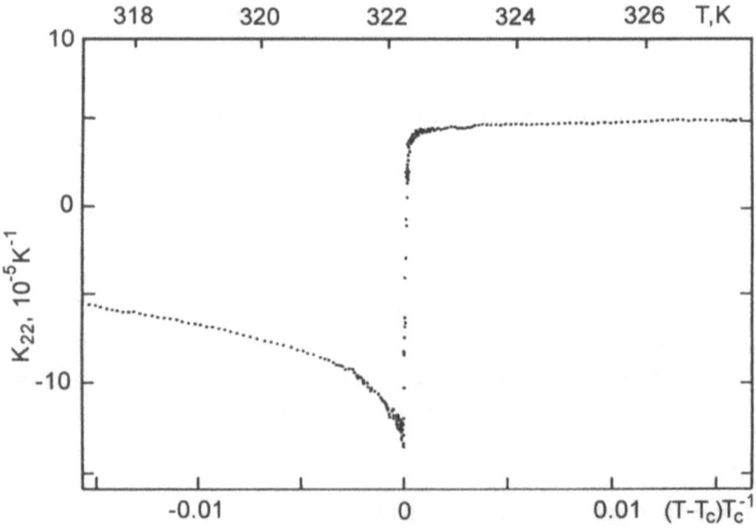

Fig. 3.4. Temperature dependence of the volume thermal expansion coefficient of TGS along the axis 2 in the vicinity of the Curie point [5].

It can be seen that at the phase transition point both coefficients experience a jump similar to the heat capacity jump (Figs. 3.4 and 3.5). A question may arise in connection with the dependence of the jumps on the volume of the crystal v. It should, however, be kept in mind that the thermodynamic potential of the crystal is proportional to the volume, which is why the coefficients in the expansion of the potential α and β may be considered to be also proportional to the volume; hence, the jumps of $\zeta_{E,T}$ and K_E are independent of the volume. It should also be noted that (3.12) and (3.13) describe the volume effects, whereas Figs. 3.4 and 3.5 give the temperature plots of definite components of the tensor of the thermal expansion coefficients and of the matrix of the elastic compliance. A more detailed account of crystal anisotropy will be given in Chap. 5.

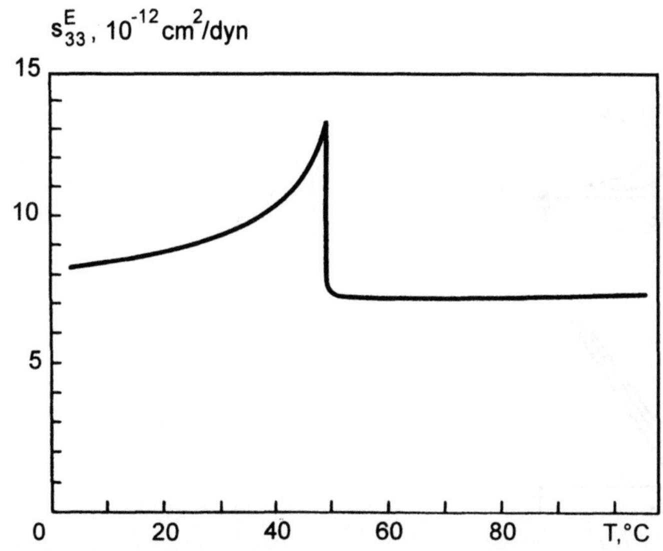

Fig. 3.5. Temperature dependence of elastic compliance s^E_{33} (longitudinal compression-extension along the z axis perpendicular to the polar axis 2) of TGS.

The curves in Figs. 3.4 and 3.5 show a characteristic discontinuous change of the parameter considered at the phase transition point, which is accompanied by certain additional singularities. Namely, in the vicinity of the point T_c the linear thermal expansion coefficient and the modulus of elastic compliance undergo anomalous changes, which exceed in magnitude those predicted by the theory expounded. These changes are presumably associated with the defects or with fluctuations of the order parameter and the possible singularity of the thermodynamic potential as a function of T at the point $T = T_c$. These effects will be discussed in Sect. 6.10.

From (3.4) it follows that the equilibrium value of the order parameter and, hence, of the spontaneous polarization, depends on the electric field applied. A crystal placed in the capacitor, in which the potential difference between the plates is nonzero and kept constant, will not undergo a symmetry change at $T = T_c$; the application of an external electric field results in a polarization of the crystal even in a symmetric phase and in a blurring of the phase transition.

To analyze the changes that occur in the temperature dependence of the thermodynamic characteristics of the crystal, let us fix the value of E_2 in (3.3). Taking E_2 as a parameter, we plot the dependence $\eta_0(T-T_c, E_2)$. The result of this graphical representation is given in Fig. 3.6a. Substituting then the corresponding values of $\eta_0(T-T_c, E_2)$ into (3.2), we obtain the results shown in Fig. 3.6b and 3.6c with account taken, respectively, of (3.7) and (3.9) and (3.10).

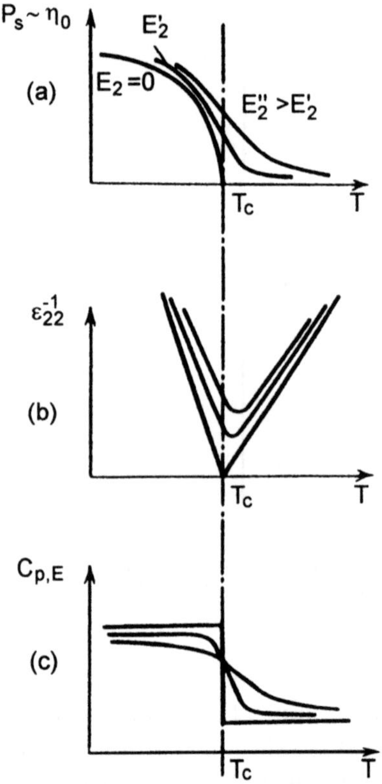

Fig. 3.6. (**a**) Temperature dependence of the spontaneous polarization $P_s \sim \eta$, (**b**) reciprocal dielectric constant ε_{22}^{-1} and (**c**) heat capacity $C_{p,E}$ in an external electric field $E_2 = 0$, E_2', $E_2'' \neq 0$ ($E_2'' > E_2'$).

On the other hand, by fixing the temperature $T - T_c$, we obtain from (3.3) the dependence of the equilibrium value of the order parameter and, hence, of the polarization on the electric field at the given temperature.

In the symmetric phase ($T > T_c$) this dependence is unambiguous, but near T_c it is essentially nonlinear (Fig. 3.7). In the nonsymmetric phase ($T < T_c$) (3.3) at not too large E_2 has three solutions, η_{01}, η_{02}, η_{03}

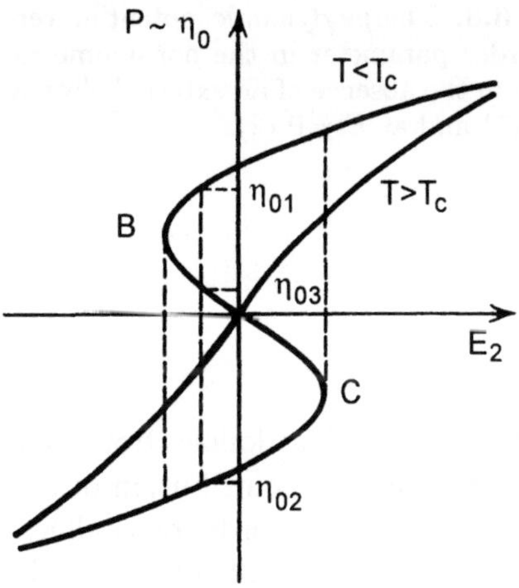

Fig. 3.7. The equilibrium value of the order parameter (spontaneous polarization) versus electric field (portion BC corresponds to unstable states).

(Fig. 3.7), one of which (η_{03}) corresponds to thermodynamically unstable states ($\partial^2\Phi/\partial\eta^2 < 0$, the BC line) and is not realized, and of the other two the one which corresponds to the lower value of thermodynamic potential, i.e., to its absolute minimum, is the equilibrium state.

With rather fast (as compared to the characteristic relaxation time of the system) changes in the electric field the crystal can be "retained" in a state with the value of the order parameter corresponding to the relative rather than the absolute minimum of Φ (Fig. 3.8). In this case there occur hysteresis phenomena characteristic of the ferroelectric phase (see Fig. 1.28). In real crystals the change of the direction of spontaneous polarization induced by an external electric field occurs not simultaneously throughout the volume of the crystal but by way of the origination and growth of regions with a "favorable" (with respect to the electric field) direction of the vector $\boldsymbol{P}_{\mathrm{s}}$. The very possibility of switching of polarization in experimentally attainable fields is in fact associated with the occurrence of a phase transition in these substances. Indeed, near T_{c} the potential barrier, which separates states with different polarization directions, is low and can be removed in this case even by a low electric field.

It should be stressed once again that the treatment of the properties of ferroelectric crystals given above refers to a temperature region sufficiently close to T_{c}. But if we are to describe the properties of crystals in a wide temperature range, it will be necessary to take account of some additional terms in the expansion of both the thermodynamic potential in η and of the coefficients at various powers of η in $T - T_{\mathrm{c}}$. This is occasionally done in describing experimental data. For example, account is taken of the term $-\frac{1}{2}\chi_0 E_2^2$ in (2.35). The polarization of the crystal will then be determined by the formula $P_2 = a\eta + \chi_0 E_2$, the second term of which on the right-hand side reflects the possibility of a contribution being introduced to the polarization not only by the order parameter but by other degrees of freedom as well. In this case, the experiment can be described much more accurately, but this is

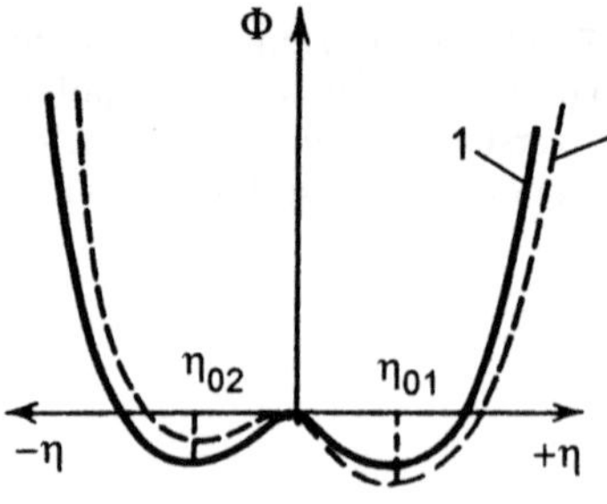

Fig. 3.8. Thermodynamic potential versus order parameter in the nonsymmetric phase in the absence of an external electric field (1) and at $E \neq 0$ (2).

not to be regarded as a refinement of the theory. Indeed, as follows from (3.3) and (3.7), $\eta_0 \propto E_2/(T - T_c)$ and, therefore, the relative contribution of this second term diminishes upon approach to T_c. But the same order of smallness in $T - T_c$ is also exhibited by terms that appear when account is taken of the temperature dependence of the coefficient a, the second term in the expansion of A in $T - T_c$, etc. Thus, taking account of the term proportional to E_2^2 and neglecting other terms having the same order of smallness is incorrect in general and must be regarded as inappropriate to the accuracy of the theory. At the same time, the importance of comparing the orders of smallnes of various terms is not usually emphasized. We can see that in the above treatment all terms of the same order of smallness in $T - T_c$ are taken into account in the thermodynamic potential. Indeed, A, $\eta_0^2 \propto |T - T_c|$; hence, the terms $A\eta_0^2$ and $B\eta_0^4$ are of the same order of magnitude and are proportional to $|T - T_c|^2$. But the term proportional to η_0^6 will have a higher order of smallness.

3.2 Anomalies of Electrical Properties (Multicomponent Order Parameter)

We will now be concerned with dielectric anomalies in proper ferroelectric phase transitions described by a multicomponent order parameter. We restrict our consideration to the case where a spontaneous polarization arises in a plane perpendicular to the axis 4 in a crystal having a $4/mmm$ symmetry (see Chap. 2). As has already been shown, here two phases of symmetry $mm2$ can appear. In order to describe dielectric properties, let us introduce into the thermodynamic potential (2.3) a mixed invaraint $a(\eta_1 E_1 + \eta_2 E_2)$, which takes into account that the components of the order parameter exhibit transformation properties, such as those shown by the corresponding components of the electric field. So, the expression for the thermodynamic potential takes on the following form:

$$\Phi = \Phi_0 + \frac{1}{2}\alpha(T - T_c)(\eta_1^2 + \eta_2^2) + \frac{1}{4}\beta_1(\eta_1^2 + \eta_2^2)^2 + \frac{1}{2}\beta_2\eta_1^2\eta_2^2 - a(\eta_1 E_1 + \eta_2 E_2). \tag{3.14}$$

The equilibrium values of the components of the order parameter are now determined by the equations

$$\frac{\partial \Phi}{\partial \eta_1} = \alpha(T - T_c)\eta_{01} + \beta_1 \eta_{01}(\eta_{01}^2 + \eta_{02}^2) + \beta_2 \eta_{01}\eta_{02}^2 - aE_1 = 0 \quad (3.15)$$

$$\frac{\partial \Phi}{\partial \eta_2} = \alpha(T - T_c)\eta_{02} + \beta_1 \eta_{02}(\eta_{01}^2 + \eta_{02}^2) + \beta_2 \eta_{01}^2\eta_{02} - aE_2 = 0. \quad (3.16)$$

The components of the dielectric susceptibility tensor χ_{11}, χ_{22} and χ_{12} can be calculated by analogy with the case of the one-component order parameter:

$$\chi_{ij} = \frac{\partial P_i}{\partial E_j} \quad (3.17)$$

$$\chi_{11} = \frac{\partial P_1}{\partial E_1} + \frac{\partial P_1}{\partial \eta_1}\frac{\partial \eta_1}{\partial E_1} + \frac{\partial P_1}{\partial \eta_2}\frac{\partial \eta_2}{\partial E_1} \quad (3.18)$$

$$\chi_{22} = \frac{\partial P_2}{\partial E_2} + \frac{\partial P_2}{\partial \eta_1}\frac{\partial \eta_1}{\partial E_2} + \frac{\partial P_2}{\partial \eta_2}\frac{\partial \eta_2}{\partial E_2} \quad (3.19)$$

$$\chi_{12} = \frac{\partial P_1}{\partial E_2} + \frac{\partial P_1}{\partial \eta_1}\frac{\partial \eta_1}{\partial E_2} + \frac{\partial P_1}{\partial \eta_2}\frac{\partial \eta_2}{\partial E_2}. \quad (3.20)$$

Since

$$P_1 = -\frac{\partial \Phi}{\partial E_1} = a\eta_1 \quad \text{and} \quad P_2 = -\frac{\partial \Phi}{\partial E_2} = a\eta_2 \quad (3.21)$$

we obtain

$$\chi_{11} = a\frac{\partial \eta_{01}}{\partial E_1}, \quad \chi_{22} = a\frac{\partial \eta_{02}}{\partial E_2}, \quad \chi_{12} = a\frac{\partial \eta_{01}}{\partial E_2}. \quad (3.22)$$

Using (3.15) and (3.16), we calculate the partial derivatives in (3.22):

$$\begin{aligned}\frac{\partial \eta_{01}}{\partial E_1} &= a[\alpha(T - T_c) + \beta_1(3\eta_{02}^2 + \eta_{01}^2) + \beta_2\eta_{01}^2] \\ &\quad \times \{[\alpha(T - T_c) + \beta_1(3\eta_{01}^2 + \eta_{02}^2) + \beta_2\eta_{02}^2] \\ &\quad \times [\alpha(T - T_c) + \beta_1(3\eta_{02}^2 + \eta_{01}^2) + \beta_2\eta_{01}^2] \\ &\quad - 4\eta_{01}^2\eta_{02}^2(\beta_1 + \beta_2)^2\}^{-1}\end{aligned} \quad (3.23)$$

$$\begin{aligned}\frac{\partial \eta_{02}}{\partial E_2} &= a[\alpha(T - T_c) + \beta_1(3\eta_{01}^2 + \eta_{02}^2) + \beta_2\eta_{02}^2] \\ &\quad \times \{[\alpha(T - T_c) + \beta_1(3\eta_{01}^2 + \eta_{02}^2) + \beta_2\eta_{02}^2] \\ &\quad \times [\alpha(T - T_c) + \beta_1(3\eta_{02}^2 + \eta_{01}^2) + \beta_2\eta_{01}^2] \\ &\quad - 4\eta_{01}^2\eta_{02}^2(\beta_1 + \beta_2)^2\}^{-1}\end{aligned} \quad (3.24)$$

$$\begin{aligned}\frac{\partial \eta_{01}}{\partial E_2} &= 2a\eta_{01}\eta_{02}(\beta_1 + \beta_2)\{[\alpha(T - T_c) + \beta_1(3\eta_{01}^2 + \eta_{02}^2) \\ &\quad + \beta_2\eta_{02}^2][\alpha(T - T_c) + \beta_1(3\eta_{01}^2 + \eta_{02}^2) + \beta_2\eta_{01}^2] \\ &\quad - 4\eta_{01}^2\eta_{02}^2(\beta_1 + \beta_2)^2\}^{-1}.\end{aligned} \quad (3.25)$$

We denote the denominator in (3.23)–(3.25) by Δ; then, for the components of the dielectric constant tensor we have

$$\varepsilon_{11} = 4\pi a^2[\alpha(T - T_c) + \beta_1(3\eta_{02}^2 + \eta_{01}^2) + \beta_2\eta_{01}^2]/\Delta \tag{3.26}$$

$$\varepsilon_{22} = 4\pi a^2[\alpha(T - T_c) + \beta_1(3\eta_{01}^2 + \eta_{02}^2) + \beta_2\eta_{02}^2]/\Delta \tag{3.27}$$

$$\varepsilon_{12} = -4\pi a^2\eta_{01}\eta_{02}(\beta_1 + \beta_2)/\Delta. \tag{3.28}$$

It is clear that in the symmetric phase ($T > T_c$) $\eta_{01} = \eta_{02} = 0$ and, hence, $\varepsilon_{11} = \varepsilon_{22}$ and $\varepsilon_{12} = 0$; the dielectric constant is isotropic in a plane perpendicular to the axis 4, being changed as $T \to T_c$ by the Curie–Weiss law:

$$\varepsilon_{11} = \varepsilon_{22} = \frac{4\pi a^2}{\alpha(T - T_c)}. \tag{3.29}$$

As indicated in Chap. 2, in the case under consideration two nonsymmetric ferroelectric phases are possible: phase I, for which $\eta_{01} = 0$, $\eta_{02}^2 = -\alpha(T - T_c)/\beta_1$ (or $\eta_{02} = 0$, $\eta_{01}^2 = -\alpha(T - T_c)/\beta_1$), and phase II, in which case $\eta_{01}^2 = \eta_{02}^2 = -\alpha(T - T_c)/(2\beta_1 + \beta_2)$.

Let us consider the manner in which the dielectric properties of the crystal are changed upon transition into phase I; we assume that the following solution is realized:

$$\eta_{01} = 0, \qquad \eta_{02}^2 = -\alpha(T - T_c)/\beta_1. \tag{3.30}$$

Substitution of (3.30) into (3.26), (3.27) and (3.28) yields for the dielectric constant as $E_2 \to 0$:

$$\varepsilon_{11} = \frac{4\pi a^2}{\alpha(T - T_c) + \beta_1(3\eta_{01}^2 + \eta_{02}^2) + \beta_2\eta_{02}^2} = -\frac{4\pi a^2\beta_1/\beta_2}{\alpha(T - T_c)} \tag{3.31}$$

$$\varepsilon_{22} = \frac{4\pi a^2}{\alpha(T - T_c) + \beta_1(\eta_{01}^2 + 3\eta_{02}^2) + \beta_2\eta_{01}^2} = -\frac{2\pi a^2}{\alpha(T - T_c)} \tag{3.32}$$

$$\varepsilon_{12} = 0. \tag{3.33}$$

It can be seen, first of all, that $\varepsilon_{11} \neq \varepsilon_{22}$, i.e., the dielectric constant displays an anisotropy associated with the lowering of the crystal's symmetry. Here, both along the direction of spontaneous polarization and in a direction perpendicular to it, the dielectric constant increases by the Curie–Weiss law as $T \to T_c$, but the constants have different values.

Upon transition into phase II ($\eta_{01}^2 = \eta_{02}^2 = \eta_0^2$) the components of the dielectric constant tensor assume the following form:

$$\varepsilon_{11} = \varepsilon_{22} = 4\pi a^2[\alpha(T - T_c) + \eta_0^2(4\beta_1 + \beta_2)]/\Delta \tag{3.34}$$

$$\varepsilon_{12} = -4\pi a^2\eta_0^2(\beta_1 + \beta_2)/\Delta. \tag{3.35}$$

Thus, the axes of the coordinate system chosen, which is natural for a tetragonal crystal, are no longer the major axes of the dielectric constant tensor in the case considered: the lowering of the symmetry into the group $mm2$ is now associated with the rotation of the major tensor axes by an angle of $\theta = \pi/4$ relative to the original coordinate system. The transformation to the coordinate system coupled with the major tensor axes can be accomplished by rotation of the coordinate system about the axis 4, whose position is not changed upon phase transition. The angle of rotation is defined as

$$\tan(2\theta) = 2\varepsilon_{12}/(\varepsilon_{22} - \varepsilon_{11}). \tag{3.36}$$

Since in phase II $\varepsilon_{22} = \varepsilon_{11}$, we have $\tan(2\theta) = \infty$ and $\theta = \pi/4$

The analysis of the dielectric properties for the case of a three-component order parameter is made in a manner quite analogous to the case considered, but with a different thermodynamic potential:

$$\begin{aligned}\Phi = \Phi_0 &+ \frac{1}{2}\alpha(T - T_c)(\eta_1^2 + \eta_2^2 + \eta_3^2) + \frac{1}{4}\beta_1(\eta_1^2 + \eta_2^2 + \eta_3^2)^2 \\ &+ \frac{1}{2}\beta_2(\eta_1^2\eta_2^2 + \eta_1^2\eta_3^2 + \eta_2^2\eta_3^2) - a(\eta_1 E_1 + \eta_2 E_2 + \eta_3 E_3).\end{aligned} \tag{3.37}$$

The potential (3.37) may be used to describe dielectric anomalies observed in ferroelectric transitions from a phase of symmetry $m3m$. As pointed out earlier, such a symmetry is exhibited by $BaTiO_3$ crystals. The analysis of the dielectric properties of these crystals is somewhat complicated because the transition from the cubic to a nonsymmetric tetragonal phase is a first-order transition close to second-order transitions.

3.3 First-Order Phase Transitions Close to Second-Order Transitions

Experiment shows that the great majority of known ferroelectric phase transformations are accompanied by a jump of the order parameter at the transition temperature, though in some cases it is small (as compared to a possible maximum jump for a given compound) (Fig. 3.9). The type of phase transition in a crystal depends, generally speaking, on external conditions. In a number of crystals, when the hydrostatic pressure is changed, the phase transition accompanied by a jump of spontaneous polarization at the transition temperature can change its parameters in a continuous way, assuming the features characteristic of second-order transitions. An analogous change in the character of phase transitions can also occur in isomorphous substitutions of ions with retention of the phase structure and symmetry. It has already been pointed out (see Chap. 1) that first-order phase transitions accompanied by a relatively small jump of the order parameter at $T = T_c$ and by a change of symmetry from a group to a subgroup, may be classified as belonging to a special class of first-order transitions close to second-order ones. They may be

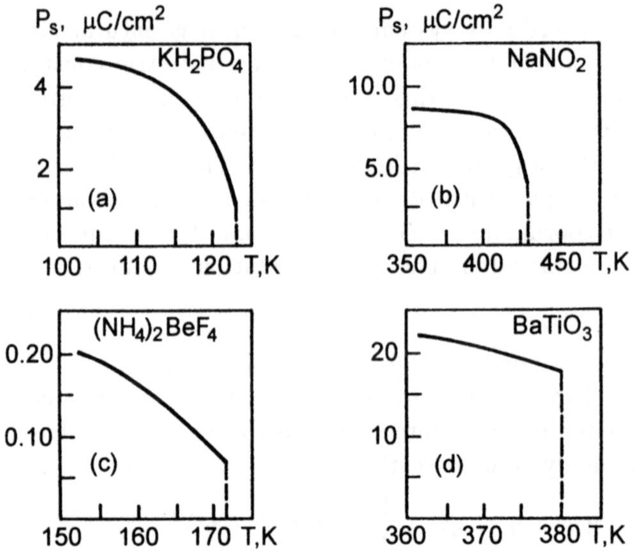

Fig. 3.9. Temperature dependence of spontaneous polarization for a number of crystals exhibiting a discontinuous variation of order parameter at the phase transition point.

described within the framework of the Landau theory but with account taken of higher-order terms with respect to the order parameter in the expansion of the thermodynamic potential.

On the p–T diagram the curve of second-order transition $A(p,T) = 0$, which separates phases of different symmetry, may change at a certain point p^*, T^* to a curve of first-order transitions. Below we will show that this will occur if the coefficient $B(p,T)$ at η^4 changes its sign. The point whose coordinates on the p–T diagram are determined by the equations

$$A(p^*,T^*) = 0 \quad \text{and} \quad B(p^*,T^*) = 0 \tag{3.38}$$

will be called the tricritical point[1] in what follows.

Let us now show that if in the expansion of $\Phi(p,T,\eta)$ the coefficient at the fourth-order term is $B(p,T) < 0$, then the phase transition is a first-order transition, i.e., an order parameter appears at the transition temperature in a jumpwise manner and there is observed a temperature hysteresis of the phase transition.

If $B(p,T) < 0$, for a crystal to be stable at sufficiently large η it is necessary to take into account the expansion term $C(p,T)\eta^6$ in expression (2.30), and then expression (2.33) will become

[1] In the literature, use is also made of the critical Curie point or the critical point of first-order transitions.

$$\Phi(p,T,\eta) = \Phi_0(p,T) + \frac{1}{2}\alpha(T - T_c)\eta^2 + \frac{1}{4}\beta\eta^4 + \frac{1}{6}\gamma\eta^6 \tag{3.39}$$

where $\gamma = 6C$; here α, $\gamma > 0$, $\beta < 0$.

The equilibrium value of the order parameter is found, as before, from the condition

$$\frac{\partial\Phi}{\partial\eta} = \eta_0[\alpha(T - T_c) + \beta\eta_0^2 + \gamma\eta_0^4] = 0. \tag{3.40}$$

In the nonsymmetric phase we have

$$\eta_0^2 = \frac{-\beta \pm \sqrt{\beta^2 - 4\alpha(T - T_c)\gamma}}{2\gamma} = -\frac{\beta}{2\gamma}\left\{1 \mp \left[1 - \frac{4\alpha(T - T_c)\gamma}{\beta^2}\right]^{1/2}\right\}. \tag{3.41}$$

We see that with $\beta < 0$ all the five roots of (3.40) are real in a certain temperature range. From the above discussion this temperature interval is bounded by a temperature which is determined from the condition of the equality of the radicand in expression (3.41) to zero. At this temperature

$$4\alpha\gamma(T_c^* - T_c)/\beta^2 = 1 \tag{3.42}$$

that is, $T_c^* = T_c + \beta^2/4\alpha\gamma > T_c$.

At the temperature $T < T_c$, (3.41) has only three roots. We are now going to consider in more detail how the form of the thermodynamic potential is changed with a change of temperature.

If the crystal is in the symmetric phase, the dependence $\Phi(\eta)$ with $T > T_c^*$ exhibits one minimum at $\eta = 0$. At $T = T_c^*$ this dependence displays points of inflection and upon further cooling there appear additional mimima (Fig. 3.10). The crystal, however, can remain in the symmetric phase unless the minimum at $\eta = 0$ disappears: this will occur at the temperature $T = T_c$ at which the state with $\eta = 0$ will become unstable and the crystal will pass over in a jumpwise manner into a state with the order parameter being of the order of $\eta_0^2(T_c) = -\beta/\gamma$.

Thus, the temperature T_c corresponds to the limit of the temperature range where the symmetric phase can exist. On the contrary, if the crystal is heated in the nonsymmetric phase, the latter can exist up to the temperature T_c. (The value of the order parameter at this temperature is $\eta_0^2 = -\beta/2\gamma$.) The difference $T_c^* - T_c$ determines the possible maximum temperature hysteresis of the first-order phase transition (Fig. 3.11).[2] In thermodynamics the phase transition point is the temperature T_0 at which the following condition is fulfilled:

$$\Phi(\eta = 0) = \Phi(\eta = \eta_{03}, \eta_{04}).$$

[2] Note that in real crystals the temperarures T_c and T_c^* are never attained even at considerable rates of temperature change: the experimentally observed hysteresis is always lower than the theoretical one since the nucleation of a new phase occurs on crystal defects and surfaces.

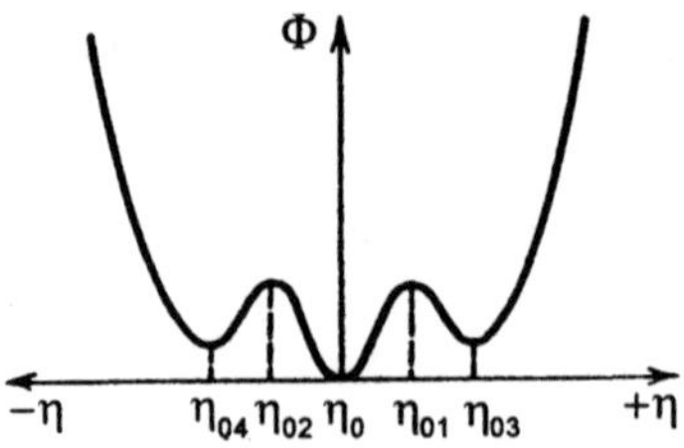

Fig. 3.10. Thermodynamic potential as a function of the order parameter in the region of temperatures where (3.40) has five real roots.

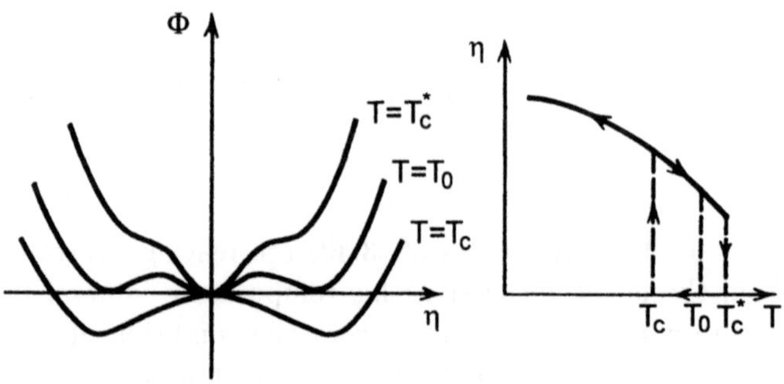

Fig. 3.11. Dependence of thermodynamic potential on order parameter and temperature dependence of order parameter in the region of a first-order phase transition close to a second-order transition.

Under real experimental conditions (not exactly equilibrium conditions) the first-order transition point can be determined with an accuracy of up to the value of the temperature hysteresis.

Evidently, upon approach to the tricritical point the temperature hysteresis of the phase transition drops to zero:

$$\frac{T_c^* - T_c}{T_c} = \frac{\beta^2}{4\alpha\gamma T_c} \to 0 \qquad \text{for} \quad \beta \to 0.$$

At the tricritical point the uncertainty in the position of the phase transition temperature disappears and the dependence $\eta_0(T)$ becomes continuous. At the first-order ferroelectric transition point the dielectric constant of the crystal does not tend to infinity but, obeying the Curie–Weiss law in the symmetric phase, it decreases discontinuously upon transition into the polar phase. Indeed, in this case, by analogy with (3.8) we have for the dielectric constant along the polar axis:

$$\varepsilon_{22}^{p,T} = 4\pi\chi_{22}^{p,T} = 4\pi a \frac{\partial \eta_0}{\partial E_2} \tag{3.43}$$

$$\frac{\partial \eta_0}{\partial E_2} = \frac{a}{\alpha(T - T_c) + 3\beta\eta_0^2 + 5\gamma\eta_0^4}. \tag{3.44}$$

Therefore, in the symmetric phase, where $\eta_0 = 0$,

$$\varepsilon_{22}^{p,T} = \frac{4\pi a^2}{\alpha(T - T_c)}. \tag{3.45}$$

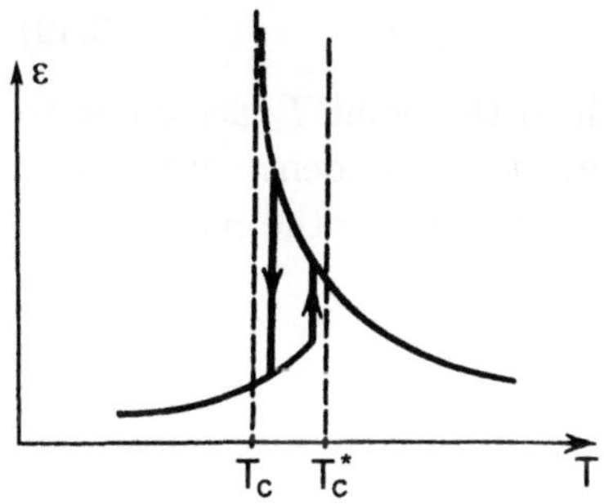

Fig. 3.12. Dielectric constant as a function of temperature in a first-order ferroelectric phase transition close to second-order transitions.

Since the temperature T_c is practically never attained upon a first-order transition, at a certain temperature $T > T_c$ there occurs a "breakaway" into the nonsymmetric phase, which is accompanied by a sharp decrease of $\varepsilon_{22}^{p,T}$ ($\eta_0^2 \propto -\beta/\gamma$) (Fig. 3.12).

An analogous "breakaway", but to the opposite side, occurs upon heating at a temperature somewhat lower than T_c^*.

It is precisely in such a manner that the dielectric constant of $BaTiO_3$ crystals, for example, depends on temperature near the transition from the cubic to the tetragonal phase (see Fig. 1.19).

For the dielectric properties of crystals in nonsymmetric phases to be analyzed in detail, we have to take into account sixth-degree invariants in the thermodynamic potential:

$$(\eta_1^2 + \eta_2^2 + \eta_3^2)^3 \tag{3.46}$$

$$\eta_1^4(\eta_2^2 + \eta_3^2) + \eta_2^4(\eta_1^2 + \eta_3^2) + \eta_3^4(\eta_1^2 + \eta_2^2) \tag{3.47}$$

$$\eta_1^2\eta_2^2\eta_3^2 . \tag{3.48}$$

This complication, however, is not critically essential: the most important features of the dielectric properties of $BaTiO_3$ crystals both upon transitions from the cubic to the tetragonal phase and in transitions between nonsymmetrical phases are described by this approach.

3.4 The Tricritical Point

As has already been pointed out, at the tricritical point the features of a first-order phase transition disappear, but a continuous transition of this type has certain features which distinguish it from ordinary second-order phase transitions. If we move to the tricritical point along the line $B(p,T) = 0$, then from the condition $\partial\Phi/\partial\eta = 0$ we obtain

$$\alpha(T - T_c) + \gamma\eta_0^4 = 0$$

that is

$$\eta_0 = \pm[-\alpha(T - T_c)/\gamma]^{1/4}. \tag{3.49}$$

A decrease in order parameter upon approach to the point T_c according to the law $(T_c - T)^{1/4}$ results in a strong temperature dependence of the heat capacity in the nonsymmetric phase: from (3.39) we have for the entropy with $T < T_c$:

$$S = S^0 + \alpha^{3/2}(T_c - T)^{1/2}/(2\gamma^{1/2}) \tag{3.50}$$

whence we get for the heat capacity

$$C_{p,E} = C^0_{p,E} + \frac{\alpha^{3/2}T}{4\gamma^{1/2}}(T_c - T)^{-1/2}. \tag{3.51}$$

Such a temperature dependence is displayed, with a high degree of accuracy, by the heat capacity of KH_2PO_4 crystals in the polar phase (Fig. 3.13) not too close to the transition point. The true tricritical point in these crystals is attained at a pressure of about 2 kbar; at normal pressure a small amount of latent heat is evolved at the transition point.

Figure 3.14 shows how the length of the KH_2PO_4 crystal is changed along the polar axis at various pressures. It can be seen that the jumps of the crystal volume and the temperature hysteresis of the phase transition decrease with increasing pressure and disappear at the tricritical point.

The existence of the tricritical point has been reliably established for a number of ferroelectric crystals (KH_2PO_4, SbSI, etc.); one may presume that the finding of tricritical points for other ferroelectric crystals with first-order phase transitions close to second-order transitions is associated with experimental difficulties, say with the necessity of attaining very high or negative pressures.

Let us now inquire into the origin of the term "tricritical point". It is known that the critical point on a p–T diagram is the point at which the curve

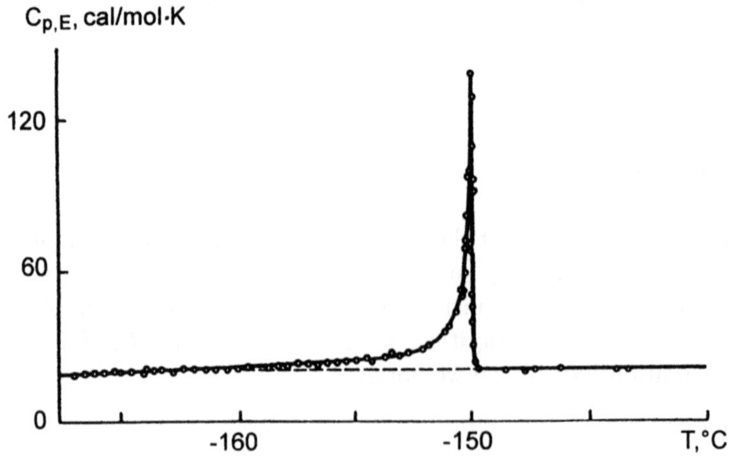

Fig. 3.13. Temperature dependence of the heat capacity of a KH_2PO_4 crystal [6].

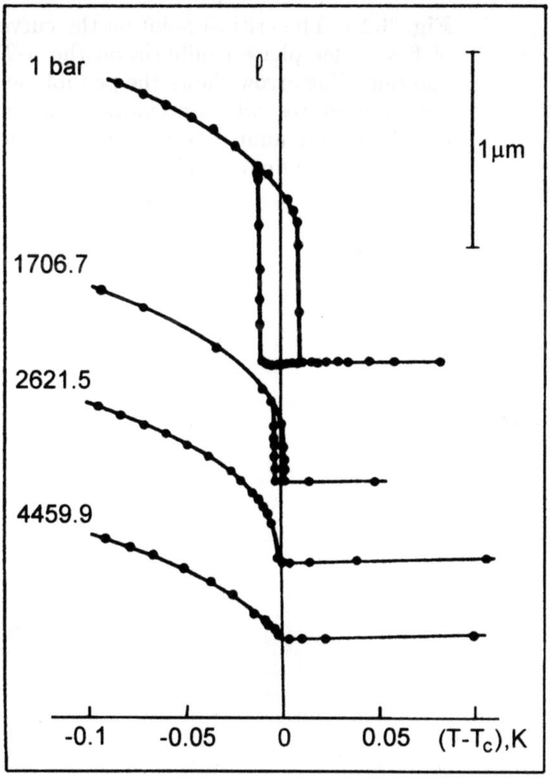

Fig. 3.14. Temperature dependence of the length of a KH_2PO_4 crystal in the region of the phase transition at various hydrostatic pressures [7].

of first-order phase transitions is terminated; in the presence of a critical point between any two states of a substance there can be accomplished a continuous transition – without separation into phases (the circular arrow in Fig. 3.15). Naturally, for phases of different symmetry there can be no critical point since it is difficult to visualize the "gradual" appearance of a symmetry element. However, for phases which are different only quantitatively, say for a liquid and a gaseous phase, which have different densities, there does exist a critical point. Figure 3.16 presents the isotherms of a liquid and a gas in the p–v plane; at $T = T_1 < T_{cr}$ there are portions on the isotherms corresponding to a metastable overheated liquid and overcooled vapors; the isotherms are discontinued at points c and d at which $\partial p/\partial v = 0$ and the conditions of thermodynamic stability are violated. If the overcooled vapor can be brought to point d at $T = T_2 < T_{cr}$, the system will pass over in a jumpwise manner into the liquid state. On the contrary, if the overheated liquid is found to be at point c, a jumpwise transition into the gaseous state will occur. With an equilibrium change of state the phase transition is represented by the line ab, which corresponds to the separation into two phases. At the critical point the

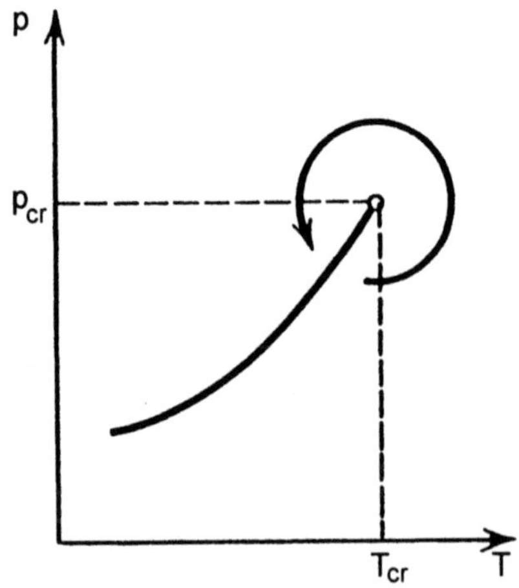

Fig. 3.15. The critical point on the curve of first-order phase equilibria on the p–T diagram. The arrow shows the possible sequence of states, which provides a continuous (without jumps) transition from the gaseous to the liquid state.

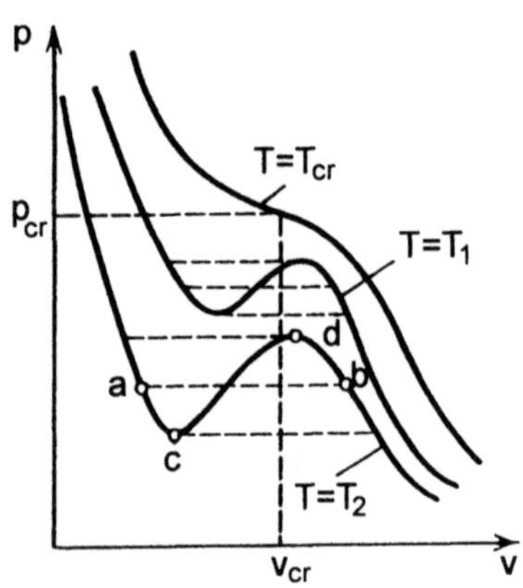

Fig. 3.16. The pressure–volume isotherm for the liquid–gas system $T = T_{cr}$, $T = T_1 < T_{cr}$ and $T = T_2$, $T_2 < T_1$.

jump of the volume upon liquid–gas transition disappears and the transition takes place continuously.

It is easy to discover a formal analogy between this picture of a gradual transformation of a first-order phase transition into a continuous one at the critical point in the liquid–gas system and phenomena that occur in a ferroelectric crystal in the nonsymmetriccal phase upon application of an electric field (Fig. 3.17). In this case we may regard the states of a crystal with oppositely oriented spontaneous polarization vectors as two phases. At $E = 0$ under equilibrium conditions the crystal is split into regions with different directions of the vector $\boldsymbol{P}$, this being an analogy with a two-phase state; the "phase transition" from the "upward-polarization" state ($+P_s$) to the "downward-polarization" state ($-P_s$) is represented by the line ab in Fig. 3.17. As the Curie point of the second-order transition T_c is approached, the portion ab diminishes (cf. Fig. 3.16) and vanishes at the point $T = T_c$ proper. It may be said that at this point the difference between the "phases" disappears and that the point of the second-order ferroelectric transition is

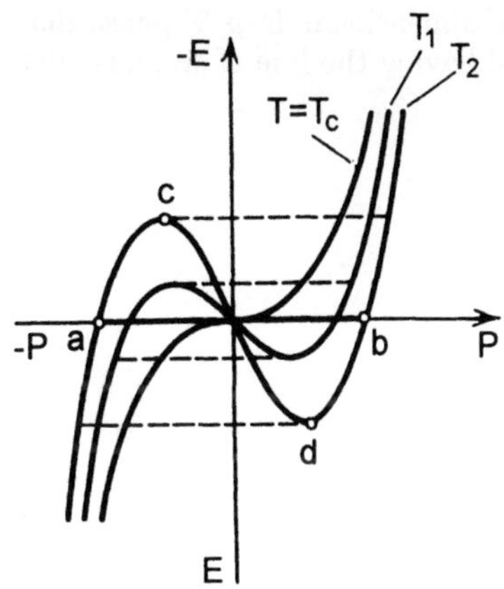

Fig. 3.17. Electric field-polarization isotherms for a ferroelectric crystal undergoing a second-order phase transition at $T = T_c$, $T = T_1 < T_c$ and $T = T_2$, $T_2 < T_1$.

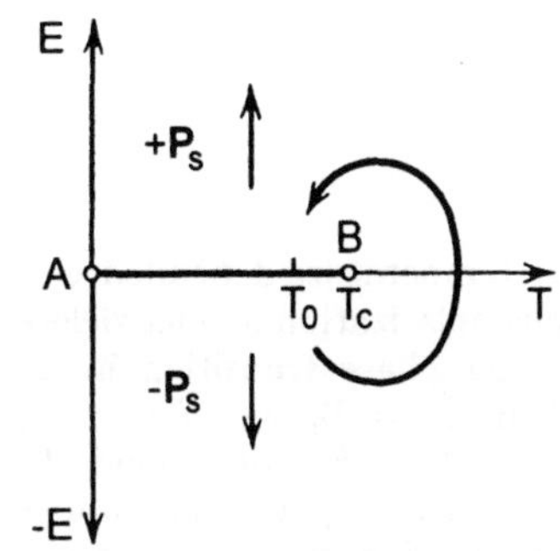

Fig. 3.18. The critical point of domain reorientation in a ferroelectric on the E–T diagram. The circular arrow indicates the succession of states, which provides a continuous transition from a phase polarized in a "negative" direction to a phase polarized in a "positive" direction.

the critical point on the E–T diagram (Fig. 3.18, cf. Fig. 3.15). The encircling of the critical point along the trajectory shown by a circular arrow makes it possible to avoid a discontinuous change of polarization (crossing the AB line) during the transition from the $-P_s$ state at T_c to the $+P_s$ state.

In fact, the diagram shown in Fig. 3.18 is the projection of the curve of second-order transitions on the E–T plane in the phase space E–p–T depicted in Fig. 3.19 at a certain fixed pressure p.

Let us now see how this diagram is changed if on the phase transition curve there is a point where the coefficient $\beta(p, T)$ changes sign, so that the line of second-order transitions turns continuously into a first-order transition curve. To do this, it is necessary to elucidate the effect of an electric field on a first-order phase transition. The temperature dependence of the equilibrium spontaneous polarization in various electric fields is given in Fig. 3.20; the positive and negative values of P_s correspond to domains of different orientations. We see that the jump of spontaneous polarization diminishes at the transition point in the electric field. At a certain critical value of the field E_{cr} the polarization jump disappears and in fields higher than E_{cr} the change of polarization and other thermodynamic parameters of the crystal as a function of temperature occurs in a continuous manner. In the field $E = E_{cr}$ there exists a temperature at which $dP/dT \to \infty$. At this temperature the heat

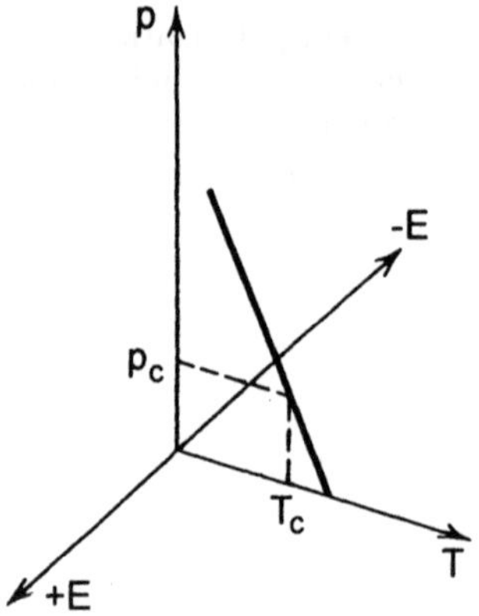

Fig. 3.19. Three-dimensional E–p–T phase diagram for a crystal having the line of second-order phase transitions.

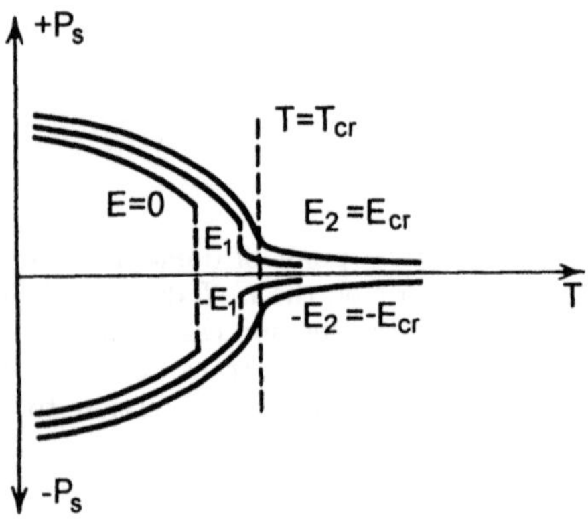

Fig. 3.20. Temperature dependence of spontaneous polarization in a ferroelectric first-order phase transition in an electric field $E = 0$, E_1, $E_2 \neq 0$; $E_2 > E_1$; the field E_2 corresponds to the electrical critical point. The upper and lower parts of the figure correspond to fields of different signs.

capacity of the crystal $C_{p,E}$ also tends to infinity. The point (E_{cr}, T_{cr}) on the E–T diagram is an electrical analog of the critical point.

Thus, in the E–p–T space, apart from the line of first-order phase transitions, there will also appear two lines of electrical critical points corresponding to different signs of E_{cr}; these lines are shown in Fig. 3.21 and converge at the tricritical point. On the other hand, this point is approached by the line of second-order transitions, which, as has been pointed out earlier, is also the line of critical points. So, three lines of critical points converge at the tricritical point.

In conclusion, it should be noted that, as a rule, ferroelectric phase transitions, both first- and second-order, appear to be close to the tricritical point. As we have found, from the standpoint based on the expansion of the thermodynamic potential in a power series in the order parameter, it means that the quantity $\beta^2/4\alpha\gamma$, which characterizes the magnitude of the temperature hysteresis for first-order transitions and the temperature region where $\eta_0^2 \propto (T_c - T)$ for second-order ones, is small as compared with T_c. As a result, all thermodynamic quantities that undergo a discontinuous change at $T = T_c$ in the nonsymmetric phase prove to be temperature dependent (see Fig. 3.3). For example, for the heat capacity we have

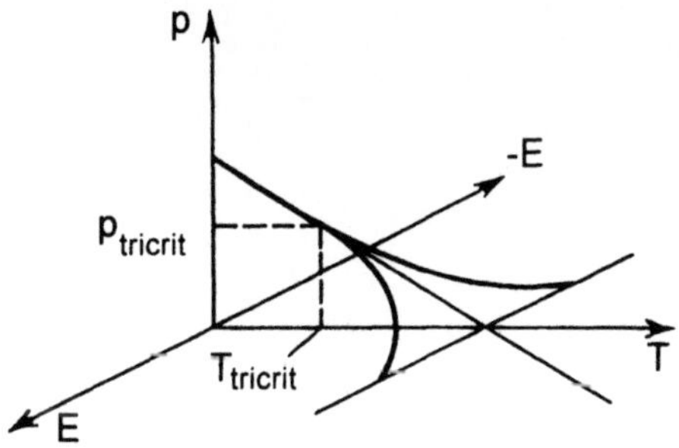

Fig. 3.21. The E–p–T phase diagram for a crystal exhibiting a tricritical point.

$$C_{p,E} = \frac{1}{2}\alpha T \frac{\partial \eta_0^2}{\partial T} \tag{3.52}$$

where η_0^2 is determined by (3.41). Differentiating with respect to temperature, we obtain near the phase transition temperature

$$C_{p,T} = C_0 + \frac{\alpha^2 T}{2\sqrt{\beta^2 - 4\alpha\gamma(T - T_c)}}. \tag{3.53}$$

From the last equation it is easy to obtain limiting cases: the second equation of (3.10) at $\gamma = 0$ and (3.51) at $\beta = 0$; besides, it can conveniently be used to determine the coefficients in the expansion of the thermodynamic potential on the basis of calorimetric data by representing them in the form

$$\left(\frac{C_{p,E} - C^0}{T}\right) = \left(\frac{C_{p,E} - C^0}{T}\right)_{T=T_c} - \frac{16\gamma}{\alpha^3}(T - T_c). \tag{3.54}$$

The first term on the right-hand side of this equation represents the inverse square of the heat capacity jump at the phase transition point (see the second equation of (3.9); the available experimental data confirm the presence of a temperature region in which the quantity $[(C_{p,E} - C^0)/T]^{-2}$ depends linearly on $T - T_c$. It is clear that for the coefficient α to be determined, we have to use data of dielectric measurements – the Curie–Weiss constant and to check the possibility of neglecting the temperature dependence of the other coefficients of the thermodynamic potential.

4. Dielectric Anomalies in Structural Nonferroelectric and Improper Ferroelectric Phase Transitions

4.1 Nonferroelectric Phase Transitions: Dielectric Anomalies

As was shown in the preceding chapter, the Landau theory allows one to successfully account for anomalies of the basic thermodynamic quantities for ferroelectric crystals in second-order transitions and also in first-order transitions close to second-order ones. While applying this theory to ferroelectrics, we presumed that the order parameter has the same transformation properties as the component of the electric field vector. As a result, it turns out that the equilibrium value of order parameter is proportional to the electric polarization of the crystal. However, there may exist ferroelectric phase transitions in which such a proportionality is not observed.

As a matter of fact, the phase transition is called ferroelectric if a spontaneous polarization occurs in the nonsymmetric phase. But it does not yet mean that it is proportional to the order parameter – in a phase transition a number of "parameters" may appear, whose symmetry properties do not describe in full measure symmetry changes that occur at the transition point. For example, in $BaTiO_3$ crystals upon transition from the cubic to the tetragonal phase, apart from a spontaneous polarization, there also occurs a spontaneous deformation. The phase transition parameter, however, is proportional exactly to the polarization of the crystal rather than to its deformation: if we simply deform the cubic phase, this will produce a tetragonal phase of symmetry $4/mmm$ and not $4mm$, which is displayed by the tetragonal phase of $BaTiO_3$. We say that a spontaneous deformation takes place in barium titanate as a result of a spontaneous polarization, being a second-order effect (the components of the deformation tensor u_{ij} and the polarization vector P_k are interconnected by a relation of the type $u_{ij} \propto P_k^2$ (see Chap. 5).

Naturally, a spontaneous polarization may also occur in a crystal as a second-order effect, accompanying a more complicated change of the crystal structure of a substance. In this case, the transition parameter will not be a quantity proportional to the polarization: if the original phase of the crystal is polarized in the required direction, the resulting phase will exhibit a different, higher symmetry than the one that appears in the phase transition.

In other words, the transition will not pass into a higher polar subgroup of the symmetry group of the original symmetric phase.

Ferroelectric phase transitions, for which the order parameter is proportional not to polarization but to another physical quantity having other transformation properties, are called *improper ferroelectric phase transitions* and substances with such transitions are referred to as *improper ferroelectrics*. Such ferroelectrics differ in a number of physical properties from ordinary, proper ferroelectrics considered in the preceding chapters. An example of an improper ferroelectric is gadolinium molybdate, $Gd_2(MoO_4)_3$. The characteristic features of improper ferroelectrics are a very slight change of the dielectric constant near the transition point and its unusual temperature dependence and also a low spontaneous polarization (Fig. 4.1). In all improper ferroelectrics known to date there is observed a multiple change of the unit cell volume upon transition into a nonsymmetric phase.[1]

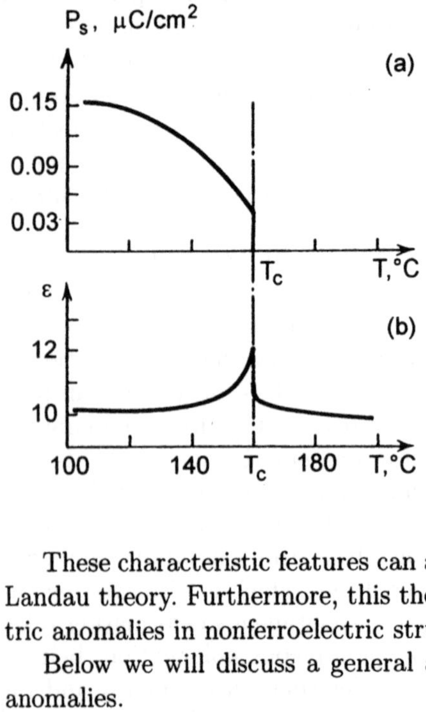

Fig. 4.1. (**a**) Temperature dependence of the spontaneous polarization and (**b**) dielectric constant of a $Gd_2(MoO_4)_3$ crystal along the polar axis [8]. Change of the space symmetry $D^3_{2d}(P\bar{4}2m) \rightarrow C^8_{2v}(Pba2)$.

These characteristic features can all be analyzed within the context of the Landau theory. Furthermore, this theory also enables one to describe dielectric anomalies in nonferroelectric structural phase transitions in crystals.

Below we will discuss a general approach to the treatment of dielectric anomalies.

We have already pointed out that in phenomenological theory the concrete physical meaning of the order parameter is of no significance; of im-

[1] From a theoretical viewpoint there are possible improper phase transitions without change of the number of atoms in the unit cell; for example, with a symmetry change $\bar{6}m2 \rightarrow 2$ (the unit-cell parameter is the axial vector in the basis plane) the phase transition is accompanied by a secondary electric polarization along the axis 2. No examples of this kind have, however, been experimentally observed so far.

portance are only its transformation properties, i.e., the manner in which it is transformed under symmetry operations of the symmetric phase. These transformation properties determine the form of the cross-order parameter-electric field terms in the thermodynamic potential. It is the form of these terms that determines the character of anomalies. The case where the order parameter is transformed as a vector component and the thermodynamic potential contains the term ηE_i has been considered above; this is a case of ordinary or proper ferroelectrics.

We will now be concerned with cross-terms of the types $\eta^2 E_i$, $\eta E_i E_j$, $\eta^2 E_i E_j$. Let us establish under what conditions they are permissible in the thermodynamic potential. So, terms of the type $\eta^2 E_i$ are allowable only if the crystallographic group of the symmetric phase corresponds to one of the pyroelectric classes, leaving the vector component E_i invariant under all transformations; this case is of no interest to us: now the substance is spontaneously polarized in the symmetric phase. Terms of the type $\eta E_i E_j$ are different from zero only with certain transformation properties of the order parameter η, namely, it is necessary that the parameter η be transformed as the product of the vector components.

Terms of the type $\eta^2 E_i E_j$ are present in the thermodynamic potential in all cases since the quantity η^2 is always invariant (η can turn only into η or $-\eta$) and quantities of the type $E_i E_j$ can always be used to form an invariant combination. For a tetragonal crystal, for example, the cross-terms have the form $\eta^2(E_1^2 + E_2^2)$ for a field lying in a plane perpendicular to the axis 4 and $\eta^2 E_3^2$ for a field directed along it. Let us consider the latter case in more detail. The parameter η may have any physical meaning here, i.e., it characterizes any nonferroelectric phase transition. The thermodynamic potential will assume the following form:

$$\begin{aligned} &\Phi(p,T,\eta) \\ &= \Phi_0(p,T) + \frac{1}{2}\alpha(T-T_c)\eta^2 + \frac{1}{4}\beta\eta^4 - \frac{1}{2}a\eta^2 E_3^2 - \frac{1}{2}\chi_{33}^0 E_3^2 \end{aligned} \tag{4.1}$$

where the invariant E_3^2 which is quadric with respect to the field has also been taken into account. It is easy to see that, in contrast to the case dealt with in Chap. 3, here this invariant is a term of the same order of smallness as the first mixed invariant $\eta^2 E_3^2$: the point is that the proportionality coefficient between η and E_3 remains finite as $T \to T_c$.

The equilibrium value of the order parameter is determined by the condition

$$\frac{\partial \Phi}{\partial \eta} = \eta_0[\alpha(T-T_c) + \beta\eta_0^2 - a\eta_0 E_3^2] = 0 \tag{4.2}$$

and the polarization is given by

$$P_3 = -\frac{\partial \Phi}{\partial E_3} = (a\eta_0^2 + \chi_{33}^0)E_3. \tag{4.3}$$

In the case under consideration the polarization along the z axis is nonzero only if an external electric field is applied in the same direction. This is what implies that the phase transition accompanied by the appearance of an order parameter

$$\eta_0^2 = -[\alpha(T - T_c) - aE_3^2]/\beta \tag{4.4}$$

is not ferroelectric. An example of such a distortion of the crystal lattice is given in Fig. 4.2. An orthorhombic crystal of symmetry mmm experiences at the phase-transition point a distortion involving a uniform shear deformation of the entire crystal and of each of the unit cells about the z axis; here, all symmetry elements disappear, except the axis 2 along z, the plane of symmetry perpendicular to it and the inversion center. It is obvious that in this case the order parameter coincides in its transformation properties with the corresponding component of the strain tensor u_6; below is given a table of transformations of the quantities u_6 and E_3 under the action of the symmetry elements of the group mmm (Table 4.1).

Table 4.1. Transformations of u_6 and E_3 under the action of the symmetry elements of the group mmm.

mmm	1	2_x	2_y	2_z	m_{xy}	m_{xz}	m_{yz}	$\bar{1}$
$\eta \propto u_6$	η	$-\eta$	$-\eta$	η	η	$-\eta$	$-\eta$	η
E_3	E_3	$-E_3$	$-E_3$	E_3	$-E_3$	E_3	E_3	$-E_3$

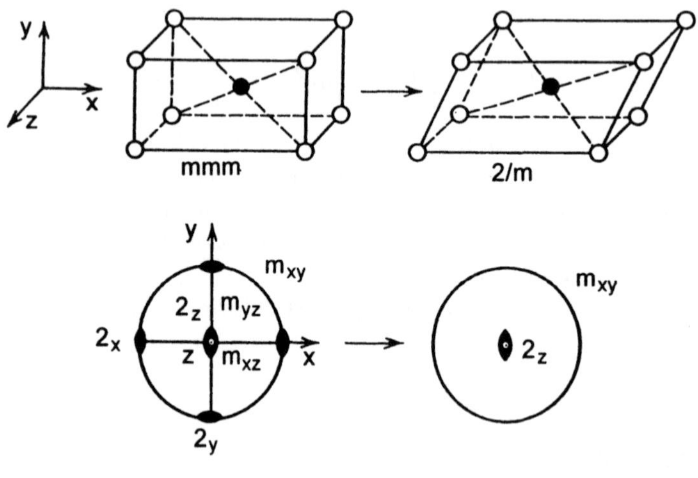

Fig. 4.2. Schematic representation of the distortion of the crystal lattice with a symmetry change $mmm \to 2/m$ described by a thermodynamic potential of the type (4.1).

It can be seen that only the combination $\eta^2 E_3^2$, which has been taken into account in the potential (4.1), is invariant. Let us analyze anomalies of the dielectric properties of a crystal in such phase transitions. From (4.3) we have

$$\varepsilon_{33} = 4\pi \frac{\partial P_3}{\partial E_3} = 4\pi a \eta_0^2 + \varepsilon_{33}^0, \qquad \varepsilon_{33}^0 = 4\pi\chi_{33}^0. \tag{4.5}$$

Since $\eta_0^2 = -\alpha(T - T_c)/\beta(E_3 \sim 0)$ with $T < T_c$, we obtain

$$\varepsilon_{33} = \begin{cases} \varepsilon_{33}^2 & \text{for} \quad T > T_c \\ \varepsilon_{33}^0 - 4\pi a\alpha(T - T_c)/\beta & \text{for} \quad T < T_c. \end{cases} \tag{4.6}$$

As can be seen from (4.6), at the point of the nonferroelectric second-order phase transition the dielectric constant experiences a kink (Fig. 4.3).

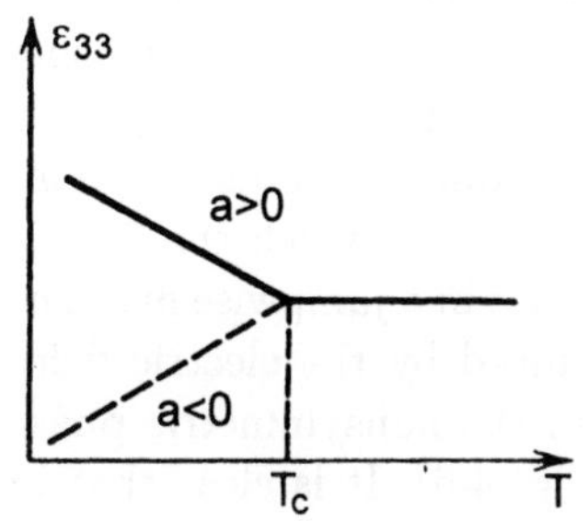

Fig. 4.3. Temperature dependence of the dielectric constant in a second-order nonferroelectric phase transition described by the one-component order parameter.

From (4.4) it also follows that the electric field can shift the phase-transition point:

$$\eta_0^2 = -\alpha(T - T_c')/\beta, \qquad \text{where} \quad T_c' = T_c + \frac{a}{\alpha} E_3^2.$$

It is essential that in this case the dependence $\eta_0(T)$ is shifted as a whole under the action of the electric field (Fig. 4.4). The direction of the shift depends on the sign of the constant a. In the case of a first-order phase transition, when the order parameter undergoes a jump, the dielectric constant

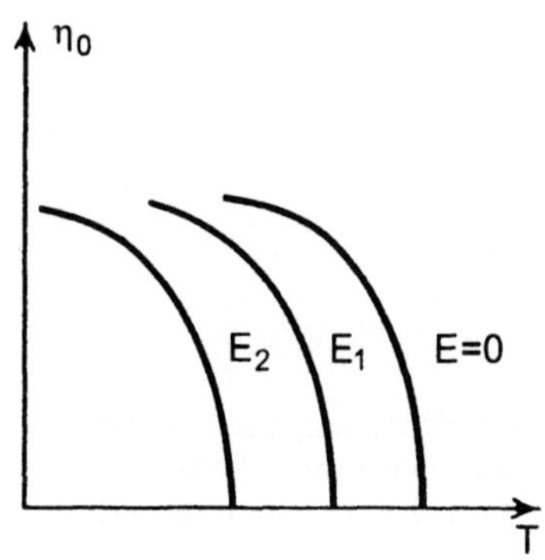

Fig. 4.4. Temperature dependence of the order parameter in a phase transition of the nonferroelectric type at various values of the external electric field $(a > 0)$ E_1, $E_2 \neq 0$; $E_2 > E_1$.

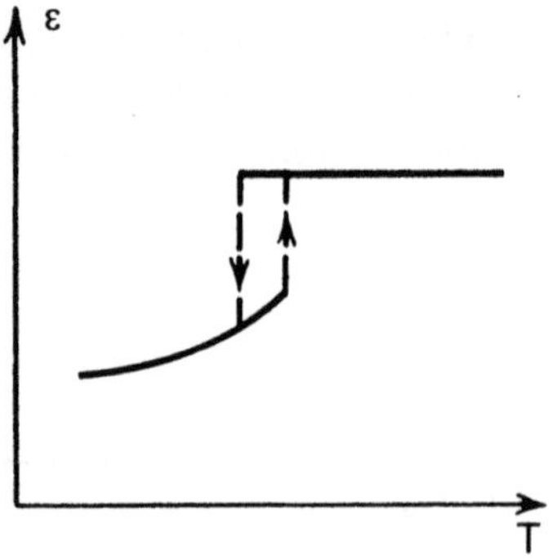

Fig. 4.5. Temperature dependence of the dielectric constant near the point of a first-order nonferroelectric phase transition close to second-order transitions.

also changes discontinuously. From (4.5) and (3.41) it follows that with $a < 0$ the temperature dependence of the dielectric constant has the form shown in Fig. 4.5. Since with $a < 0$ the transition temperature falls off with increasing field, upon application of a sufficiently large electric field it may be expected that in the nonsymmetric phase near the transition point the transition temperature will eventually coincide with the temperature at which the field is applied and the field-induced polarization will increase in a jumpwise manner: there will occur a first-order phase transition induced by the electric field. The dependence of polarization on electric field in the nonsymmetric phase will have the form of a double hysteresis loop (Fig. 4.6). It is clear that in moving away from the transition point the petals of the loop are shifted toward higher electric fields. The hysteresis character of the dependence $P(E)$ is evidently of the same nature as the temperature hysteresis of the transition point.

The anomaly of ε and the presence of a double hysteresis loop on the $P(E)$ plot in a nonsymmetric phase are sometimes considered to be the features of a certain group of substances called *antiferroelectrics.* These specific features are often associated with the character of a structural reconstruction in a phase transition, with a so-called "antipolar" ordering (see Fig. 1.6). From our treatment it follows that with a phenomenological approach the characteristic

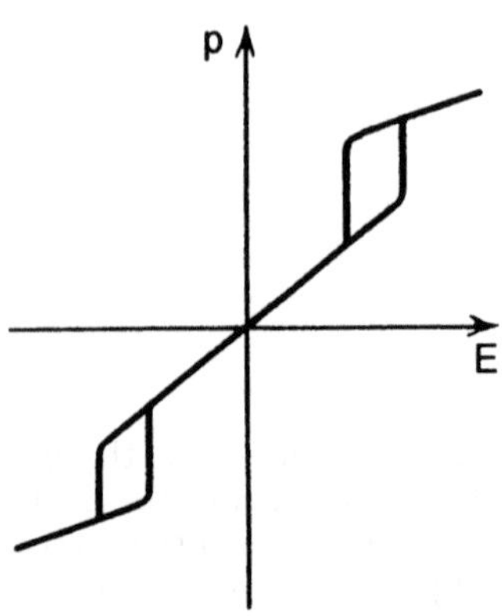

Fig. 4.6. Plot of polarization against electric field in the nonsymmetric phase for a first-order nonferroelectric phase transition close to second-order transitions.

anomalies of ε and the double hysteresis loops in a nonsymmetrical phase appear without reference to the specific structure of a substance; it will suffice only to assume that one of the constants of the thermodynamic potential has a certain sign.

One should bear in mind that there are examples of crystals which display the following interesting properties: upon transition from a symmetrical to a nonsymmetric phase the dielectric constant of a substance behaves just as in the case of proper ferroelectrics, i.e., it obeys the Curie–Weiss law. In a nonsymmetrical phase, which may also be a centrosymmetric phase, the application of a relatively low external electric field brings about a discontinuous increase of electric polarization, this being an indication of the field-induced transition into a new phase, which is different from both phases realized in the absence of an electric field. Here, a new phase appears and disappears at different values of electric field, so that the $P(E)$ plot exhibits a double hysteresis loop similar to that depicted in Fig. 4.6.

It is natural to presume that the properties similar to those described above may be associated with the fact that the thermodynamic potential of a nonsymmetric phase is only slightly lower than that of the ferroelectric phase; in an electric field, when it appears to be higher than a certain critical value, the ferroelectric state becomes more stable than the initial nonsymmetric state. The first-order transition that occurs in this case involves a discontinuous change of the crystal polarization.

A typical example of crystals that exhibit such properties is lead zirconate, $PbZrO_3$, the projection of whose structure is shown in Fig. 1.20. The temperature dependence of the dielectric constant of lead zirconate is given in Fig. 4.7 and the polarization vs. electric field plot is shown in Fig. 4.8. It can be seen that the anomalous hysteresis loops display a tendency toward saturation, just as in the case of normal loops for proper ferroelectrics.

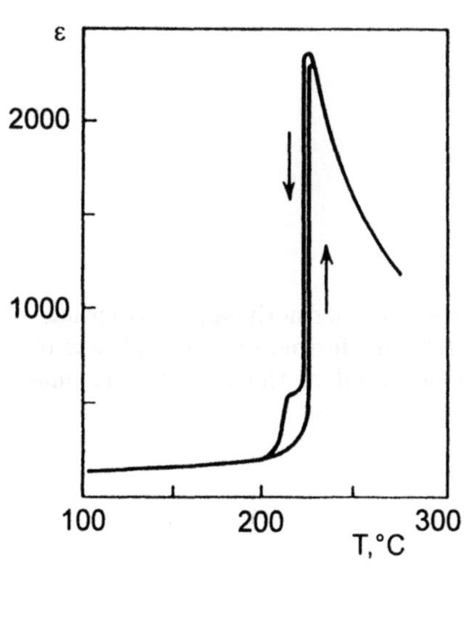

Fig. 4.7. Temperature dependence of the dielectric constant of a $PbZrO_3$ crystal ($T_c = 230\,K$) [9].

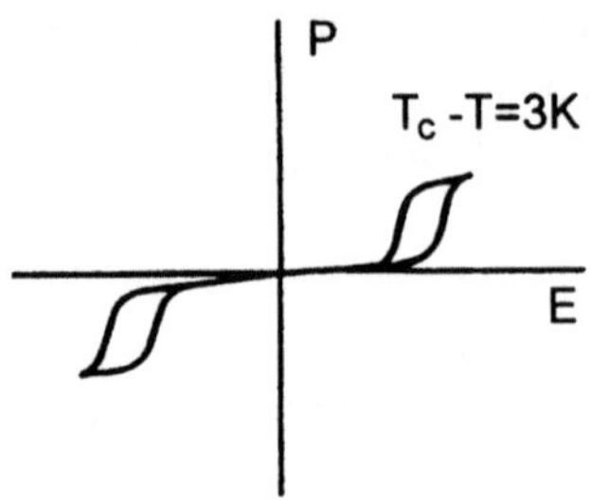

Fig. 4.8. Plot of polarization against electric field for a $PbZrO_3$ crystal in the nonsymmetric phase [9].

The sequence of phase transitions in a crystal of lead zirconate can be qualitatively understood if we assume that the dependence of the specific thermodynamic potentials (per unit volume) of all three phases looks like the one shown in Fig. 4.9. One can see that in the absence of an electric field ($E = 0$) the phase sequence $1 \to 2$ is realized with a decrease of temperature. With $E > E_{cr}$ in a certain neighborhood of T_c the specific thermodynamic potential of the ferroelectric phase is found to be lowest and, accordingly, this phase appears to be stable. The phase sequence in the field is $1 \to 3$.

Note that the term "antiferroelectric" is often used in reference to the experimental situation described above, which is characteristic of $PbZrO_3$ crystals and which involves a well-pronounced anomaly of the dielectric constant described, just as in the case of ferroelectrics, by the Curie–Weiss law, double hysteresis loops on the $P(E)$ plot in a nonsymmetric phase and the presence of superstructure in a nonsymmetric phase.

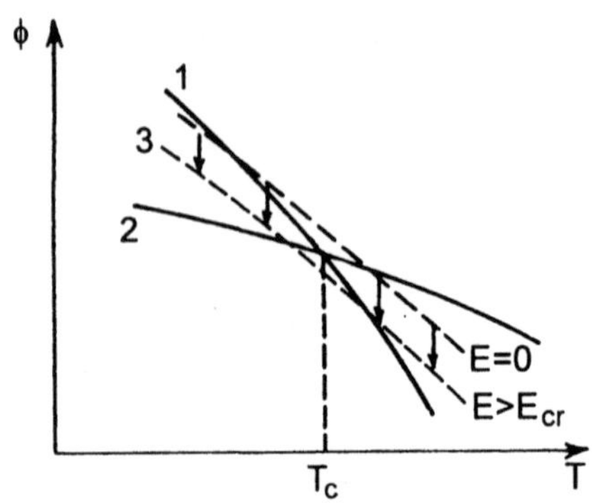

Fig. 4.9. Schematic representation of the specific thermodynamic potentials of the symmetric (*1*), nonpolar nonsymmetric (*2*) and ferroelectric (*3*) phases of a $PbZrO_3$ crystal. The dashed line shows the potential of the ferroelectric phase, which is decreased by the electric field.

4.2 Improper Ferroelectric Phase Transitions: Dielectric Anomalies

We will now analyze the properties of improper ferroelectrics. It is clear that in order to describe ferroelectrics, including improper ones, it is necessary that the thermodynamic potential contain an order parameter-electric field cross-term linear in the electric field. In all other cases the polarization is proportional to any power of the electric field and vanishes as soon as the field is removed.

Of all the variants with a one-component order parameter considered above none satisfies this requirement (except for the case of a proper ferroelectric which is of no interest to us here).

We will consider the case where the phase transition is described by an order parameter with two components η_1 and η_2. An invariant linear in the electric field exists in those cases when any combination of η_1 and η_2 (say, $\eta_1\eta_2$, $\eta_1^2-\eta_2^2$, $\eta_1\eta_2^2$, etc.) has the same transformation properties in a nonpolar symmetric phase as one of the components of the electric field vector (say, E_3).

The properties of improper ferroelectrics will be considered for the particular case of a second-order phase transition with a symmetry change $2/m \rightarrow m$ and a doubling of the unit-cell volume, so that the change of the space group (Fig. 4.10) is given by

$$C_{2h}^5(P2_1/c) \longrightarrow C_s^2(P_c). \tag{4.7}$$

The new aspect that appears in the treatment of such phase transitions is that the phase transition parameter may be a multicomponent quantity not only for transitions from sufficiently symmetric phases but also for transitions in monoclinic and rhombic crystals that involve a change of the number of

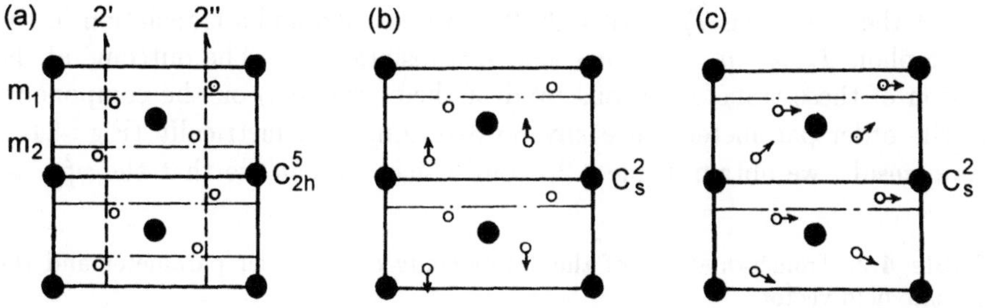

Fig. 4.10. (**a**) Schematic representation of the distortion of the unit cell of an improper ferroelectric in a phase transition with a symmetry change $C_{2h}^5(P2_1/c) \rightarrow C_s^2(P_c)$ (the arrows indicate the directions of displacements of atoms): the original symmetric phase ($\eta = 0$), the space group C_{2h}^5; (**b**) the distorted structure ($\eta \neq 0$) of symmetry C_s^2 (one glide reflection plane disappeared and the translation a_1 along the axis 2 was doubled); (**c**) the nonsymmetric phase distorted additionally by a uniform displacement of atoms of one of the sublattices parallel to the plane of symmetry (the same group of space symmetry C_s^2).

atoms in the unit cell. To illustrate, we use a phase transition with a symmetry change (4.7).

Figure 4.10a is a schematic representation of a unit cell of symmetry C_{2h}^5 and of its distortion described by an order parameter proportional to displacements of atoms (Fig. 4.10b). With $\eta \neq 0$ some of the planes of glide reflection (every other one) and twofold screw axes disappear; the unit-cell volume is doubled and the symmetry is lowered to C_s^2. Note that the application of an electric field along the x axis lying in the plane of symmetry will result in an additional distortion of the structure, which however is not accompanied by a symmetry change in our model (Fig. 4.10c).

We are now going to deal with the transformation properties of the order parameter. The components of the order parameter (η_1 and η_2) will be in this case the displacements of the upper and lower pairs of atoms in each unit cell along the screw axes, respectively (Fig. 4.11). Since the number of atoms in the unit cell is changed upon transition, i.e., the unit-cell parameter is doubled, we have to deal not with a point but with a space symmetry of the symmetric phase. It is known that, apart from rotations and reflections, the space group also includes, as symmetry operations, translations and combinations (the product) of rotations and reflections with translations. It should be noted that, as can be seen from Fig. 4.10, in the phase transition considered the translation of the vector $\boldsymbol{a}_1(T_{\boldsymbol{a}_1})$ ceases to be a symmetry element in a nonsymmetric phase and the translation to doubled periods is retained. What has been said holds for all translations, odd and even with respect to $\boldsymbol{a}_1$ translations, respectively; we may therefore ignore all even translations along $\boldsymbol{a}_1$ in the composition of the space group, assuming them to be equivalent to an identity transformation 1; all odd translations may be represented by a single element $T_{\boldsymbol{a}_1}$ (assuming, by virtue of what has been said above, that $T_{\boldsymbol{a}_1} \cdot T_{\boldsymbol{a}_1} = 1$). As a result, instead of a space group having an infinite number of symmetry elements, we will deal with a finite group formed by the multiplication of the following elements: the rotation about the screw axis $T_{\boldsymbol{a}_1/2} \cdot 2 = 2', 2''$, a reflection and a translation in the glide plane $T_{\boldsymbol{a}_2/2} \cdot m = m_1, m_2$, and a translation $T_{\boldsymbol{a}_1}$. The outcome of the action of these transformations (and of their products) on the components of the order parameter can easily be determined geometrically (Fig. 4.11). As a result, we obtain Table 4.2, from which one can see that the $\eta_1^2 - \eta_2^2$

Table 4.2. Transformations of the components of the order parameter and the electric field vector.

$P2_1/c$	1	m	$2', 2''$	T_{a_1}	$\bar{1}$
η_1	η_1	$-\eta_1$	η_2	$-\eta_1$	$-\eta_2$
η_2	η_2	η_2	η_1	$-\eta_2$	$-\eta_1$
E_3	E_3	E_3	$-E_3$	E_3	$-E_3$

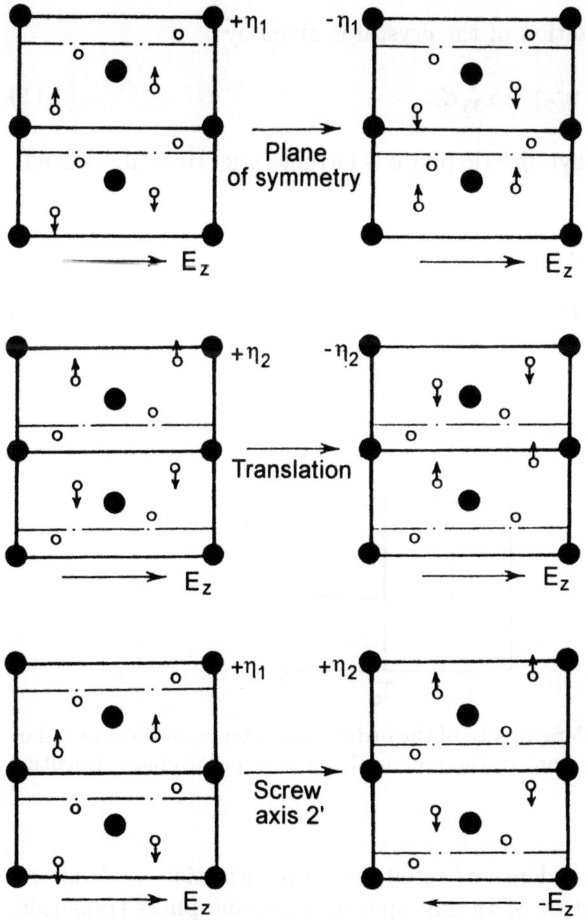

Fig. 4.11. Transformation of the components of the order parameter under symmetry operations of the group C_{2h}^5 lost upon transformation to the group C_s^2.

combination is transformed as the component of the electric field E_3 and, therefore, the thermodynamic potential of the crystal has the following form:

$$\Phi(p,T,\eta_1,\eta_2,E_3) = \Phi_0(p,T) + \frac{1}{2}\alpha(T-T_c)(\eta_1^2+\eta_2^2)$$

$$+ \boxed{\begin{array}{c}\text{fourth-degree}\\ \text{invariants}\end{array}} - a(\eta_1^2-\eta_2^2)E_3 - \frac{1}{2}\chi_{33}^0 E_3^2. \quad (4.8)$$

It has been shown in Chap. 2 that a thermodynamic potential of this type (at $E_3 = 0$) describes phase transitions to the following nonsymmetric phases:

$$(1) \quad \eta_{01}^2 \neq 0, \quad \eta_{02} = 0 \quad \text{or} \quad \eta_{01} = 0, \quad \eta_{02}^2 \neq 0 \qquad (4.9)$$

$$(2) \quad \eta_{01}^2 = \eta_{02}^2 \neq 0. \qquad (4.10)$$

The equilibrium polarization of the crystal is given by

$$P_3 = -\frac{\partial \Phi}{\partial E_3} = a(\eta_{01}^2 - \eta_{02}^2) + \chi_{33}^0 E_3. \tag{4.11}$$

Thus, only the first nonsymmetric phase is ferroelectric. Here, if, as usual,

$\eta_{01} \propto (T_c - T)^{1/2}$

then the spontaneous polarization is

$$P_s \propto \eta_{01}^2 - \eta_{02}^2 \sim T_c - T \tag{4.12}$$

(see Fig. 4.12a).

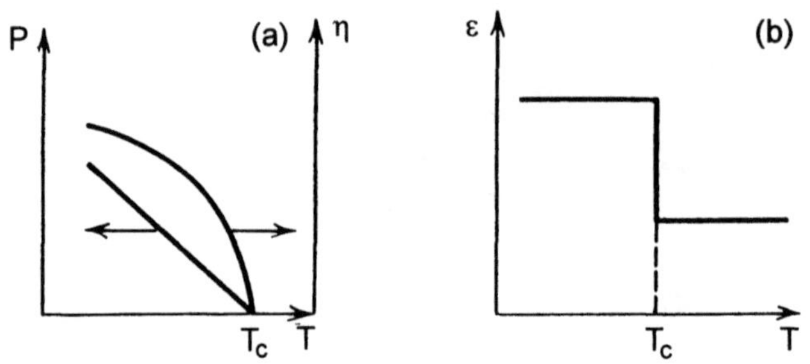

Fig. 4.12. (**a**) Temperature dependence of the order parameter, spontaneous polarization and (**b**) dielectric constant in the region of a second-order phase transition in an improper ferroelectric.

The unusual linear dependence of spontaneous polarization on temperature may serve as an indication of an improper second-order phase transition. To analyze the temperature dependence of the dielectric constant near the transition point, we will consider only the case of $\eta_{01} \neq 0$, $\eta_{02} = 0$ and the thermodynamic potential

$$\Phi = \Phi_0 + \frac{1}{2}\alpha(T - T_c) + \frac{1}{4}\beta\eta^4 - a\eta^2 E_3 - \frac{1}{2}\chi_{33}^0 E_3^2. \tag{4.13}$$

The equation that describes the equilibrium value of the order parameter will have the form

$$\frac{\partial \Phi}{\partial \eta} = \eta_0[\alpha(T - T_c) + \beta\eta_0^2 - 2aE_3] = 0. \tag{4.14}$$

Differentiating (4.14) with respect to E_3, we get

$$\frac{\partial \eta_0}{\partial E_3} = \frac{2a\eta}{\alpha(T - T_c) + 3\beta\eta_0^2 - 2aE_3}$$

and since

$$\chi_{33} = \frac{\partial P_3}{\partial E_3} = 2a\eta_0 \frac{\partial \eta_0}{\partial E_3} + \chi_{33}^0$$

we obtain with $E_3 \to 0$

$$\varepsilon_{33} = \begin{cases} \varepsilon_{33}^0 & \text{for} \quad T > T_c \\ \varepsilon_{33}^0 + 4\pi a^2/\beta & \text{for} \quad T > T_c . \end{cases} \tag{4.15}$$

Thus, the dielectric constant of an improper ferroelectric must experience at the second-order transition point a discontinuous change similar to a heat capacity change (Fig. 4.12b).

There is still another important feature of improper ferrolectric phase transitions. Equation (4.14) shows that $\eta_0^2 = -[\alpha(T - T_c) - 2aE_3]/\beta$, i. e., in an external electric field there occurs only a shift of the phase-transition point but not its "smearing", as was the case with proper ferroelectrics. This is not surprising since in this particular case the symmetry of the low-temperature phase is such that the electric field does not remove the difference in phase symmetry. To put it differently, in this case the symmetry of the polar phase is not the highest subgroup of the symmetry group common to the symmetric phase and the polar vector (cf. Chap. 2).

We have considered the specific properties of an improper ferroelectric for the case where the thermodynamic potential contains an invariant linear in the electric field and quadratic in the order parameter. But if the mixed invariant contains higher powers of the order parameter, the character of anomalies in an improper ferroelectric phase transition is changed: instead of the jump of the dielectric constant for a second-order transition, it undergoes a kink (just as in the case of a nonferroelectric phase transition) and the spontaneous polarization is proportional not to the first but to a higher power of $T_c - T$.

As pointed out above, phase transitions in ferroelectrics, including improper ones, are accompanied, as a rule, by a small jump of the order parameter, i.e., are first-order transitions close to second-order ones. This is clearly seen in Fig. 4.1, which shows the temperature dependence of the spontaneous polarization and dielectric constant of a crystal of gadolinium molybdate.

Therefore, in order to compare the theory described above with experiment, it is necessary to introduce sixth-degree invariants into the potential (4.8) in accordance with the considerations given in Chap. 3.[2] Here, the dielectric constant in a nonsymmetric phase becomes temperature dependent and the order parameter and polarization undergo a discontinuous change, this being characteristic of first-order transitions (Fig. 4.13). In contrast to ordinary ferroelectric transitions of this type, the dielectric constant remains unchanged in a symmetric phase and the spontaneous polarization depends on temperature in the same way as P_s^2 in proper ferroelectrics. The shift of the first-order transition point under the influence of an electric field results in this case in the appearance of double hysteresis loops.

[2] For crystals of gadolinium molybdate the symmetry change at the phase-transition point is such that the thermodynamic potential is somewhat different from (4.8), which, however, does not affect the character of anomalies.

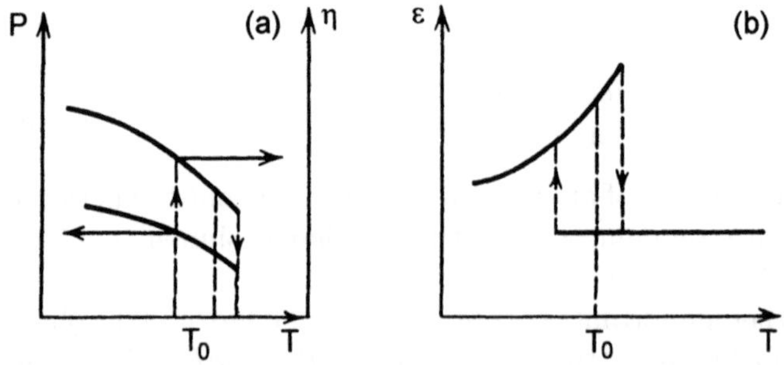

Fig. 4.13. (a) Temperature dependence of the order parameter and spontaneous polarization and (b) dielectric constant in the region of a first-order phase transition close to second-order transitions in an improper ferroelectric.

The phenomenological theory of structural phase transformations based on Landau's ideas yields a qualitative (and, in a number of cases, quantitative up to some constants) description of anomalies observed near the phase-transition points. The determining factor here is the change of the crystal's symmetry in a phase transition and the transformation properties of the order parameter.

Experiment, however, shows that even in cases where the order parameter is proportional to polarization, i.e., the transition occurs into a higher polar subgroup of the symmetry group of a symmetric phase, the temperature dependence of the dielectric constant is different for different crystals. This difference is more quantitative than qualitative and refers to the size of the temperature region near T_c, where the Curie–Weiss law is fulfilled, and also to the value of the Curie–Weiss constant itself.

We have already seen above that dielectric anomalies are determined quantitatively by a set of constants that appear in the expression for the thermodynamic potential of the system: α, β, γ and a. In particular, the value of the Curie–Weiss constant depends on the coefficients α and a; the latter, which determines the relationship between the equilibrium value of the order parameter and the electric polarization, may, evidently, vary within wide limits for different crystals. It is essential that the order parameter η is an "internal" variable associated directly with certain displacements of atoms; naturally, not all of these displacements make the same contribution to polarization and dielectric constant, which are measured experimentally.

It seems, therefore, to be quite natural that the Curie–Weiss constant and also the size of the temperature region, where the Curie–Weiss law is satisfied, can vary within wide limits. It should be noted that ferroelectrics with a small Curie–Weiss constant and a narrow region where the Curie–Weiss law is fulfilled, are usually classified in the literature as belonging to a group of *pseudoproper ferroelectrics*. It is clear that this group of substances has nothing in common with improper ferroelectrics.

However, the term "pseudoproper ferroelectric" is usually used with the following connotation. It is presumed that, in spite of the fact that the symmetry change in a crystal at $T = T_c$ corresponds to an ordinary (proper) ferroelectric phase transition and that the latter is accompanied by a spontaneous polarization, this polarization is not a true or "primary" phase transition parameter; its appearance is treated as a "secondary" effect induced by another, "primary" phase-transition parameter, which has the same transformation properties as the corresponding component of the polarization vector. For this case we may consider, for example, the thermodynamic potential in the form

$$\Phi = \Phi_0 + \frac{1}{2}A\eta^2 + \frac{1}{2}\beta\eta^4 + \frac{1}{2}b\xi^2 - d\eta\xi, \tag{4.16}$$

where η is the "primary" order parameter ($A = \alpha(T - T_c)$) and $\xi \propto P_i$ is the "secondary" one (b, d = const). It is also possible to take into account mixed invariants of a higher order. Simultaneous solution of the equations that determine the equilibrium value of the two parameters

$$\frac{\partial \Phi}{\partial \eta} = 0, \qquad \frac{\partial \Phi}{\partial \xi} = 0 \tag{4.17}$$

yields different versions of the $\eta_0(T)$ and $\xi_0(T)$ plots; the latter may appear to be rather "exotic" and the temperature dependence of spontaneous polarization may be brought into agreement with experiment by choosing appropriate coefficients. Such a procedure could hardly be accepted as an explanation of the experimentally observed "exoticism": if the physical meaning of the "primary" order parameter is not defined distinctly enough and the parameter itself cannot be found by experiment, the number of coefficients that are physically free significantly increases. In fact, we are dealing here with a phenomenological model with a certain set of parameters being varied.

At the same time, in a number of special cases the concept of pseudoproper ferroelectricity may prove rather fruitful and may extend the possibilities of the theory to a certain extent. It is clear that the introduction of a few interrelated order parameters into the thermodynamic potential does not in itself contradict the Landau theory. If these parameters having the same transformation properties can be determined experimentally and if we can find out with certainty which of the coefficients in the expansion of the thermodynamic potential tend to zero at $T = T_c$, the division into "primary" and "secondary" is also substantiated. In what follows (Chap. 5) we will see that in crystals, say KH_2PO_4 (symmetry $\bar{4}2m$ for $T > T_c$), the order parameter may be either the vector component along the $\bar{4}(z)$ axis or the component of the shear strain tensor u_{12} about that axis. In this case, we may assume that $\eta \propto P$ and $\xi \propto u_{12}$. It is well known from experiment that the inverse dielectric constant of this crystal along the z axis at u_{12} = const (the coefficient of P_z^2) vanishes at $T = T_c$ and that at the same time the elastic compliance of the crystal at P_z = const is practically temperature independent. It is therefore natural to assume that the polarization in the expansion of the thermodynamic potential is the primary order parameter; this assumption has been used as the basis for the development of the phenomenological theory of a phase transition in KH_2PO_4 crystals. A different situation is also possible, in principle, in which the component of the elastic strain tensor will be the primary component, but this inference may be based only on the experimentally verified constancy of the quantity ε_{ii}^{-1} at a constant value of deformation with T being approximately equal to T_c.

Summing up what has been said above, it may be pointed out that the introduction of several order parameters into the thermodynamic potential and the classification of a phase transition to a ferroelectric state as a pseudoproper transition may be fruitful if there are weighty grounds for distinguishing the "primary" parameter, the coefficient of the square of which in the expansion of the thermodynamic potential vanishes at $T = T_c$. The introduction of "speculative" phase-transition parameters of unknown physical nature brings about an increase of the number of unknown coefficients in the theory, which would become a model theory. Here it should be borne in mind that the solution of the problem of a complete, detailed description of the diversity of phenomena observed in phase transitions in a sufficiently wide temperature range can be found only within the framework of a strictly microscopic approach.

At a later time (see Chap. 12) we will see that the positive content of the "pseudoproper" concept can be realized in the analysis of certain types of phase transitions that are accompanied by a dipolar ordering in liquid crystals.

5. Anomalies of Elastic and Electromechanical Characteristics of Crystals in Second-Order Phase Transitions

5.1 One-Component Order Parameter: Elastic Properties of an Isotropic Liquid

In the preceding chapters we have seen that, according to a simple treatment based on the Landau theory, the temperature dependence of the dielectric constant (dielectric permittivity) experiences anomalies of three types in second-order phase transitions. If the transition is a proper ferroelectric transition, i.e., if the order parameter exhibits the transformation properties of a component (or components) of the polarization vector, then one or more components of the permittivity tensor goes to infinity at $T = T_c$, obeying the Curie–Weiss law in a certain vicinity of this point. If the transition is an improper ferroelectric transition, i.e., if the transformation properties of the order parameter differ from those for polarization but have a certain specific form (see Chap. 4), then the dielectric constant remains slightly temperature dependent in both phases, but one or more components of it increase discontinuously upon transition into the polar phase. Finally, in the most general case, when the order parameter has no specific transformation at all with respect to an electric field, the temperature dependence of the dielectric constant undergoes only a kink at $T = T_c$.[1] In order to establish these differences, we did not need to know either the physical meaning of the order parameter or the character of interactions leading to a phase transition; it was sufficient to take into account the symmetry properties of a crystal and the order parameter.

In this chapter we will use this approach for the analysis of anomalies of the elastic and electromechanical characteristics of a crystal. This example will be used to show that the Landau theory provides a general method of determination of anomalies of any thermodynamic quantities. True, as has been pointed out above, this theory is valid only in a limited temperature range, but this region is rather wide for ferroelectrics. The corrections to the Landau theory will be discussed in other chapters.

---

[1] It is presumed that the symmetric phase is nonpyroelectric. Otherwise certain tensor components ε_{ij} experience jumps upon transition into a nonsymmetric phase. This case was not considered in Chap. 4.

Just as in Chap. 4, the starting point will be the expression for the incomplete thermodynamic potential, which depends on the order parameter and some other variables. In Chap. 4 the electric field was used as such a variable. Here, account will be taken, in addition, of the dependence of the incomplete thermodynamic potential on the components of the elastic stress tensor σ_{ik}. In fact, this was done when we took into account the dependence of the incomplete thermodynamic potential on pressure p. Such a description, however, is not complete since the stressed state and deformations of a solid are described by symmetric tensors of rank two: the stress tensor σ_{ij} and the strain tensor u_{ik}. The elastic equation of state, which relates u_{ik} to σ_{ik}, has the form

$$u_{ik} = -\frac{\partial \Phi}{\partial \sigma_{ik}} \tag{5.1}$$

and is a direct mechanical analog of (2.4).

It has been shown earlier that at the second-order transition point the compressibility undergoes a jump (see (3.12)) . However, the volume compressibility, which characterizes the "compliance" of the system's volume with respect to an external pressure, does not fully characterize its elastic properties if the system is anisotropic. If the mechanically stressed state is created by a hydrostatic pressure p, then

$$\sigma_{ik} = \begin{vmatrix} -p & 0 & 0 \\ 0 & -p & 0 \\ 0 & 0 & -p \end{vmatrix}$$

or $\sigma_{ik} = -p\delta_{ik}$, where δ_{ik} is the Kronecker delta. It is evident that $\sigma_{ii} = \sigma_{11} + \sigma_{22} + \sigma_{33} = -3p$ and it is this quantity which will be used here instead of p. We will first consider a general approach to the analysis of changes of elastic properties in the region of phase transitions on the simplest example of a phase transition in an isotropic system – a liquid; we will assume that the transition is described by a one-component order parameter.[2]

The part of the thermodynamic potential that depends on pressure may evidently contain any powers of p, i.e.,

$$\Phi_{\text{el}} = -\frac{g}{3}\sigma_{ii} - \frac{\zeta}{18}\sigma_{ii}^2 - \dots, \tag{5.2}$$

[2] One may consider a real physical system possessing this property – liquid helium, which undergoes at a temperature of 2.3 K a second-order transition into a superfluid state. It should be kept in mind, however, that the phase transition parameter in liquid helium is a certain complex quantity; its phase is changed under certain transformations, the meaning of which is not discussed here. An important point here is that these transformations are not associated with the change of the orientation of the coordinates in space, which is why the order parameter in helium is not a tensor quantity. The thermodynamic potential includes only the square of the modulus of the order parameter η and the thermodynamic potential therefore has the same form as in the case of the one-component order parameter.

where g and ζ are certain coefficients which depend on temperature. Using (5.1), we find from (5.2) that the nondiagonal components of the tensor u_{ik} are equal to zero and the diagonal ones are equal to one another. This is an indication that no static shear deformations occur in the liquid. In other words, the deformations of the liquid are described by a single quantity. We will choose the quantity $u_{ii} = u_{11}+u_{22}+u_{33}$ which is the sum of the diagonal components of the strain tensor. Taking account of (5.1), we obtain

$$u_{ii} = g + \frac{\zeta}{3}\sigma_{ii}. \tag{5.3}$$

Since $u_{ii} = \Delta v/v$, where Δv is the change of the volume upon deformation, it can be seen that the coefficient ζ in (5.3) has the meaning of isothermal compressibility and that the quantity g bears a relation to the thermal expansion of the liquid, since the thermal expansion coefficient is given by

$$K = \left(\frac{\partial u_{ii}}{\partial T}\right)_{\sigma_{ii}=0} = \left(\frac{\partial g}{\partial T}\right)_{\sigma_{ii}=0} \tag{5.4}$$

Now we will consider a phase transition in a liquid and will write an expression for the incomplete thermodynamic potential as a function of η and σ_{ii}. If (just as in the case of helium) the quantity η is not a tensor quantity, i.e., it does not change under rotations of the coordinate system but may change sign under other transformations, say under translations,[3] then the cross-terms containing η and σ_{ii} (mixed invariants) will reduce to the product of "pure" invariants – the powers of the quantities η^2 and σ_{ii}. In other words, the incomplete thermodynamic potential will have the following form:

$$\begin{aligned}\Phi(T, \sigma_{ii}, \eta) = \Phi_0(T) + A\eta^2 + \frac{1}{4}\beta\eta^4 \\ - \frac{1}{3}g\sigma_{ii} - \frac{1}{18}\sigma_{ii}^2 - \frac{1}{2}r\eta^2\sigma_{ii} + \dots\end{aligned} \tag{5.5}$$

where the three dots represents higher-order terms; r is a constant.

In accordance with what was said in Chap. 2, we are to find the value of η corresponding to the minimal value of Φ and to substitute this value into (5.5). However, instead of writing down the minimum condition, we may use the second formula of (2.34) for the equilibrium Φ, replacing $A = (1/2)\alpha(T-T_c)$ by $A - (1/2)r\sigma_{ii}$ and adding to Φ_0 the fourth and fifth terms in (5.5). Then, below the phase transition point, i.e., with $A - l/2r\sigma_{11} < 0$, we have

$$\eta_0^2 = -[\alpha(T-T_c) - r\sigma_{ii}]/\beta \tag{5.6}$$

[3] The one-component order parameter that describes a phase transition from an isotropic liquid cannot be a tensor quantity. Indeed, the tensorial order parameter for a liquid is always a multicomponent one since under an arbitrary transformation of the coordinate system one of the tensor components is expressed in terms of the others; in other words, the tensor components form the basis of multidimensional representations in the case of an isotropic medium.

$$\begin{aligned}\Phi &= \Phi_0(T) - \frac{[\alpha(T-T_c) - r\sigma_{ii}]^2}{4\beta} - \frac{1}{3}g\sigma_{ii} - \frac{1}{18}\zeta\sigma_{ii}^2 \\ &= \Phi_0(T) - \frac{1}{4}\frac{\alpha^2(T-T_c)^2}{\beta} - \left(\frac{g}{3} - \frac{\alpha(T-T_c)r}{2\beta}\right)\sigma_{ii} \\ &\quad - \left(\frac{\zeta}{18} + \frac{r^2}{4\beta}\right)\sigma_{ii}^2. \end{aligned} \tag{5.7}$$

From (5.5) it can be seen that the linear expansion coefficient defined by relation (5.4) corresponds to the paraelectric phase ($\eta_0 = 0$). In going to a nonsymmetric phase it decreases by a factor of $\Delta K = 3\alpha r/(2\beta)$ and the compressibility increases by $\Delta\zeta = 9r^2/(2\beta)$.

The phase transition temperature depends on σ_{ii}:

$$\alpha(T_c - T_c^0) - r\sigma_{ii} = 0$$

(here T_c^0 is the transition temperature at $\sigma_{ii} = 0$) and

$$\frac{dT_c}{d\sigma_{ii}} = -\frac{1}{3}\frac{dT_c}{dp} = \frac{r}{\alpha}. \tag{5.8}$$

Taking into account the heat capacity jump at the phase transition point given by the second equation of (3.10) and the fact that the four quantities being measured (ΔC, ΔK, $\Delta\zeta$ and dT_c/dp) are expressed in terms of the three parameters of the theory α, β and r, we come to the conclusion that these quantities are interconnected by relations that follow from theory. We see that

$$\Delta\zeta = \Delta K \frac{dT_c}{dp} \tag{5.9}$$

and

$$\Delta C_p = \frac{T_c \Delta K}{dT_c/dp} = \frac{T_c \Delta\zeta}{(dT_c/dp)^2}. \tag{5.10}$$

Relations (5.9) and (5.10) were derived by Ehrenfest long before Landau worked out his theory. They follow directly from the continuity of the entropy and volume of the system during a second-order phase transition. However, here the presence of jumps had to be postulated.

Let us discuss the physical factor responsible for an increase in the compressibility of the system upon transition to a nonsymmetric phase.[4] To do this, we write down the equation of state (5.1), which follows from (5.5) and (5.7):

$$u_{ii} = -\frac{\partial\Phi}{\partial\sigma_{ii}} = g + \frac{1}{3}\zeta\sigma_{ii} + \frac{3}{2}r\eta_0^2, \tag{5.11}$$

[4] The jump of the linear expansion coefficient may be either positive or negative, depending on the sign of the coefficient r.

where η_0 is the equilibrium value of the order parameter with $\sigma_{ii} \neq 0$, which is determined by the equation

$$\frac{\partial \Phi}{\partial \eta} = \alpha(T - T_c^0)\eta_0 + \beta\eta_0^3 - r\sigma_{ii}\eta_0 = 0. \tag{5.12}$$

Let us simplify (5.11) and (5.12) by reducing them to linear relations and by considering the quantities η_0 and u_{ii}, which differ little from their equilibrium values η_{00} and $u_{ii,0}$ at $\sigma_{ii} = 0$. This corresponds to small values of σ_{ii}. In other words, we put $\Delta u_{ii} = u_{ii} - u_{ii,0}$ and $\Delta\eta = \eta_0 - \eta_{00}$ and retain in (5.11) and (5.12) terms of first order with respect to small values of u_{ii}, $\Delta\eta$ and σ_{ii}. For $T > T_c$ we have

$$\Delta u_{ii} = \frac{1}{3}\zeta\sigma_{ii} \tag{5.13}$$

and for $T < T_c$ we obtain

$$\Delta u_{ii} = \frac{1}{3}\zeta\sigma_{ii} + 3r\eta_0\Delta\eta\,, \tag{5.14}$$

where the quantity η_0 is defined by (5.12):

$$\Delta\eta = -\eta_0\frac{r\sigma_{ii}}{2\alpha(T - T_c)}. \tag{5.15}$$

One can see that, in contrast to the symmetric phase, in the nonsymmetric phase the application of a mechanical stress leads not only to the appearance of the corresponding strain u_{ii} but also to a change of η. The change of η in turn causes an additional deformation (the second term in (5.14)).

From (5.14) we see that if the deformation is accomplished at a fixed value of η, the compressibilities of the symmetric and nonsymmetric phases will coincide. To put it differently, the compressibility jump during the transition is equal to the difference between the compressibility providing $\partial\Phi/\partial\eta = 0$ (this is said to be compliance at a constant force coupled to η) and the compressibility at constant η. Proceeding from the general thermodynamic considerations, we can show that the former compressibility is greater than the latter; we will first explain this by using a simple model.

The equations given above describe two coupled quantities, u_{ii} and η. In this sense they may be regarded as being analogous to a system of equations for two coupled oscillators, say for two balls coupled by a spring (Fig. 5.1). It is evident that the force exerted by a certain fixed quantity will shift ball 1 to a greater distance if ball 2 is free than in the case when the latter is fixed, i.e., the compliance of the system with respect to the shift of ball 1 is lower if the position of ball 2 is fixed. This example is an illustration of one of the general propositions of thermodynamics – Le Chatelier's principle, which may be formulated as follows: if the system is subjected to an external force, which brings it out of the equilibrium state, the approach of the system to equilibrium reduces the result of this external influence. In our case, the external force is the shift of ball 1. Suppose that this shift is made so rapidly that ball 2 has no time to change its position. A stress will appear in springs 1 and 2, which will decrease when ball 2 is brought into motion. In an analogous manner we may visualize that the strain of the nonsymmetric phase first occurs at a fixed value of the order parameter. The stress that arises is the same as the one which would have occurred with the same strain of the symmetric phase. After the fixation is removed we "release" η, "allowing" it to relax. According to Le Chatelier's principle, the stress will decrease in this case. Thus, the elastic compliance at constant η is lower than the equilibrium elastic compliance; in other words, the positive jump of elastic compliance upon transition into a nonsymmetric phase is a general thermodynamic regularity.

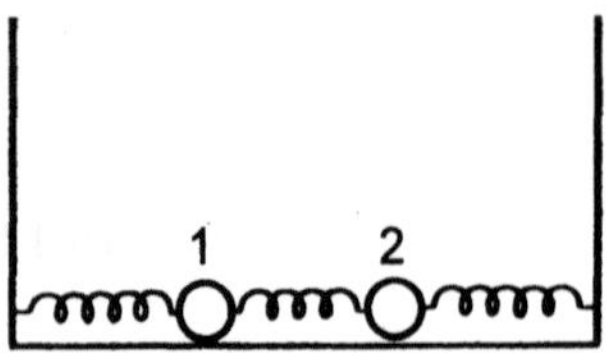

Fig. 5.1. A model of two coupled oscillators.

We emphasize that what has been said above refers to a second-order phase transition. In a first-order transition the compliance jump may, of course, have either sign since in such transitions all the constants of a crystal, including the compliance at constant η, undergo a jump.

5.2 One-Component Order Parameter: Elastic Properties of an Anisotropic Crystal

Let us now consider a phase transition in an anisotropic crystal, whose elastic properties are described by the elastic compliance matrix s_{ijkl}:

$$u_{ij} = s_{ijkl}\sigma_{kl}. \tag{5.16}$$

The symmetry of the tensor s_{ijkl} upon permutation of the index pairs and of the indices within these pairs is responsible for the fact that in a triclinic crystal there are 21 independent elastic coefficients and their number decreases with increasing number of symmetry elements of the crystal.

In a second-order phase transition the crystal symmetry is lowered and there appear new elastic coefficients; the appearance of an order parameter is accompanied by a spontaneous strain of the crystal and the elastic moduli undergo various types of anomalies.

The qualitative character of anomalies will be considered on a planar model of a rhombic crystal (Fig. 5.2). Figure 5.2 also shows the change of the crystal symmetry during a phase transition.

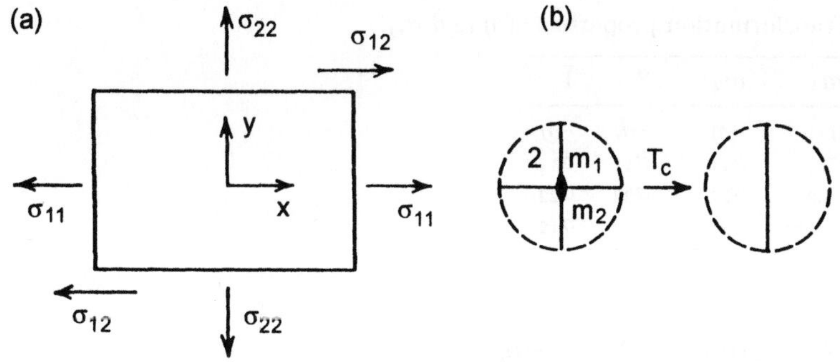

Fig. 5.2. (**a**) Orientation of the components of the mechanical stress tensor and (**b**) change of the symmetry of a two-dimensional model of a rhombic crystal in the phase transition.

The elastic properties are described by the matrix

	σ_{11}	σ_{22}	σ_{12}
u_{11}	s_1	s_3	0
u_{22}	s_3	s_2	0
u_{12}	0	0	s_4

and by the following equations:

$$u_{11} = s_1\sigma_{11} + s_3\sigma_{22} \tag{5.17}$$

$$u_{22} = s_3\sigma_{11} + s_2\sigma_{22} \tag{5.18}$$

$$u_{12} = s_4\sigma_{12}. \tag{5.19}$$

Note that the thermal expansion of the rhombic crystal will be anisotropic in the xy plane; the tensor of the thermal expansion coefficients is

$$K_{ij} = \begin{vmatrix} K_{11} & 0 \\ 0 & K_{22} \end{vmatrix} .$$

We will assume that the phase transition parameter η is a one-component vector quantity, which transforms as the coordinate y (the appearance of such a distortion results in a lowering of the symmetry shown in Fig. 5.2).

Below is given a table which describes the transformation properties of the quantities η and σ_{ij} under transformations of the symmetry of the initial group (Table 5.1).

It can be seen that the components of the mechanical stress tensor σ_{11} and σ_{22} are the invariants of the symmetry group of the model; the combinations σ_{12}^2, $\eta^2\sigma_{11}$, $\eta^2\sigma_{22}$, and $\eta^2\sigma_{12}^2$ are also invariants. Therefore, the thermodynamic potential assumes the form

Table 5.1. Transformation properties of η and σ_{ij}.

	1	m_1	m_2	2	$\bar{1}$
η	η	η	$-\eta$	$-\eta$	$-\eta$
σ_{11}	σ_{11}	σ_{11}	σ_{11}	σ_{11}	σ_{11}
σ_{22}	σ_{22}	σ_{22}	σ_{22}	σ_{22}	σ_{22}
σ_{12}	σ_{12}	$-\sigma_{12}$	$-\sigma_{12}$	σ_{12}	σ_{12}

$$\begin{aligned}\Phi = \Phi_0(T) &+ \frac{1}{2}\alpha(T-T_c)\eta^2 + \frac{1}{4}\beta\eta^4 \\ &- \frac{1}{2}r_1\eta^2\sigma_{11} - \frac{1}{2}r_2\eta^2\sigma_{22} - \frac{1}{2}a_1\eta^2\sigma_{12}^2 - g_1\sigma_{11} - g_2\sigma_{22} \\ &- \frac{1}{2}s_1^0\sigma_{11}^2 - \frac{1}{2}s_2^0\sigma_{22}^2 - s_3^0\sigma_{11}\sigma_{22} - \frac{1}{2}s_4^0\sigma_{12}^2. \end{aligned} \tag{5.20}$$

Here, as before, r_1, r_2, a_1, g_1, g_2, s_1^0, s_2^0, s_3^0 and s_4^0 are constants, the meaning of which will be clear from what follows.

With $\sigma_{ij} \neq 0$ the equilibrium values of the order parameter are determined by the condition

$$\frac{\partial\Phi}{\partial\eta} = \left(\tilde{\alpha} + \beta\eta_0^2\right)\eta_0 = 0,$$

where $\tilde{\alpha} = \alpha(T-T_c) - r_1\sigma_{11} - r_2\sigma_{22} - a_1\sigma_{12}^2$. Just as in the case of an isotropic liquid, the phase transition temperature appears to be a function of the components of the mechanical stress tensor.

In the region of temperatures $T > T_c$, $\eta_0 = 0$; from (5.20) it follows that the quantities g_i have the meaning of the strain tensor components and that s_i^0 have the meaning of the components of the elastic compliance tensor of the symmetric phase.

Substituting into (5.20) the equilibrium value of the order parameter

$$\eta_0^2 = -\frac{\tilde{\alpha}}{\beta}, \tag{5.21}$$

we obtain the thermodynamic potential of the nonsymmetric phase:

$$\Phi = \Phi_0 - \frac{1}{4}\frac{\tilde{\alpha}^2}{\beta} + \Phi(\sigma_{ij}), \tag{5.22}$$

where

$$\Phi(\sigma_{ij}) = -g_1\sigma_{11} - g_2\sigma_{22} - \frac{1}{2}s_1^0\sigma_{11}^2 - \frac{1}{2}s_2^0\sigma_{22}^2 - s_3^0\sigma_{11}\sigma_{22} - \frac{1}{2}s_4^0\sigma_{12}^2.$$

With $T < T_c$ we have for small σ_{ij}:

$$\begin{aligned} u_{11} &= -\frac{\partial\Phi}{\partial\sigma_{11}} \\ &= \left[g_1 - \frac{\alpha r_1}{2\beta}(T-T_c)\right] + \left(s_1^0 + \frac{r_1^2}{2\beta}\right)\sigma_{11} + \left(s_3^0 + \frac{r_1 r_2}{2\beta}\right)\sigma_{22} \end{aligned} \tag{5.23}$$

$$u_{22} = -\frac{\partial \Phi}{\partial \sigma_{22}}$$
$$= \left[g_2 - \frac{\alpha r_2}{2\beta}(T - T_c)\right] + \left(s_3^0 + \frac{r_1 r_2}{2\beta}\right)\sigma_{11} + \left(s_2^0 + \frac{r_2^2}{2\beta}\right)\sigma_{22} \quad (5.24)$$

$$u_{12} = -\frac{\partial \Phi}{\partial \sigma_{12}} = \left(s_4^0 + \frac{\alpha a_1 (T - T_c)}{2\beta}\right)\sigma_{12}. \quad (5.25)$$

In (5.23)–(5.25) the smallness of σ_{ij} implies the neglect of terms quadric in the components of the mechanical stress tensor.

Let us define the components of the thermal expansion tensor in the nonsymmetric phase:

$$K_{11}^{T<T_c} = \frac{\partial u_{11}}{\partial T} = K_{11}^{T>T_c} - \frac{\alpha r_1}{2\beta} \quad (5.26)$$

$$K_{22}^{T<T_c} = \frac{\partial u_{22}}{\partial T} = K_{22}^{T>T_c} - \frac{\alpha r_2}{2\beta} \quad (5.27)$$

where $K_{ii}^{T>T_c} = \partial g_i / \partial T$.

A discontinuous change of the thermal expansion coefficient K_{ij} during the transition from the symmetric into the nonsymmetric phase is typical when the "cross-term" order parameter – components of σ_{ij} tensor has the form $\eta^2 \sigma_{ij}$. In this case the elastic compliance coefficients also undergo jumps at $T = T_c$:

$$s_1 = \frac{\partial u_{11}}{\partial \sigma_{11}} = \begin{cases} s_1^0 & \text{for} \quad T > T_c \\ s_1^0 + r_1^2/(2\beta) & \text{for} \quad T < T_c \end{cases} \quad (5.28)$$

$$s_2 = \frac{\partial u_{22}}{\partial \sigma_{22}} = \begin{cases} s_2^0 & \text{for} \quad T > T_c \\ s_2^0 + r_2^2/(2\beta) & \text{for} \quad T < T_c \end{cases} \quad (5.29)$$

$$s_3 = \frac{\partial u_{11}}{\partial \sigma_{22}} = \begin{cases} s_3^0 & \text{for} \quad T > T_c \\ s_3^0 + r_1 r_2/(2\beta) & \text{for} \quad T < T_c. \end{cases} \quad (5.30)$$

It is interesting that an invariant of the type $\eta^2 \sigma_{ij}^2$ leads to a different temperature dependence of the elastic compliance coefficient; from (5.25) it follows that in our model the elastic compliance coefficient, which characterizes the "compliance" of the model with respect to shear strains, exhibits an anomaly of the "kink" type (Fig. 5.3):

$$s_4 = \frac{\partial u_{12}}{\partial \sigma_{12}} = \begin{cases} s_4^0 & \text{for} \quad T > T_c, \\ s_4^0 - \alpha a_1 (T - T_c)/(2\beta) & \text{for} \quad T < T_c. \end{cases} \quad (5.31)$$

Though we have considered a seemingly abstract model, in fact the results obtained are rather general for all cases where there is a one-component order parameter. As a matter of fact, this also includes uniaxial proper ferroelectrics having a center of inversion in the symmetric phase (triglycine sulphate; see, for example, Figs. 3.4 and 3.5).

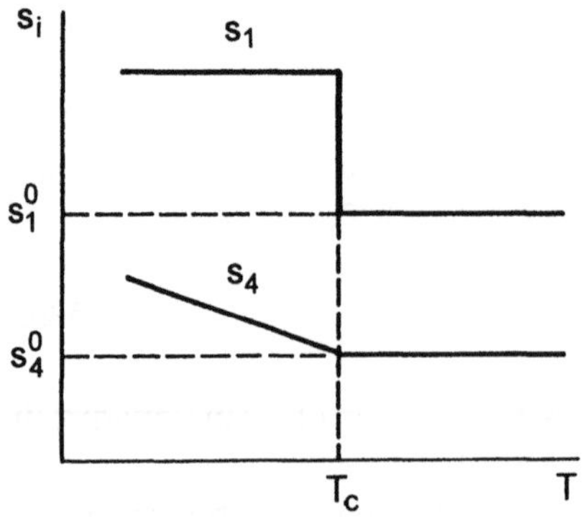

Fig. 5.3. Temperature dependence of the coefficients of elastic compliance of the planar model shown in Fig. 5.2.

Comparison of experimental data on the elastic and thermal properties of uniaxial ferroelectrics with the inferences of the theory expounded above may be made, as has been pointed out at the beginning of Chap. 3, only within a certain vicinity of the phase transition point. Confirming the presence of jumps of the corresponding thermodynamic quantities at $T = T_c$, experiment shows at the same time that they exihibit a rather strong temperature dependence (a λ-shaped form) in the nonsymmetric phase. We have not obtained such a temperature dependence in the theory because in the expansion of the thermodynamic potential in powers of the order parameter we took into account only terms of fourth order in the order parameter. Considering terms of higher order in η in the expansion of the thermodynamic potential (see (3.39)), one can describe the temperature dependence of the parameters η_0, C_p, s_{ijkl} and K_{ij} over a wider temperature range (see, for example, (3.53)). Furthermore, it can be shown that the temperature dependences of the anomalous parts of the quantities C_p and of the components of the elastic compliance tensor and thermal expansion, which experience a jump at the phase transition point, are similar and are interconnected in the nonsymmetric phase by relations analogous to Ehrenfest's relations (3.10) and (3.11):

$$\frac{\Delta C_p}{T} = \frac{\Delta K}{\mathrm{d}T_c/\mathrm{d}p} + \text{const}, \tag{5.32}$$

$$\Delta K = \frac{\Delta \zeta}{\mathrm{d}T_c/\mathrm{d}p} + \text{const}, \tag{5.33}$$

where $\mathrm{d}T_c/\mathrm{d}p$ is the rate of variation of the phase transition temperature under the action of hydrostatic pressure.

For the case where the external mechanical force is described not by hydrostatic pressure but by the set of components of the mechanical stress tensor σ_{ij}, relations (5.32) and (5.33) may be generalized:[5]

[5] In the literature expressions (5.32) and (5.33) are known as Pippard equations and expressions (5.34) and (5.35) as Pippard–Janovec equations.

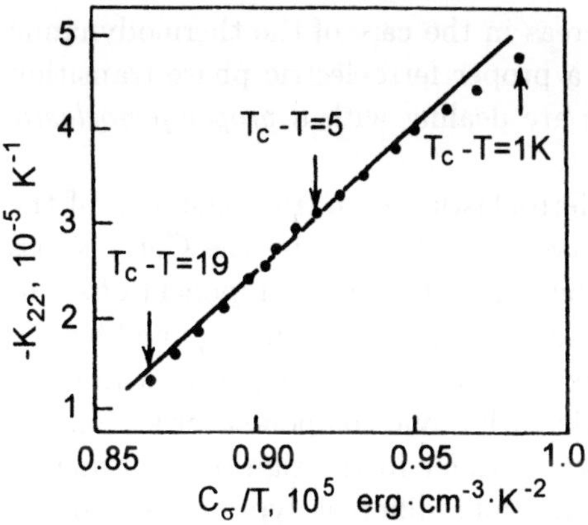

Fig. 5.4. Heat capacity referred to temperature plotted as a function of the thermal linear expansion coefficient for a TGS crystal [10].

$$\frac{\Delta C_\sigma}{T} = \frac{\Delta K_{ij}}{\mathrm{d}T_\mathrm{c}/\mathrm{d}\sigma_{ij}} + \mathrm{const}, \tag{5.34}$$

$$\Delta s_{ijkl} = \frac{\partial T_\mathrm{c}}{\partial \sigma_{ij}} \Delta K_{kl} + \mathrm{const}. \tag{5.35}$$

The proportionality of the quantities contained in these equations is nicely confirmed by experiment for many structural phase transitions, including ferroelectric ones (Fig. 5.4). From relations (5.34) and (5.35) it follows that in order to elucidate the type of phase transition and the character of anomalies of various thermodynamic characteristics, it is sufficient to obtain data on a single parameter, say the temperature dependence of the heat capacity.

5.3 Ferroelectric-Ferroelastics: One-Component Order Parameter with Transformation Properties of the Component of a Second-Rank Tensor and the Polar Vector

Let us now consider the case where the order parameter displays the transformation properties of the component of a tensor of rank two. The thermodynamic potential will assume the following form:

$$\Phi = \Phi_0(T) + \frac{1}{2}\alpha(T - T_c)\eta^2 + \frac{1}{4}\beta\eta^4 - a\eta\sigma_{ij}. \tag{5.36}$$

Here we have taken into account only one component (σ_{ij}) of the mechanical stress tensor, the one which exhibits the transformation properties of the order parameter since the inclusion of other invariants would be a repetition of the cases considered above.

Comparing (5.36) with (2.35), we see that they are completely equivalent, including the replacement of σ_{ij} by E_2. Therefore, all conclusions made on the basis of the thermodynamic potential (2.35) are valid in the case of (5.36). The difference is only in terminology: whereas in the case of the thermodynamic potential (2.35) we were speaking of a proper ferroelectric phase transition at $T = T_c$, in this particular case we are dealing with *a proper ferroelastic transition.*

Thus, near the point of a proper ferroelastic phase transition one of the elastic compliances of a crystal increases with $T \to T_c$ by the Curie–Weiss law and in the nonsymmetric phase there appears a new component of spontaneous strain; there is observed a hysteresis dependence $u_{ij}(\sigma_{ij})$ and ferroelastic domains are formed. With respect to the latter we are to bear in mind the following: whereas in a ferroelectric with a one-component order parameter there are two types of domain which differ in the sign of spontaneous polarization, in a ferroelastic there exist four different types of domain at least. This is associated with the specific feature of spontaneous strain. As is known, the strain tensor may be expressed in terms of a component of the displacement vector $\boldsymbol{u}$:

$$u_{ik} = \frac{1}{2}\left(\frac{\partial u_i}{\partial x_k} + \frac{\partial u_k}{\partial x_i}\right).$$

Thus, the homogeneous spontaneous strain corresponding to the component u_{ik} of a given sign can be realized in two ways: the displacement along the i axis proportionately to x_k, i.e., $u_i \propto x_k$ but $u_k = 0$ or $u_k \propto x_i$, in which case $u_i = 0$. So, to each sign of the order parameter there correspond two types of domain (Fig. 5.5).

Special attention should be focused on the case where the order parameter exhibits the transformation properties of the component of a tensor of rank two and of the vector component simultaneously. Hence, we will be dealing here with *ferroelectric-ferroelastics.* With respect to symmetry such crystals are characterized by the absence of an inversion center above the phase transition point. So, we are dealing here with a ferroelectric which exhibits a

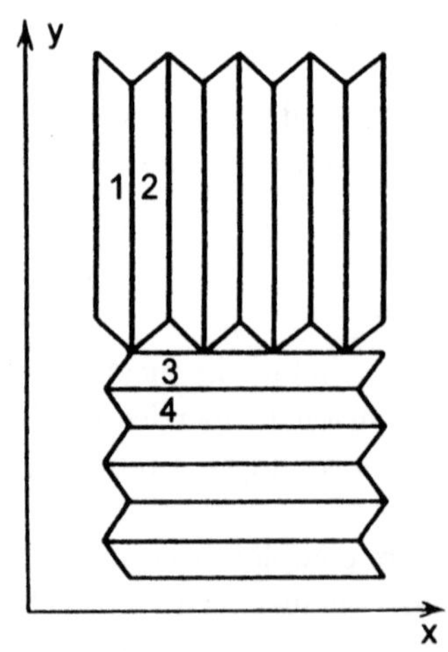

Fig. 5.5. Four different types of domain in the case where the order parameter has the transformation properties of a second-rank tensor.

Table 5.2. The action of symmetry operations of the group 222.

	1	2_x	2_y	2_z
E_1, $P_1 \propto x$	x	x	$-x$	$-x$
σ_{23}, $u_{23} \propto yz$	yz	yz	$-yz$	$-yz$

piezoelectric effect in a paraelectric phase. Examples of such ferroelectrics are crystals of Rochelle salt (symmetry change 222 $\rightarrow$ 2) and potassium dihydrogen phosphate ($\bar{4}2m \rightarrow mm2$). It is easy to see that the action of symmetry operations of the group 222 (Fig. 5.6) on the x-component of the vector (E_1, P_1) and the tensor component (σ_{23}, u_{23}) is the same (Table 5.2).

Since the same transformation properties are also displayed in the case under consideration by the order parameter, the thermodynamic potential will take the form

$$\Phi = \Phi_0(T) + \frac{1}{2}\alpha(T - T_c)\eta^2 + \frac{1}{4}\beta\eta^4 - a_1\eta E - a_2\eta\sigma, \tag{5.37}$$

where, for the sake of brevity, σ represents the corresponding tensor component σ_{ij} and E stands for the component of the vector $\boldsymbol{E}$, which is transformed as η. If in (5.37) we put $\sigma = 0$ or $E = 0$, we will obtain a thermodynamic potential of the form (2.35) or (5.36). Hence, in the phase transition considered both the dielectric permittivity at $\sigma = 0$ ($\varepsilon_{11}^{\sigma}$) and the elastic compliance at $E = 0$ (s_{2323}^{E}) will change in accordance with the Curie–Weiss law.

It is evident that in this case

$$P_{01} = -\frac{\partial \Phi}{\partial E} = a_1\eta_0, \tag{5.38}$$

$$u_{23}^0 = -\frac{\partial \Phi}{\partial \sigma} = a_2\eta_0. \tag{5.39}$$

Thus, it turns out that in crystals, say those of Rochelle salt, below the phase transition point there also arise an electric polarization along the x axis and a spontaneous shear strain u_{23}^0, these two quantities being interrelated as follows:

$$u_{23}^0 = -\frac{a_2}{a_1}P_{01}. \tag{5.40}$$

Let us now consider a clamped crystal, i.e., a crystal with a constant strain u_{23} equal to zero. From the proportionality between u and η (see (5.39)) it may seem that the clamping of the crystal would eliminate the phase transition. Recall, however, that the letter η is used here to designate the volume-averaged value of the order parameter, which is why our treatment only shows that in a clamped crystal there cannot occur a phase transition with the formation of a homogeneous, i.e., monodomain, nonsymmetric phase. It can be shown that at a temperature practically equal to T_c, the transition takes place in a clamped crystal, but only into a multidomain state.

Note that the elastic compliance coefficients, which are not associated with a shear strain, experience anomalies of the "jump" or "kink" type, such anomalies being usual for second-order phase transitions.

5.4 Temperature Dependences of "Morphic" Moduli of Elasticity

The lowering of the crystal symmetry at the phase transition point brings about a change – an increase in the number of independent elastic compliance coefficients; new ones appear, which are called *morphic coefficients.* In particular, in crystals of Rochelle salt upon transition into the polar phase to nine coefficients of the group 222 there are added four morphic coefficients, which are enclosed in boxes in the matrix given below for the group 2 (the twofold axis along the x axis in Fig. 5.6):

$$s_{ijkl} = \begin{vmatrix} s_{1111} & s_{1122} & s_{1133} & \boxed{s_{1123}} & 0 & 0 \\ s_{1122} & s_{2222} & s_{2233} & \boxed{s_{2223}} & 0 & 0 \\ s_{1133} & s_{2233} & s_{3333} & \boxed{s_{3323}} & 0 & 0 \\ \boxed{s_{1123}} & \boxed{s_{2223}} & \boxed{s_{3323}} & s_{2323} & 0 & 0 \\ 0 & 0 & 0 & 0 & s_{1313} & \boxed{s_{1312}} \\ 0 & 0 & 0 & 0 & \boxed{s_{1312}} & s_{1212} \end{vmatrix}$$

Of the four morphic coefficients, three of them – s_{1123}, s_{2223} and s_{3323} – relate the longitudinal strains u_{11}, u_{22} and u_{33} to the shear stress σ_{23}, which exhibits the transformation properties of the order parameter, and the remaining one, s_{1312}, determines the elastic compliance to "shear" (strain u_{13}) under the action of a mechanical shear stress σ_{12}.

Let us consider the first group for the coefficient s_{1123}. Considering that the quantity σ_{11} by itself is an invariant of the group, we have

$$\Phi = \Phi_0 + \frac{1}{2}\alpha(T - T_c)\eta^2 + \frac{1}{4}\beta\eta^4 - a\eta\sigma_{23} - \frac{1}{2}b\sigma_{11}\eta^2, \tag{5.41}$$

where a and b are constant quantities, which are independent of T.

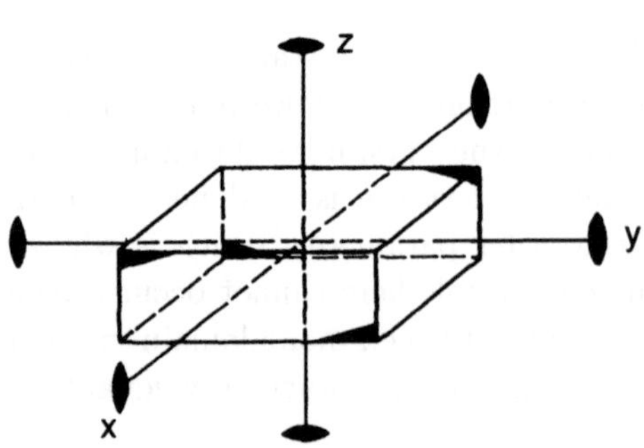

Fig. 5.6. Symmetry elements and the disposition of coordinates for the symmetry group 222.

The temperature dependence of the quantity s_{1123} can be obtained from the relation

$$s_{1123} = -\frac{\partial^2 \Phi(\eta_0)}{\partial\sigma_{11}\partial\sigma_{23}}, \tag{5.42}$$

where η_0 is the equilibrium value of the order parameter, which is determined by the condition

$$\frac{\partial\Phi}{\partial\eta} = \alpha(T - T_c)\eta_0 + \beta\eta_0^3 - a\sigma_{23} - b\sigma_{11}\eta_0 = 0. \tag{5.43}$$

It can be seen that $\eta_0 = \eta_0(\sigma_{11}, \sigma_{23})$; substitution of the equilibrium value of the order parameter into (5.41) and differentiation with respect to σ_{11} yield

$$u_{11} = -\frac{\partial\Phi(\eta_0)}{\partial\sigma_{11}} = \frac{1}{2}b\eta_0^2, \tag{5.44}$$

and, hence,

$$s_{1123} = \frac{\partial u_{11}}{\partial\sigma_{23}} = b\eta_0\frac{\partial\eta_0}{\partial\sigma_{23}}. \tag{5.45}$$

Differentiating (5.43) with respect to σ_{23}, we get

$$\frac{\partial\eta_0}{\partial\sigma_{23}} = \frac{a}{\alpha(T - T_c) + 3\beta\eta_0^2 - b\sigma_{11}}. \tag{5.46}$$

Assuming that σ_{12}, $\sigma_{23} \to 0$, we obtain for the polar phase, where $\eta_0^2 = -[\alpha(T - T_c)]/\beta$,

$$s_{1123} = \frac{ab}{2\sqrt{\alpha\beta(T_c - T)}}.$$

Thus, upon approach to the Curie point from the side of the nonsymmetric phase the given morphic modulus increases by the law $s_{1123} \propto (T_c - T)^{-1/2}$ and then vanishes at $T = T_c$: as follows from (5.44), at $\eta_0 = 0$ (symmetric phase) the mechanical stress σ_{23} does not cause a strain u_{11} (Fig. 5.7).

Concluding this section, it should be noted that the above-considered variants of the temperature dependence of the elastic compliance coefficients do not cover all possibilities. An analogous consideration of other possible invariants of the order parameter with a mechanical stress tensor component ($\eta^2\sigma_{13}^2$, $\eta^2\sigma_{12}^2$, $\eta\sigma_{12}\sigma_{13}$) shows that the corresponding coefficients experience either a kink (s_{1212}, s_{1313}) or a proportionality to η_0 and a decrease upon approach to T_c from the side of the ferroelectric phase by the law of $(T_c-T)^{1/2}$ (the morphic coefficient s_{1312}).

The results obtained may be generalized as follows. In order to find the temperature dependence of the elastic compliance of a crystal of arbitrary symmetry, one has to establish the form of the thermodynamic potential of a crystal undergoing a phase transition at $T = T_c$ with the known transformation properties of the order parameter and being acted upon by a mechanical

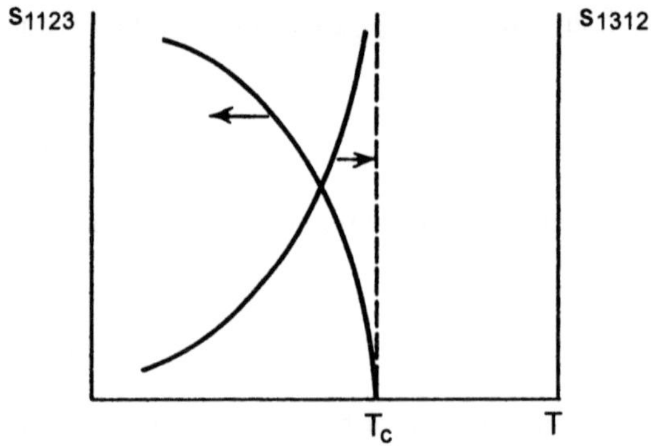

Fig. 5.7. Temperature dependence of the morphic coefficients of elastic compliance s_{1123} and s_{1312} of a crystal of Rochelle salt.

stress specified by the tensor components σ_{ij}. Writing down the tensor components in matrix notation, we have, in general form, for the one-component order parameter:

$$\begin{aligned}\Phi = \Phi_0 &+ \frac{1}{2}\alpha(T-T_c)\eta^2 + \frac{1}{4}\beta\eta^4 - \frac{1}{2}s_{ij}^0\sigma_i\sigma_j \\ &- a_i\eta\sigma_j - b_{ij}\eta\sigma_i\sigma_j - \frac{1}{2}c_i\sigma_i\eta^2 - \frac{1}{2}f_{ij}\sigma_i\sigma_j\eta^2 - \dots .\end{aligned}$$

As can be seen, the thermodynamic potential covers all possible (and symmetry-allowed) types of order parameter-mechanical stress invariants. Operating further, as before, we obtain

$$s_{ij} = \begin{cases} s_{ij}^0 + \frac{a_i a_j}{\alpha(T-T_c)} & \text{for} \quad T > T_c \\ s_{ij}^0 + \frac{c_i c_j}{2\beta} + \frac{a_i a_j}{2\alpha(T-T_c)} + \frac{a_i c_j + a_j c_i}{2}[\alpha\beta(T_c - T)]^{-1/2} & \\ \quad + b_{ij}\left[\frac{\alpha(T_c-T)}{\beta}\right]^{1/2} + \frac{\alpha f_{ij}}{\beta}(T_c - T) & \text{for} \quad T < T_c . \end{cases} \tag{5.47}$$

This formula includes all possible anomalies of the elastic compliance coefficients – a jump, an increase in both phases by the Curie–Weiss law, a kink for certain coefficients different from zero in the symmetric phase; and vanishing or tending to infinity upon approach to T_c from the side of the nonsymmetric phase for morphic coefficients.

5.5 Two-Component Order Parameter: Elastic Properties of Crystals

We will now discuss the character of anomalies of the elastic properties of crystals, the phase transition in which is described by a two-component order parameter. We will limit ourselves to the case of a proper ferroelectric,

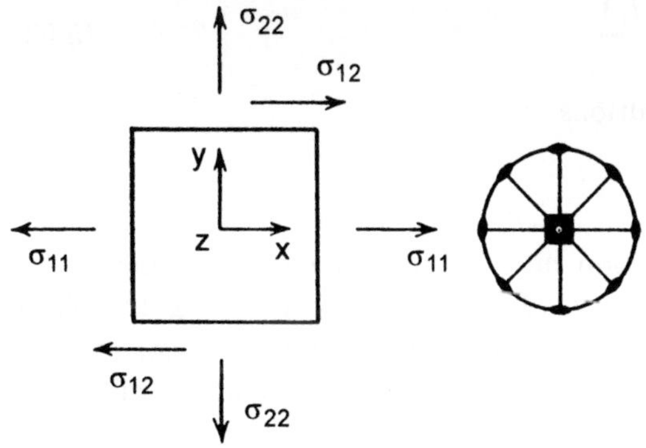

Fig. 5.8. Orientation of the components of the mechanical stress tensor relative to the symmetry elements of a crystal belonging to the group $4/mmm$.

assuming that the crystal has a symmetry group $4/mmm$ (Fig. 5.8) in the symmetric phase and a spontaneous polarization arises in a direction perpendicular to the fourfold axis, i.e., the components of the order parameter η_1 and η_2 exhibit the same transformation properties as the components E_1 and E_2 of the electric field strength vector.

We write the thermodynamic potential as a function of the components of the order parameter and the mechanical stress tensor:

$$\begin{aligned}\Phi = \Phi_0 &+ \frac{1}{2}\alpha(T-T_\mathrm{c})(\eta_1^2+\eta_2^2) + \frac{1}{4}\beta_1(\eta_1^2+\eta_2^2)^2 + \frac{1}{2}\beta_2\eta_1^2\eta_2^2 \\ &- g(\sigma_{11}+\sigma_{22}) - r_1(\eta_1^2+\eta_2^2)(\sigma_{11}+\sigma_{22}) - r_2(\eta_1^2\sigma_{22}+\eta_2^2\sigma_{11}) \qquad (5.48) \\ &- r_3\eta_1\eta_2\sigma_{12} - \frac{1}{2}s_1(\sigma_{11}^2+\sigma_{22}^2 - \frac{1}{2}s_2\sigma_{12}^2 - s_3\sigma_{11}\sigma_{22}.\end{aligned}$$

For the sake of simplicity, we have restricted our consideration, as before (see Sect. 5.2), to stresses and strains in the xy plane. It is not difficult, however, to see that by virtue of a symmetry higher than in the cases considered earlier, the thermal expansion coefficients are identical along the axes x and y and there are only three independent elastic compliance coefficients in the xy plane.

Using (5.1), we obtain for the strain tensor components:

$$u_{11} = g + r_1(\eta_{01}^2+\eta_{02}^2) + r_2\eta_{02}^2 + s_1\sigma_{11} + s_3\sigma_{22}, \qquad (5.49)$$

$$u_{22} = g + r_1(\eta_{01}^2+\eta_{02}^2) + r_2\eta_{01}^2 + s_1\sigma_{22} + s_3\sigma_{11}, \qquad (5.50)$$

$$u_{12} = r_3\eta_{01}\eta_{02} + s_2\sigma_{12}. \qquad (5.51)$$

It has been shown in Chap. 2 that a thermodynamic potential of the type (5.48) at $\sigma_{ij} = 0$ describes second-order phase transitions into two different nonsymmetric phases. One of these phases has four solutions of the types

$$\eta_{01} = 0, \qquad \eta_{02}^2 = -\frac{\alpha(T-T_\mathrm{c})}{\beta_1} \qquad (5.52)$$

or

$$\eta_{02} = 0, \qquad \eta_{01}^2 = -\frac{\alpha(T - T_c)}{\beta_1} \tag{5.53}$$

and the second also has four solutions:

$$\eta_{01}^2 = \eta_{02}^2 = -\frac{\alpha(T - T_c)}{2\beta_1 + \beta_2}. \tag{5.54}$$

We will first consider the transition into a structure corresponding to the solution (5.52). As can be seen from (5.49)–(5.51), in the nonsymmetric phase ($\eta_{01} \neq 0$, $\eta_{02} = 0$) the equilibrium components of the strain tensor u_{11}^0 and u_{22}^0 at $\sigma_{ij} = 0$ (spontaneous polarization) are no longer equal to each other:

$$u_{11}^0 = g + r_1\eta_{01}^2, \qquad u_{22}^0 = g + (r_1 + r_2)\eta_{01}^2,$$

that is, there appears an anisotropy of the thermal expansion in the xy plane. The intersection of the thermal expansion ellipsoid by the xy plane becomes an ellipse with the major and minor axes being parallel to the axes x and y. Upon transition into another nonsymmetric phase ($\eta_{01}^2 = \eta_{02}^2$, solution (5.54)) $u_{11}^0 = u_{22}^0$, as can be seen from (5.49) and (5.50), but there appears a spontaneous strain u_{12}^0 (see (5.51)). In other words, in this case too there appears an anisotropy of thermal expansion in the xy plane, but the major and minor axes of the ellipse are parallel to the bisectrices of the angles between the axes x and y. It is also easy to find jumps in the components of the thermal expansion tensor since $K_{ij} = \partial u_{ij}/\partial T$.

Note that in both cases there occurs a spontaneous strain $u_{11} = u_{22}$ or u_{12}, which is symmetry-forbidden at $T > T_c$. Therefore, a proper ferroelectric with a multicomponent order parameter may be regarded as an *improper ferroelastic.*

Turning to the coefficients of elastic compliance, let us first consider only the strains u_{11} and u_{22}, i.e., in the equations $\partial\Phi/\partial\eta_1 = 0$ and $\partial\Phi/\partial\eta_2 = 0$ we put $\sigma_{12} = 0$. We have

$$\eta_{01}[\alpha(T - T_c) + \beta_1(\eta_{01}^2 + \eta_{02}^2) + \beta_2\eta_{02}^2 \\ -2r_1(\sigma_{11} + \sigma_{22}) - 2r_2\sigma_{22}] = 0, \tag{5.55}$$

$$\eta_{02}[\alpha(T - T_c) + \beta_1(\eta_{01}^2 + \eta_{02}^2) + \beta_2\eta_{01}^2 \\ -2r_1(\sigma_{11} + \sigma_{22}) - 2r_2\sigma_{11}] = 0. \tag{5.56}$$

We will now be concerned with a phase corresponding to the solution (5.52). As can be seen from (5.55) and (5.56), upon superposition of the stresses σ_{11} and σ_{22} the character of the solution does not change, i.e., we can put $\eta_{01} \neq 0$ and $\eta_{02} = 0$ and use (5.55) for the determination of η_{01} (as a function of stresses). We have

$$\alpha(T - T_c) + \beta_1\eta_{01}^2 - 2r_1(\sigma_{11} + \sigma_{22}) - 2r_2\sigma_{22} = 0, \tag{5.57}$$

that is, the effect of imposition of external stresses was reduced to the replacement of the coefficient $\alpha(T - T_c)$ by $\alpha(T - T_c) - 2r_1(\sigma_{11} + \sigma_{22}) - 2r_2\sigma_{22}$

in the equation for the order parameter. We have already encountered this situation in considering the one-component order parameter. Operating in an analogous way, we find jumps of the elastic compliance coefficients; in tensor notation

$$\Delta s_{1111} = \frac{2r_1^2}{\beta_1}, \quad \Delta s_{2222} = \frac{2(r_1 + r_2)^2}{\beta_1}, \quad \Delta s_{1122} = \frac{2r_1(r_1 + r_2)}{\beta_1}. \tag{5.58}$$

Thus, the lowering of the crystal symmetry is reflected also in the set of its elastic compliances: at $T > T_c$ the equality $s_1 = s_2$ was valid, which is violated for the nonsymmetric phase.

For a nonsymmetric phase corresponding to the solution (5.54) $\eta_{01} \neq 0$ and $\eta_{02} \neq 0$, which is why from (5.55) and (5.56) it follows that the expressions enclosed in square brackets in these formulas are equal to zero. Subtracting and adding these expressions, we find

$$\begin{aligned} \alpha(T - T_c) + (\beta_1 + \frac{1}{2}\beta_2)(\eta_{01}^2 + \eta_{02}^2) &= (2r_1 + r_2)(\sigma_{11} + \sigma_{22}) \\ &= (2r_1 + r_2)\sigma_1, \end{aligned} \tag{5.59}$$

$$\beta_2(\eta_{01}^2 - \eta_{02}^2) = 2r_2(\sigma_{11} - \sigma_{22}) = 2r_2\sigma_2. \tag{5.60}$$

(Here we have used the notation $\sigma_1 = \sigma_{11} + \sigma_{22}$ and $\sigma_2 = \sigma_{11} - \sigma_{22}$.) The same operations can be done with (5.49) and (5.50). As a result, we obtain

$$u_1 = u_{11} + u_{22} = 2g + (2r_1 + r_2)(\eta_{01}^2 + \eta_{02}^2) + (s_1 + s_3)\sigma_1, \tag{5.61}$$

$$u_2 = u_{11} - u_{22} = -r_2(\eta_{01}^2 - \eta_{02}^2) + (s_1 - s_3)\sigma_1. \tag{5.62}$$

From (5.59)–(5.62) it follows that the equilibrium value of the components of the order parameter is a function of the components of the mechanical stress tensor; the corresponding variation of the quantity $\eta_{01}^2 + \eta_{02}^2$, as follows from (5.59), is

$$\Delta(\eta_{01}^2 + \eta_{02}^2) = \frac{2r_1 + r_2}{\beta_1 + \frac{1}{2}\beta_2}\sigma_1. \tag{5.63}$$

Analogously, from (5.61) we have

$$\begin{aligned} \Delta u_1 &= (2r_1 + r_2)\Delta(\eta_{01}^2 + \eta_{02}^2) + (s_1 + s_3)\sigma_1 \\ &= \frac{(2r_1 + r_2)^2}{\beta_1 + \frac{1}{2}\beta_2}\sigma_1 + (s_1 + s_3)\sigma_1, \end{aligned} \tag{5.64}$$

and from (5.60) and (5.62) we obtain

$$\Delta u_2 = -\frac{4r^2}{\beta_2}\sigma_2 + (s_1 - s_3)\sigma_2. \tag{5.65}$$

Returning to the components of the stress and strain tensors and establishing a linear relationship between them, we further find the jumps of the elastic compliance coefficients in the nonsymmetric phase; in tensor notation we have

$$\Delta s_{1111} = \frac{(2r_1 + r_2)^2}{2\beta_1 + \beta_2} - \frac{2r_2^2}{\beta_2}, \tag{5.66}$$

$$\Delta s_{1111} = \Delta s_{2222}, \tag{5.67}$$

$$\Delta s_{1122} = \frac{(2r_1 + r_2)^2}{2\beta_1 + \beta_2} + \frac{2r_2^2}{\beta_2}. \tag{5.68}$$

From (5.66)–(5.68) we cannot yet conclude that the number of independent elastic compliances is not changed upon transition into a nonsymmetric phase since we have not considered all the components of the stress tensor. Thus, if the component σ_{12} is taken into account, then in the square brackets of (5.55) there will also appear the term $-r_3\eta_2\sigma_{12}$ and to (5.56) there will be added the term $-r_3\eta_1\sigma_{12}$. In a phase corresponding to the solution (5.52) the equilibrium values of the order parameter components corresponding to the case $\sigma_{11} = \sigma_{22} = 0$, $\sigma_{12} \neq 0$ can be determined from (5.55) and (5.56) with account taken of the following considerations. Since we are interested only in linear elastic compliances, η_{01} may be assumed to be equal to its value in the absence of an external mechanical stress; $\eta_{02} = 0$ at $\sigma_{12} = 0$ and the dependence of η_{02} on σ_{12} can then be found from (5.56) with the added terms. So, we have

$$\eta_{02}[\alpha(T - T_c) + \beta_1(\eta_{01}^2 + \eta_{02}^2) + \beta_2\eta_{01}^2] - r_3\eta_{01}\sigma_{12} = 0$$

or with account taken of the fact that $\eta_{01}^2 \gg \eta_{02}^2$,

$$\eta_{02} = \frac{r_3\sigma_{12}\eta_{01}}{\alpha(T - T_c) + \eta_{01}^2(\beta_1 + \beta_2)} = \frac{r_3}{\eta_{01}\beta_2}\sigma_{12}. \tag{5.69}$$

Further we have (see (5.51))

$$u_{12} = r_3\eta_{01}\eta_{02} + s_2\sigma_{12} = \left(\frac{r_3^2}{\beta_2} + s_2\right)\sigma_{12}. \tag{5.70}$$

Thus, in this case

$$\Delta s_{1212} = \frac{r_3^2}{\beta_2}. \tag{5.71}$$

We will now turn to the solution (5.54). For the sake of definiteness, we assume that in the absence of mechanical stresses $\eta_{01} = \eta_{02}$. From (5.51) we now have

$$\begin{aligned} \Delta u_{12} = u_{12} - u_{12}^0 &= r_3\eta_{01}\Delta\eta_{02} + r_3\eta_{02}\Delta\eta_{01} + s_2\sigma_{12} \\ &= r_3\eta_{01}\Delta(\eta_{01} + \eta_{02}) + s_2\sigma_{12}, \end{aligned} \tag{5.72}$$

where $\Delta\eta_{01}$ and $\Delta\eta_{02}$ are increments of the order parameter components upon application of stresses. These increments may appear when either the stresses σ_{12} or the stresses σ_{11}, σ_{22} are applied. From formula (5.63) it can be seen that this is what actually occurs. Indeed, from this formula it follows that

$$\begin{aligned} 2\eta_{01}\Delta\eta_{01} + 2\eta_{02}\Delta\eta_{02} &= 2\eta_{01}\Delta(\eta_{01} + \eta_{02}) \\ &= \frac{(2r_1 + r_2)(\sigma_{11} + \sigma_{22})}{\beta_1 + \frac{1}{2}\beta_2}. \end{aligned} \tag{5.73}$$

Using (5.72) and (5.73), we find that in a nonsymmetric phase corresponding to the solution (5.54) there appear elastic compliances

$$s_{1112} = s_{2212} = \frac{r_3(2r_1 + r_2)}{2\beta_1 + \beta_2}. \tag{5.74}$$

Recall that, as was shown in Chap. 2, the phase under consideration exhibits a group of point symmetry $mm2$. It is not difficult to see that coefficients of the type (5.74) are missing in the sets of elastic compliances of this group that are usually used. The point is that the components of the elastic compliance matrix are given here in the coordinate system coupled to the symmetric phase. For the solution (5.52) such a choice corresponds to the usual choice of the crystallographic coordinate system, but this is not the case for the solution (5.54). In order to pass over to the ordinary system, we have to perform a rotation in the xy plane by 45° (see Chap. 3). In the new system of coordinates x', y' the components $s_{1'1'1'2'}$ and $s_{2'2'1'2'}$ are equal to zero, but $s_{1'1'1'1'} \neq s_{2'2'2'2'}$, which of course is quite natural since (see Chap. 2) to the solutions (5.52) and (5.54) there corresponds one and the same point group, whose elements are, however, oriented differently (relative to the symmetry element of the symmetric phase) for different solutions.

Let us now take into account the variation of η_{01} and η_{02} under the action of stresses σ_{12}. Assuming that $\sigma_{11} = \sigma_{22} = 0$ in (5.55) and (5.56), adding terms linear in σ_{12} and taking into account that the expressions enclosed in square brackets in (5.55) and (5.56) are equal to zero at equilibrium values of η_{01} and η_{02} corresponding to the solution (5.54), we get

$$2\beta_1\eta_{01}(\Delta\eta_{01} + \Delta\eta_{02}) + \beta_2\eta_{01}\Delta\eta_{02} = r_3\sigma_{12}, \tag{5.75}$$

$$2\beta_1\eta_{01}(\Delta\eta_{01} + \Delta\eta_{02}) + \beta_2\eta_{01}\Delta\eta_{01} = r_3\sigma_{12}. \tag{5.76}$$

Addition of (5.75) and (5.76) yields

$$\Delta(\eta_{01} + \eta_{02}) = \frac{2r_3\sigma_{12}}{(2\beta_1 + \beta_2)\eta_{01}}. \tag{5.77}$$

Substitution of (5.77) into (5.72) gives

$$\Delta s_{1212} = \frac{2r_3^2}{2\beta_1 + \beta_2}. \tag{5.78}$$

5.6 Piezoelectric Effect and Electrostriction in the Case of One-Component Order Parameter and Centrosymmetric Paraelectric Phase

We will be concerned here with the electromechanical properties of ferroelectric crystals and with the singularities of the temperature dependence of the corresponding piezoelectric and electrostriction coefficients in the region of phase transitions. Under the action of an electric field the crystal undergoes a deformation:

$$u_{ij} = d_{kij} E_k + q_{ijkl} E_k E_l . \tag{5.79}$$

Here the first term on the right-hand side describes the converse piezoelectric effect and the second describes electrostriction. A third-rank tensor d_{kij} has nonzero components only in crystals devoid of an inversion center; in centrosymmetric crystals the strain quadric in electric field is the only electromechanical effect. The converse piezoelectric effect is accompanied by a direct effect, which describes the origination of surface charges on a crystal under the action of stresses:

$$P_k = d_{kij} \sigma_{ij} . \tag{5.80}$$

Naturally, in a nonsymmetric polar phase a ferroelectric crystal belongs to one of the polar (and, hence, noncentrosymmetric) symmetry groups and must exhibit a piezoelectric effect.

If above the phase transition point the symmetry elements of a crystal include an inversion center, no piezoelectric effect will be observed; so, all piezoelectric moduli that appear in this case below the phase transition point should be classified as morphic.

In order to quantitatively analyze the behavior of the electromechanical parameters of a crystal, we will proceed from the incomplete thermodynamic potential, which depends on the order parameter and on the components of the stress tensor σ_{ij} and the vector of the electric field E_i. A detailed analysis will be made for the TGS crystal (symmetry change $2/m \to 2$). Figure 5.9 shows the orientation of the crystallophysical coordinates and of the quantities σ_{ij} and E_k relative to the crystal sample. The transformation properties of the order parameter and of the quantities σ_{11}, σ_{22}, σ_{12}, E_1 and E_2 are demonstrated in Table 5.3.

Below the transition point the crystal symmetry is lowered to the group 2 and a piezoelectric effect is observed; the transformation of the matrix of piezo moduli in such a transition may be illustrated schematically as follows:

$$\begin{Vmatrix} 0 & 0 & 0 & 0 & 0 & 0 \\ 0 & 0 & 0 & 0 & 0 & 0 \\ 0 & 0 & 0 & 0 & 0 & 0 \end{Vmatrix} \to \begin{Vmatrix} 0 & 0 & 0 & d_{123} & 0 & d_{112} \\ d_{211} & d_{222} & d_{233} & 0 & d_{231} & 0 \\ 0 & 0 & 0 & d_{323} & 0 & d_{312} \end{Vmatrix} .$$

It should be pointed out that all morphic piezo moduli may be divided into two groups – d_{2jk} (associated with a polarization along the y axis) and $d_{i\neq 2,jk}$

Table 5.3. Transformation of the order parameter and of σ_{11}, σ_{22}, σ_{12}, E_1, and E_2.

$2/m$	1	m	2	$\bar{1}$
η	η	$-\eta$	η	$-\eta$
E_1	E_1	E_1	$-E_1$	$-E_1$
E_2	E_2	$-E_2$	E_2	$-E_2$
σ_{11}	σ_{11}	σ_{11}	σ_{11}	σ_{11}
σ_{22}	σ_{22}	σ_{22}	σ_{22}	σ_{22}
σ_{12}	σ_{12}	$-\sigma_{12}$	$-\sigma_{12}$	σ_{12}

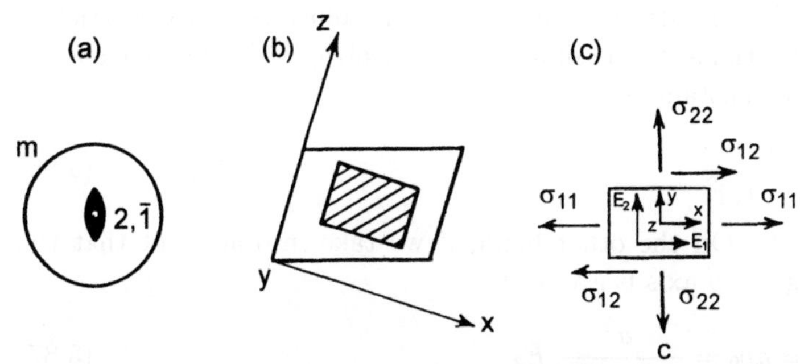

Fig. 5.9a–c. Orientation of the components of the mechanical stress tensor and of the electric field vector relative to the symmetry elements of a crystal belonging to the group $2/m$: (**a**) symmetry elements; (**b**) crystallophysical coordinates and orientation of a sample (hatched area); (**c**) orientation of the components of the tensor σ_{ij} and the vector $\boldsymbol{E}$.

(associated with a polarization along the x and z axes). For a concrete analysis we choose one piezo modulus from each group.

We write down the thermodynamic potential of a crystal as a function of η, σ_{11}, σ_{12}, and E_2 with account taken of the transformation properties of all the parameters:

$$\Phi = \Phi_0 + \frac{1}{2}\alpha(T - T_c)\eta^2 + \frac{1}{4}\beta\eta^4 - g_1\sigma_{11} - a\eta E_2 - r\eta^2\sigma_{11} - c\eta E_1\sigma_{12}. \quad (5.81)$$

It should be noted that, in accordance with Table 5.2, the products $\sigma_{11}E_2$ and $\sigma_{11}E_1$ are not invariants of the symmetric phase in this particular case; this just corresponds to the absence of the piezo effect in the symmetric phase (cf. (5.37)). Moreover, we do not take into account an invariant of the type $E_2^2\sigma_{11}$ since it has a next order of smallness as compared with $\eta^2\sigma_{11}$ (see Sect. 3.1) and also $E_1E_2\sigma_{12}$, which is small as compared with $E_1\eta\sigma_{12}$.

For a mechanically free crystal we obtain from (5.81)

$$\left(\frac{\partial\Phi}{\partial\eta}\right)_{\sigma=0} = \alpha(T - T_c)\eta_0 + \beta\eta_0^3 - aE_2 = 0, \quad (5.82)$$

$$u_{11} = -\frac{\partial \Phi}{\partial \sigma_{11}} = g_1 + r\eta_0^2. \tag{5.83}$$

It is obvious that at $T > T_c$ and small E_2

$$\eta_0 = \frac{a}{\alpha(T - T_c)} E_2, \tag{5.84}$$

and, hence,

$$u_{11} = g_1 + \frac{a^2 r}{\alpha^2 (T - T_c)^2} E_2^2. \tag{5.85}$$

It can be seen that the strain in the paraelectric phase is of electrostrictive nature and is proportional to the square of the electric field. The corresponding electrostriction constant

$$q_{1122} = \frac{a^2 r}{\alpha^2 (T - T_c)^2} \tag{5.86}$$

diverges as $T \to T_c$. On the other hand, if we take into acoount that the polarization along the y axis is given by

$$P_2 = -\frac{\partial \Phi}{\partial E_2} = a\eta_0 = \frac{a^2}{\alpha(T - T_c)} E_2, \tag{5.87}$$

then the dependence of the strain on polarization will assume the following form:

$$u_{11} = g_1 + \frac{r}{a_2} P_2^2. \tag{5.88}$$

It turns out that the relationship between the strain u_{11} and the polarization P_2 is determined by the coefficient r/a^2, which is temperature independent. The relationship between the components of the strain tensor and polarization vector is often described by the equations

$$u_{ij} = Q_{ijkl} P_k P_l; \tag{5.89}$$

in accordance with the result obtained by us:

$$Q_{1122} = r/a^2 = \text{const.} \tag{5.90}$$

These electrostriction constants are called "true" in the sense that they are temperature independent even in the vicinity of the phase transition point and in the polar phase.

At $T < T_c$ the equilibrium value of the order parameter is no longer equal to zero; now $\eta_0 = \eta_{00} + \Delta\eta$, where η_{00} is the equilibrium value of the order parameter at E_2, $\sigma_{ij} = 0$, and $\Delta\eta$ is its change with a change of one of the external parameters. From (5.82) we have

$$\Delta\eta = \frac{a}{2\alpha(T_c - T)} E_2, \tag{5.91}$$

and from (5.83) the corresponding component of the strain tensor is

$$u_{11} = r(\eta_{00} + \Delta\eta)^2 = u_{11}^0 + 2r\eta_{00}\Delta\eta \tag{5.92}$$

with an accuracy of up to terms of second order of smallness in $\Delta\eta$. Here $u_{11}^0 = r\eta_{00}^2$ is the strain of the crystal associated with the appearance of an order parameter – spontaneous strain – in the crystal. Thus, we obtain

$$\Delta u_{11} = u_{11} - u_{11}^0 = \eta_{00}\frac{ar}{\alpha(T_c - T)}E_2. \tag{5.93}$$

Comparing (5.79) with (5.93), we find that

$$d_{211} = \eta_{00}\frac{ar}{\alpha(T_c - T)} = \frac{ar}{(\alpha\beta)^{1/2}(T_c - T)^{1/2}}, \tag{5.94}$$

that is, the given piezoelectric modulus appears at the phase transition point "from infinity" and then falls off in moving away from T_c as $(T_c - T)^{1/2}$. Note that, according to (5.92) and (5.93), the linear piezoelectric effect that arises in the polar phase is intimately connected with electrostriction: the strain induced by an external electric field is, as it were, a linearized continuation of the spontaneous strain (Fig. 5.10). Two circumstances should be stressed here.

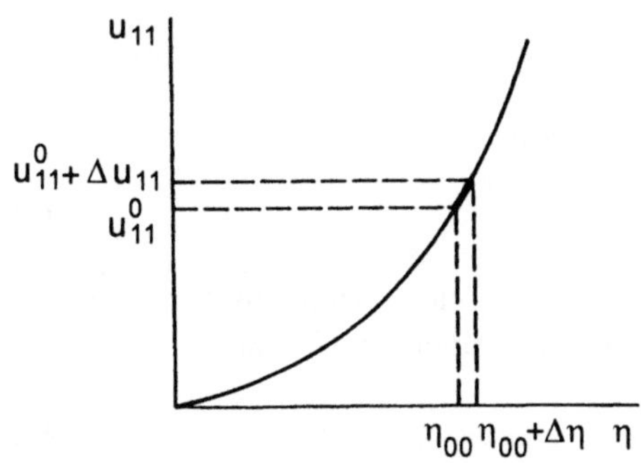

Fig. 5.10. The piezoelectric effect in ferroelectrics with a centrosymmetric phase as a "linearized" electrostriction.

First, it is remarkable that, according to (5.94), the piezoelectric modulus under consideration has anomalously large values in the polar phase near T_c, which may be of certain practical interest. Second, the piezoelectric modulus appears to be proportional to the order parameter and, hence, the sign of the modulus depends on the direction of the spontaneous polarization vector. This means, in particular, that for a significant piezoelectric effect to be realized it is necessary to have a monodomain crystal; a multidomain crystal will not be piezoelectrically active. A monodomain crystal can be produced by an external electric field. The dependence of the strain on electric field with account taken of the possibility of monodomainization and change of the direction (sign) of spontaneous polarization is shown in Fig. 5.11. It can be seen that in the case considered the strain sign is independent of the sign of the electric field if the magnitude of the latter exceeds that of the coercive field.

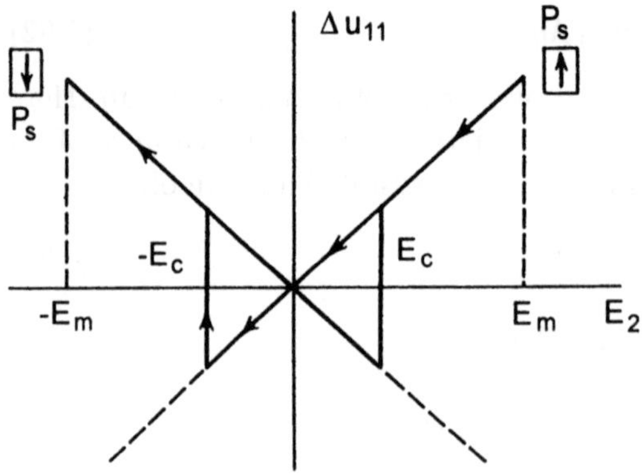

Fig. 5.11. The piezoelectric strain in the polar phase of a ferroelectric crystal with a complete cycle of variation of the electric field.

Let us now consider the piezo modulus from another group not associated with the polarization along the polar axis. From (5.81) we have

$$u_{12} = -\frac{\partial \Phi}{\partial \sigma_{12}} = c\eta_0 E_1. \tag{5.95}$$

With account taken of (5.84), we find at $T > T_c$ the typical electrostriction relationship

$$u_{12} = c\frac{a}{\alpha(T - T_c)} E_1 E_2 \tag{5.96}$$

to the temperature-dependent electrostriction constant

$$q_{1212} = \frac{ca}{\alpha(T - T_c)}. \tag{5.97}$$

In the polar phase $\eta_0 \neq 0$ and from (5.95) it follows that the temperature dependence of the piezoelectric modulus d_{112} coincides with η_0:

$$d_{112} = c\eta_0 = \frac{c\alpha^{1/2}(T_c - T)^{1/2}}{\beta^{1/2}}. \tag{5.98}$$

A noteworthy point is the similarity between the temperature dependences of the piezoelectric moduli of type d_{2ij} (coupled to the polarization along the polar axis) and type $d_{i\neq 2,jk}$ (not coupled to the polarization along the polar axis) and the temperature dependences of the morphic elastic moduli of two types (see Fig. 5.7).

5.7 Piezoelectric Effect in the Case of One-Component Order Parameter and Noncentrosymmetric Paraelectric Phase

We will now discuss the electromechanical properties of crystals which have no center of inversion in the paraelectric phase and, hence, exhibit piezoelectric properties at $T > T_c$. To the category of such crystals belong ferroelectric-ferroelastics, Rochelle salt and potassium dihydrogen phosphate.

Let us consider, as an example, a phase transition with the symmetry change 222 → 2 (Rochelle salt). Locating the coordinates in the way indicated in Fig. 5.6 and choosing the x axis along the polar axis 2 in the ferroelectric phase, we write down schematically the matrices of the piezoelectric moduli for both phases:

$$\begin{array}{c} 222 \\ \left|\begin{array}{cccccc} 0 & 0 & 0 & d_{123} & 0 & 0 \\ 0 & 0 & 0 & 0 & d_{213} & 0 \\ 0 & 0 & 0 & 0 & 0 & d_{312} \end{array}\right| \end{array} \rightarrow \begin{array}{c} 2 \\ \left|\begin{array}{cccccc} d_{111} & d_{122} & d_{133} & d_{123} & 0 & 0 \\ 0 & 0 & 0 & 0 & d_{213} & d_{212} \\ 0 & 0 & 0 & 0 & d_{313} & d_{312} \end{array}\right| \end{array} .$$

We will be concerned not with morphic piezoelectric moduli (their analysis in this case is completely analogous to the treatment for crystals with a centrosymmetric paraelectric phase given above) but with "transient" piezoelectric moduli d_{123}, d_{213} and d_{312}. Like the morphic moduli, the latter may be split into two types – those coupled to the polarization along the polar axis (d_{1jk}) and those which are not ($d_{i\neq 1,j,k}$).

In order to choose mixed invariants correctly, we draw up a table (see Table 5.4) which reflects the transformation properties of all quantities of interest to us.

Table 5.4. Transformation properties.

222	1	2_x	2_y	2_z
η	η	η	$-\eta$	$-\eta$
E_1	E_1	E_1	$-E_1$	$-E_1$
E_2	E_2	$-E_2$	E_2	$-E_2$
σ_{11}	σ_{11}	σ_{11}	σ_{11}	σ_{11}
σ_{13}	σ_{13}	$-\sigma_{13}$	σ_{13}	$-\sigma_{13}$
σ_{23}	σ_{23}	σ_{23}	$-\sigma_{23}$	$-\sigma_{23}$

We see that to the invariants linear in η and E_2, η and σ_{23} (see (5.37)) there are added $\sigma_{13}E_2$ and $\eta^2\sigma_{13}E_2$, as a result of which the thermodynamic potential assumes the following form:

$$\begin{aligned} \Phi = \Phi_0(T) &+ \frac{1}{2}\alpha(T - T_c)\eta^2 + \frac{1}{4}\beta\eta^4 \\ &- a_1\eta E_1 - a_2\eta\sigma_{23} - b\eta^2\sigma_{13}E_2 - d\sigma_{13}E_2 . \end{aligned} \tag{5.99}$$

whence, assuming for the moment that $\sigma_{13} = 0$, we have

$$u_{23} = -\frac{\partial\Phi}{\partial\sigma_{23}} = a_2\eta_0 . \tag{5.100}$$

In the presence of an electric field E_1 and a mechanical stress σ_{23} the equilibrium value of the order parameter changes; in this case at $T > T_c$, $\eta_{00} = 0$ and

$$\Delta\eta = \frac{a_1 E_1 + a_2 \sigma_{23}}{\alpha(T - T_c)}, \tag{5.101}$$

which is why in the paraelectric phase

$$\Delta u_{23} = u_{23} - u_{23}^0 = a_2 \Delta\eta = \frac{a_1 a_2}{\alpha(T - T_c)} E_1 + \frac{a_2^2}{\alpha(T - T_c)} \sigma_{23}, \tag{5.102}$$

where $u_{23}^0 = a_2 \eta_{00}$ is the spontaneous strain. The proportionality coefficient between Δu_{23} and E_1 is what we call the piezoelectric modulus:

$$d_{123} = \frac{a_1 a_2}{\alpha(T - T_c)} \quad \text{at} \quad T > T_c. \tag{5.103}$$

From (5.102) it follows that the elastic compliance s_{2323} varies in the same manner as a function of temperature.

In the ferroelectric phase $\eta_{00}^2 = -\alpha(T - T_c)/\beta$; evaluating the order parameter increment linear in E_1 and σ_{23}:

$$\Delta\eta = \frac{a_1 E_1 + a_2 \sigma_{23}}{2\alpha(T_c - T)}, \tag{5.104}$$

we find

$$\Delta u_{23} = \frac{a_1 a_2}{2\alpha(T_c - T)} E_1 + \frac{a_2^2}{2\alpha(T_c - T)} \sigma_{23}. \tag{5.105}$$

Thus, both the piezoelectric modulus d_{123} and the elastic compliance coefficient s_{2323} obey the Curie–Weiss law in both phases. One can say that these anomalies result from an anomaly in the dielectric constant (dielectric permittivity). Indeed, the "true" piezoelectric constant, which relates the strain u_{23} to the polarization P_1, is equal to a_2/a_1, i.e., is independent of temperature.

Finally let us consider the second type of "transient" piezoelectric moduli; differentiation of (5.99) with respect to σ_{13} yields

$$u_{13} = -\frac{\partial \Phi}{\partial \sigma_{13}} = (b\eta_0^2 + d)E_2. \tag{5.106}$$

It can be seen that the strain u_{13} occurs only in the presence of an external electric field E_2, i.e., a spontaneous strain of the type u_{13}^0 does not arise in this case. With $T \to T_c$ and σ_{13}, $E_2 \to 0$ we obtain

$$d_{213} = \begin{cases} d & \text{for} \quad T > T_c \\ d + \alpha b(T_c - T)/\beta & \text{for} \quad T < T_c. \end{cases} \tag{5.107}$$

This means that d_{213} experiences a kink at the phase transition point.

Summing up the analysis carried out, we may state that the piezoelectric moduli of a crystal with a second-order phase transition undergo an anomalous change at the transition point. An analogous approach can be used to consider the electromechanical properties of crystals, the phase transitions in which are described by a multicomponent order parameter; an example of this analysis has been given above for the elucidation of the character of anomalies in the elastic compliance coefficients. We presume that the reader has a sufficient knowledge of the methods of finding anomalies in various thermodynamic quantities with the aid of the Landau theory and will be able to do it on his own.

Let η be transformed as a component of a third-rank tensor. This means that in the expression for the thermodynamic potential there will appear additonal invariants, of which we are interested only in $\eta E_i \sigma_{kl}$ (it is obvious that $E_i \sigma_{kl}$ has the transformation properties of the component of a third-rank tensor). For the component u_{kl} we have at $\sigma_{kl} = 0$

$$u_{kl} \sim \eta_0 E_i . \tag{5.108}$$

Thus, a piezoelectric effect occurs in the nonsymmetric phase, i.e., there is observed a strain proportional to the electric field strength. This is one more example of the morphic effect – the appearance of new nonzero components of material tensors with the symmetry being lowered. The value of the piezoelectric constant is proportional to the equilibrium value of the order parameter η_0, i.e., to different domains there correspond different signs of the piezoelectric constant. It would seem that the piezoelectric constant for such a phase transition (the situation under consideration is realized, for example, for the transition from the cubic phase in NH_4Cl and for the α–β transition in quartz) plays the same role as the polarization for a ferroelectric transition. It is sometimes said that in the case under discussion the piezoelectric constant is an order parameter. This statement, however, is not correct. In contrast to polarization, the piezoelecrtic constant is a material rather than a field tensor. But the variables, on which the incomplete thermodynamic potential depends and which have the meaning of the degrees of freedom of the system, are field quantities. Note that the proportionality between η_0 and the value of the piezoelectric constant provides a convenient tool of measurement of the temperature dependence of η_0. As regards other anomalies, the phase transition under consideration is devoid of a specific character as compared with the case of a nontensorial order parameter.

Further, let η_0 be transformed as a certain component of a fourth-rank tensor. Such a situation is realized, for example, for the α–β transition in quartz. It may seem strange that the order parameter for quartz exhibits simultaneously the transformation properties of a third- and a fourth-rank tensor. It would seem that the difference between tensors of various ranks is just what we call transformation properties. This is valid, however, for the group of all transformations of the symmetry of isotropic space, i.e., for all rotations of the coordinate system and reflections in all planes. At the same time, while speaking of the transformation properties of η we mean the symmetry group of the symmetric phase. This group contains only a limited number of rotations and reflections, in which the components of tensors of various ranks can be transformed in the same manner.

In the case under discussion, to the cross-terms will be added the following terms:

$$\eta\sigma_{ik}\sigma_{lm}, \qquad \eta E_i E_k \sigma_{lm}. \tag{5.109}$$

Since $\sigma_{ik}\sigma_{lm}$ is not an invariant (this product is tranformed as an order parameter), it follows that the thermodynamic potential does not have a term containing only $\sigma_{kl}\sigma_{lm}$, i.e., the symmetric phase is devoid of the corresponding component of the elastic compliance tensor $s_{iklm} = 0$. In the nonsymmetric phase this component will appear with s_{iklm} being approximately equal to η_0, i.e., the new component of the elastic compliance tensor has different signs in different domains. This does not mean, of course, that the elastic properties of different domains are different. The fact that the shear compliance s_{iklm} has different signs in different domains implies only the difference in the orientation of the indicatrix surfaces of the elastic compliance tensor in different domains. In the case considered the measurement of the compliance s_{iklm} is a convenient method of investigating the temperature dependence of η_0. It is natural that if η coincides in transformation properties with a sixth-rank tensor, in the nonsymmetric phase there will appear a new constant of the tensor of elastic compliances proportional to η_0, etc.

6. Fluctuations of the Order Parameter in Phenomenological Theory

6.1 Spatially Inhomogeneous Fluctuations of the Order Parameter in the Incomplete Thermodynamic Potential

We will now be concerned with the assumptions that were made in the construction of the Landau phenomenological theory of phase transitions. The basic assumption of the theory is the assumption of the normal, "nonsingular" temperature dependence of the coefficients in the expansion of the incomplete themodynamic potential. It was assumed, for example, that the dependence $\Phi_0(p,T)$ in the vicinity of the phase transition point is the same as it is far away from it. By definition, the incomplete thermodynamic potential $\Phi_0(p,T)$ takes account of the contribution of all degrees of freedom, except η, to the complete thermodynamic potential. It was assumed implicitly that these degrees of freedom "do not feel" a phase transition. The concept that the phase transition does not "touch upon" other degrees of freedom, except η, lies also implicitly at the basis of the assumption of the "nonsingular" temperature dependence of the coefficients in the expansion of the thermodynamic potential.

We will show below that these assumptions are not correct. In the preceding chapters we have dealt with the incomplete thermodynamic potential which is dependent on one degree of freedom of a crystal – the order parameter η, which is homogeneous throughout its volume. The spatially inhomogeneous distributions of the order parameter correspond to other degrees of freedom of the crystal and their number is comparable with the total number of particles or with the total number of degrees of freedom. It turns out that the thermal motion corresponding to these degrees of freedom is determined by a phase transition, which manifests itself in a significant increase of fluctuations corresponding to those degrees of freedom. Taking this circumstance into account, it is possible to refine the phenomenological theory.

Let us consider in more detail the problem of *fluctuations of the order parameter*. Even from the form of the thermodynamic potential (2.33) it follows that near the Curie point the fluctuations must sharply increase.

This increase results from the fact that the $\Phi(\eta)$ curve exhibits a minimum at $\eta = 0$ above T_c and a maximum below T_c. Hence, at small η and $T = T_c$ the

$\Phi(\eta)$ plot becomes very weak, so that rather significant changes in the order parameter do not bring about a noticeable change in the thermodynamic potential; the probability of fluctuations increases in this case.

We will now determine the mean square deviation of the order parameter in the symmetric phase ($\eta = 0$):

$$\langle \eta^2 \rangle = \int_{-\infty}^{\infty} \eta^2 \exp\left(-\frac{\Phi(T,\eta)}{k_B T}\right) d\eta \left[\int_{-\infty}^{\infty} \exp -\frac{\Phi(T,\eta)}{k_B T} d\eta\right]^{-1} . \quad (6.1)$$

Restricting ourselves in the expression for Φ in (6.1) to the term quadric in η (see (2.33)) and substituting the values of tabular integrals, we obtain

$$\langle \eta^2 \rangle = k_B T / v\alpha(T - T_c) \quad (6.2)$$

where v is the volume of the crystal.[1]

One can see that $\langle \eta^2 \rangle \to \infty$ as $T \to T_c$. It is clear that near the Curie point there is also observed an increase in the probability of *spatially inhomogeneous* fluctuations in order parameter associated with states for which $\eta = \eta(\boldsymbol{r})$, and the order parameter varies rather slowly throughout the crystal volume. Therefore, if the local value of the order parameter $\eta(\boldsymbol{r})$ is given in the form

$$\eta(\boldsymbol{r}) = \sum_{\boldsymbol{k}} \eta_{\boldsymbol{k}} e^{i\boldsymbol{k}\cdot\boldsymbol{r}} , \quad (6.3)$$

where $\eta_{\boldsymbol{k}}$ is its Fourier component with a wave vector $\boldsymbol{k}$, then we can state that the values of $\eta_{\boldsymbol{k}}$ corresponding to small $\boldsymbol{k}$ are strongly fluctuating.

We will now discuss in more detail the meaning of the expansion (6.3). It presupposes, in fact, that the crystal is treated as a homogeneous continuum medium rather than a discrete structure. Of course, this notion does not correspond to the function $\eta(\boldsymbol{r})$ which does not change too rapidly in space: the characteristic lengths over which this function undergoes significant changes must, in any case, be much larger than the lattice constant. This means that the summation over $\boldsymbol{k}$ in (6.3) must be limited to the vectors, the moduli of which do not exceed $k_{\max} \ll \pi/a$ (a is the lattice constant). It will be reasonable to choose $k_{\max}$ so that we can take into account in expression (6.3) those degrees of freedom which "feel" a phase transition. In doing so we split all degrees of freedom into two groups, one "feeling" and the other "nonfeeling" a phase transition; the variables of the thermodynamic potentials will be degrees of freedom corresponding only to the first group.

The quantity $\eta(\boldsymbol{r})$ is, of course, real and therefore in the expansion (6.3) the Fourier components η must satisfy the condition

$$\eta_{-\boldsymbol{k}} = \eta_{\boldsymbol{k}}^* \quad (6.4)$$

(the asterisk implies a complex-conjugate value). Indeed,

[1] Unlike the preceding chapters, in this chapter the crystal is not assumed to have a unit volume and, hence, $\Phi(p,T) = v\varphi(p,T)$.

$$\eta(\boldsymbol{r}) = \sum_{\boldsymbol{k}} \eta_{\boldsymbol{k}} \mathrm{e}^{\mathrm{i}\boldsymbol{k}\cdot\boldsymbol{r}} = \eta^*(\boldsymbol{r}) = \sum_{\boldsymbol{k}} \eta^*_{\boldsymbol{k}} \mathrm{e}^{\mathrm{i}\boldsymbol{k}\cdot\boldsymbol{r}}$$

from which it follows that the equality (6.4) is valid. The quantities $\mathrm{e}^{\mathrm{i}\boldsymbol{k}\cdot\boldsymbol{r}}$ are assumed, as usual, to satisfy periodic boundary conditions; so, if the volume of a rectangular crystal sample is $v = L_1 L_2 L_3$, then the permissible values of the components of the wave vector are

$$k_x = \frac{2\pi}{L_1} n_1, \qquad k_y = \frac{2\pi}{L_2} n_2, \qquad k_z = \frac{2\pi}{L_3} n_3 , \tag{6.5}$$

where n_1, n_2, $n_3 = 0, 1, 2, \ldots$ are integers. Here the values of $\boldsymbol{k}$ are distributed in the space of k_x, k_y, k_z with a constant density equal to $v/(2\pi)^3$. Then, the total number of wave vectors appearing in (6.3) will be equal to $(4\pi/3)k^3_{\mathrm{max}} v/(2\pi)^3$, i.e., will be proportional to the crystal volume. In other words, the relative fraction of the number of degrees of freedom corresponding to weakly inhomogeneous, i.e., longwave, fluctuations η, remains finite for systems as large as desired. Note that in (6.3) the term $\eta_{k=0}$ corresponds to an order parameter which is homogeneous throughout the entire volume of the crystal.

The incomplete thermodynamic potential of the crystal will now be represented by a functional:

$$\Phi = \int_v \varphi(\eta(\boldsymbol{r}))\mathrm{d}v , \tag{6.6}$$

where $\varphi(\eta(\boldsymbol{r}))$ is the density of the thermodynamic potential.

Let us derive an expression for the density of the incomplete thermodynamic potential by proceeding from the following considerations. Suppose that the crystal is divided into a number of regions, so small that inside each of the regions we may assume that $\eta(\boldsymbol{r}) = \mathrm{const}$. The crystal energy may be represented as a sum of the volume energies of these regions and the energy of interaction between them. The expression for the density of the thermodynamic potential of each such region may be written in a form similar to what is done in the Landau theory; the interaction between the regions may be taken into account by assuming that the variation of the order parameter in the boundary region is sufficiently smooth and involves a large number of unit cells. Then, for the neighboring regions n and $n+1$ the approximate expression for the interaction energy per unit volume will look as follows:

$$\frac{c}{2d^3}[\eta(x_n) - \eta(x_{n+1})]^2 \approx \frac{1}{2}\delta\left(\frac{\mathrm{d}\eta}{\mathrm{d}x}\right)^2 ,$$

where x is a certain direction in the crystal; c and δ are constants which differ in dimensions: $\delta = c/d^2$; and d is a quantity with the dimensions of length, which is of the order of the region size. The final form of the expression for the density of the incomplete thermodynamic potential is

$$\varphi(\eta) = \varphi_0 + \frac{1}{2}A\eta^2(\boldsymbol{r}) + \frac{1}{4}\beta\eta^4(\boldsymbol{r}) + \frac{1}{2}\delta(\operatorname{grad}\eta)^2 , \tag{6.7}$$

where $A = \alpha(T - T_c)$.

Note that expression (6.7) can also be obtained on the basis of considerations associated with the possibility of expanding the thermodynamic potential into a series in powers not only of η but of the derivatives of η as well. The constant δ, as will be seen at a later time, has a direct relation to the form of the vibrational spectrum of the crystal (see Chap. 9) and is determined by the interaction of atoms which are displaced during the phase transition of the sublattices: the last term in (6.7) renders the sharp (discontinuous) changes of the order parameter energetically unfavorable.

So, the incomplete thermodynamic potential is given by

$$\Phi = \varphi_0 v + \int_v \left\{ \frac{1}{2}A\eta^2(\boldsymbol{r}) + \frac{1}{4}\beta\eta^4(\boldsymbol{r}) + \frac{1}{2}\delta(\operatorname{grad}\eta)^2 \right\} \mathrm{d}v. \tag{6.8}$$

Substituting (6.3) into this equation and taking into account that

$$\int_v \mathrm{e}^{\mathrm{i}\boldsymbol{k}_1\cdot\boldsymbol{r}}\mathrm{e}^{\mathrm{i}\boldsymbol{k}_2\cdot\boldsymbol{r}}\mathrm{d}v = \begin{cases} 0 & \text{at} \quad \boldsymbol{k}_1 + \boldsymbol{k}_2 \neq 0 \\ v & \text{at} \quad \boldsymbol{k}_1 + \boldsymbol{k}_2 = 0 \end{cases}$$

$$\int_v \mathrm{e}^{\mathrm{i}\boldsymbol{k}_1\cdot\boldsymbol{r}}\mathrm{e}^{\mathrm{i}\boldsymbol{k}_2\cdot\boldsymbol{r}}\mathrm{e}^{\mathrm{i}\boldsymbol{k}_3\cdot\boldsymbol{r}}\mathrm{e}^{\mathrm{i}\boldsymbol{k}_4\cdot\boldsymbol{r}}\mathrm{d}v = \begin{cases} 0 & \text{at} \quad \boldsymbol{k}_1 + \boldsymbol{k}_2 + \boldsymbol{k}_3 + \boldsymbol{k}_4 \neq 0 \\ v & \text{at} \quad \boldsymbol{k}_1 + \boldsymbol{k}_2 + \boldsymbol{k}_3 + \boldsymbol{k}_4 = 0 \end{cases}$$

and $\operatorname{grad}\eta = \sum_{\boldsymbol{k}} \mathrm{i}\boldsymbol{k}\eta_{\boldsymbol{k}}\mathrm{e}^{\mathrm{i}\boldsymbol{k}\cdot\boldsymbol{r}}$, we obtain, retaining only terms of fourth order in η,

$$\Phi(p, T, \eta_{\boldsymbol{k}=0}, \ldots, \eta_{\boldsymbol{k}_{\max}}) = \varphi_0 v + \frac{1}{2}v\sum_{\boldsymbol{k}}(A + \delta k^2)\eta_{\boldsymbol{k}}\eta_{-\boldsymbol{k}} + \frac{1}{4}v\beta \sum_{\boldsymbol{k}_1+\ldots+\boldsymbol{k}_4=0} \eta_{\boldsymbol{k}_1}\eta_{\boldsymbol{k}_2}\eta_{\boldsymbol{k}_3}\eta_{\boldsymbol{k}_4} . \tag{6.9}$$

Here the summation has been taken over all values of $k \leq k_{\max}$, which determine the number of variables in the incomplete thermodynamic potential. It should be noted that the coefficients φ_0, A and β in (6.9) do not coincide with the coefficients of the thermodynamic potential in Landau theory since in the derivation of (6.9) the integration was carried out over a much lower number of variables. These variables do not "feel" a phase transition, and now there is every ground for believing that these new coefficients do not display any singularities in temperature behavior near T_c. The transformation from expression (6.9) to the potential of Landau theory can be made in a simple manner if the last sum on the right-hand side of (6.9) is ignored. It means that the results obtained will be valid only in a symmetric phase and not too close to the transition point.

Taking account of the condition (6.4) and the general procedure outlined in Chap. 2, we obtain the equilibrium thermodynamic potential of the system:

$$\Phi(p,T) = \Phi_0 - k_\mathrm{B}T \times \ln \int_{-\infty}^{\infty} \exp\left\{-\frac{1}{k_\mathrm{B}T}\sum_{\boldsymbol{k}}\frac{v}{2}[\alpha(T-T_\mathrm{c})+\delta k^2](a_{\boldsymbol{k}}^2+b_{\boldsymbol{k}}^2)\right\}\prod_{\boldsymbol{k}}\mathrm{d}a_{\boldsymbol{k}}\mathrm{d}b_{\boldsymbol{k}}. \quad (6.10)$$

Here $a_{\boldsymbol{k}}$ and $b_{\boldsymbol{k}}$ are the real and imaginary parts of the complex quantities $\eta_{\boldsymbol{k}}$, respectively:

$$\eta_{\boldsymbol{k}} = a_{\boldsymbol{k}} + \mathrm{i}b_{\boldsymbol{k}}, \qquad \eta_{-\boldsymbol{k}} = a_{\boldsymbol{k}} - \mathrm{i}b_{\boldsymbol{k}}.$$

The summation in (6.10) has been taken over all $\boldsymbol{k}$ that satisfy the condition $|k| < k_\mathrm{max}$.

Assuming, as usual, that $a_{\boldsymbol{k}}^2 + b_{\boldsymbol{k}}^2 = r_{\boldsymbol{k}}^2$ and $\mathrm{d}a_{\boldsymbol{k}}\mathrm{d}b_{\boldsymbol{k}} = 2\pi r_{\boldsymbol{k}}\mathrm{d}r_{\boldsymbol{k}}$, we get

$$\int_{-\infty}^{\infty} \exp\left\{-\frac{1}{k_\mathrm{B}T}\sum_{\boldsymbol{k}}\frac{v}{2}[\alpha(T-T_\mathrm{c})+\delta k^2]r_{\boldsymbol{k}}^2\right\}2\pi\prod_{\boldsymbol{k}} r_{\boldsymbol{k}}\mathrm{d}r_{\boldsymbol{k}}$$

$$= \prod_{\boldsymbol{k}}\frac{\pi k_\mathrm{B}T}{v/2[\alpha(T-T_\mathrm{c})+\delta k^2]} \quad (6.11)$$

and, hence, (6.10) becomes

$$\Phi(p,T) = \Phi_0 + k_\mathrm{B}T\sum_{\boldsymbol{k}}\ln[\alpha(T-T_\mathrm{c})+\delta k^2] \quad (6.12)$$

where Φ_0 now includes all monotonically temperature-dependent parts of the thermodynamic potential.

The summation over $\boldsymbol{k}$ in (6.12) may be replaced by an integration since, as has already been mentioned, the permissible values of $\boldsymbol{k}$ are uniformly distributed in the space of wave vectors with a density of $v/(2\pi)^3$:

$$\Phi(p,T) = \Phi_0 + k_\mathrm{B}T\frac{v}{(2\pi)^3}\int_0^{k_\mathrm{max}}\ln[\alpha(T-T_\mathrm{c})+\delta k^2]\mathrm{d}^3k$$

$$= \Phi_0 + k_\mathrm{B}T\frac{v}{(2\pi)^3}\int_0^{\pi}\sin\theta\mathrm{d}\theta\int_0^{2\pi}\mathrm{d}\varphi\int_0^{k_\mathrm{max}}k^2\ln[\alpha(T-T_\mathrm{c})+\delta k^2]\mathrm{d}k$$

$$= \Phi_0 + k_\mathrm{B}T\frac{v}{2\pi^2}\int_0^{k_\mathrm{max}}k^2\ln[\alpha(T-T_\mathrm{c})+\delta k^2]\mathrm{d}k$$

where the volume element d^3k in the k_x, k_y, k_z space is represented in an ordinary manner in the form of $\mathrm{d}^3k = k^2\sin\theta\mathrm{d}\varphi\mathrm{d}k\mathrm{d}\theta$.

The contribution to the heat capacity of the symmetric phase due to fluctuations is

$$\Delta C_p^{\mathrm{fl}} = -\frac{T}{v}\frac{\partial^2 \Phi}{\partial T^2} = \frac{k_{\mathrm{B}} T^2 \alpha^2}{2\pi^2 \delta^2} \int_0^{k_{\max}} \frac{k^2 \mathrm{d}k}{[\alpha(T - T_{\mathrm{c}})/\delta + k^2]^2}. \tag{6.13}$$

The tabular integral on the right-hand side of (6.13) tends to zero at the lower limit, which is why we have

$$\Delta C_p^{\mathrm{fl}} = \frac{k_{\mathrm{B}} T^2 \alpha^2}{2\pi^2 \delta^2} \left[-\frac{k_{\max}}{\alpha(T - T_{\mathrm{c}})/\delta + k_{\max}^2} + \frac{1}{2}\left(\frac{\delta}{\alpha(T - T_{\mathrm{c}})}\right)^{1/2} \arctan\left(\frac{\delta k_{\max}^2}{\alpha(T - T_{\mathrm{c}})}\right)^{1/2} \right]. \tag{6.14}$$

This expression can be simplified if we assume that $\alpha(T - T_{\mathrm{c}}) \ll \delta k_{\max}^2$, i.e., if we are dealing with a region sufficiently close to the phase transition point. In this case $\arctan[\delta k_{\max}^2/\alpha(T - T_{\mathrm{c}})]^{1/2}$ may be approximately replaced by $\pi/2$ and the denominator in the first term by $k_{\max}^2$ and, introducing terms that are monotonically temperature dependent in the vicinity of T_{c} into the normal part of the heat capacity, we obtain

$$\Delta C_p^{\mathrm{fl}} \approx \frac{k_{\mathrm{B}} T^2 \alpha^{3/2}}{8\pi \delta^{3/2}} (T - T_{\mathrm{c}})^{-1/2}. \tag{6.15}$$

It is essential that the final result does not include the quantity $k_{\max}$. This means that for the theory expounded the approximateness of the division of the degrees of freedom into those "feeling" and those "not feeling" a phase transition is unimportant.

6.2 Applicability of Landau Theory to Nonferroelectric Structural Phase Transitions

From relation (6.15) for the part of the heat capacity due to fluctuations of the order parameter it can be seen that the parameter δ, which in essence determines the interaction of fluctuationally appearing regions with different η, plays an important role in the determination of anomalies in the thermodynamic characteristics of a crystal. In particular, the limiting cases $\delta \to \infty$ and $\delta \to 0$ correspond to nonphysical situations where, respectively, either the displacements of atoms are completely correlated throughout the crystal and fluctuation effects are absent, or the appearance of spatially inhomogeneous distributions of the order parameter is not accompanied by an appreciable increase of the crystal energy, and the displacements of atoms are not correlated even at distances of the order of the distance between the nearest neighbors. But, in general, the physical estimate of the parameter δ in order of magnitude can be obtained on the basis of the following considerations. It may be presumed that far from the transition point ($T - T_{\mathrm{c}} \sim T_{\mathrm{c}}$) the interaction energy of fluctuations is of the order of the volume energy $(1/2)\alpha T_{\mathrm{c}} \eta_0^2$

with variation of η from 0 to η_0 at a certain distance ξ_0 characteristic of the system:

$$\frac{1}{2}\alpha T_c \eta_0^2 \sim \frac{1}{2}\delta\frac{\eta_0^2}{\xi_0^2}$$

that is,

$$\delta \sim \alpha\xi_0^2 T_c. \tag{6.16}$$

The quantity $\xi_0 = \sqrt{\delta/\alpha T_c}$ itself may be regarded as a correlation radius of fluctuations far from the transition point, which is of the order of the unit-cell parameter or more depending on the nature of the transition. The result obtained (see (6.15)) shows that one should expect an anomalous change of heat capacity with $T \to T_c$, apart from the discontinuous change predicted by the Landau theory (Fig. 6.1). It should be kept in mind that expression (6.13) is approximate. It was derived by neglecting fourth-order terms in the expression for the thermodynamic potential (6.9). In a more exact theory it turns out that the inclusion of these terms is essential in the immediate vicinity of the Curie point. According to this theory the fluctuation contribution to the heat capacity has a weaker temperature dependence than in expression (6.15), but the discrepancy nevertheless remains. We have shown that the results following from the Landau theory (see Chap. 2) will be inapplicable in a certain neighborhood of the phase transition point. This neighborhood can be estimated, for example, with the aid of the following inequality:

$$\Delta C_p^{\mathrm{fl}} \geq \Delta C_p^{\mathrm{L}} = \alpha^2 T_c/2\beta \tag{6.17}$$

where ΔC_p^{L} is the heat capacity jump in the Landau theory. Thus, we have

$$\frac{k_B T_c^{3/2}\alpha^{3/2}}{8\pi\delta^{3/2}}\frac{T_c^{1/2}}{(T-T_c)^{1/2}} \geq \Delta C_p^{\mathrm{L}}$$

or

$$\frac{T-T_c}{T_c} \leq \frac{k_B^2}{64\pi^2\xi_0^6 \Delta C_p^{\mathrm{L}^2}}. \tag{6.18}$$

In the temperature region where the inequality (6.18) is fulfilled, the fluctuation effects are essential and the Landau theory does not describe the temperature dependence of thermodynamic parameters.

An essential point is that in the region where the Landau theory is not applicable, a special temperature dependence is exhibited not only by Φ_0 but also by the other coefficients in the expansion of the thermodynamic potential in powers of the order parameter, which is why the expansion itself appears to be invalid as $T \to T_c$.

This can be explained as follows. We transform (6.9) so as to single out in explicit form the order parameter – the only degree of freedom appearing in the Landau theory – $\eta_{k=0}$, following which we integrate the resultant expression over all variables, except $\eta_{k=0}$. So, we have

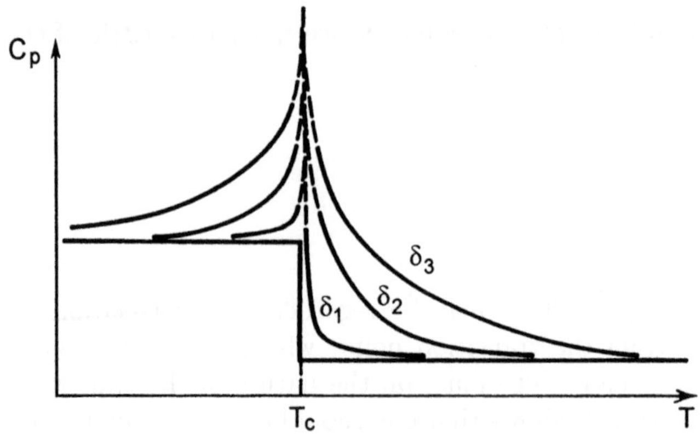

Fig. 6.1. Temperature dependence of heat capacity in the vicinity of the second-order transition point for various values of δ: $\delta_1 > \delta_2 > \delta_3$. The step corresponds to the Landau theory.

$$\begin{aligned}\frac{1}{v}\Phi(p,T,\eta_{\boldsymbol{k}=0},\ldots,\eta_{\boldsymbol{k}},\ldots) &= \varphi_0 + A\eta_{\boldsymbol{k}=0}^2 + \frac{1}{4}\beta\eta_{\boldsymbol{k}=0}^4 \\ &+\frac{1}{2}\sum_{\boldsymbol{k}}{}'(A+\delta k^2)\eta_{\boldsymbol{k}}\eta_{-\boldsymbol{k}} + \frac{1}{4}\beta\sum_{\boldsymbol{k}_1+\ldots+\boldsymbol{k}_4=0}{}'\eta_{\boldsymbol{k}_1}\eta_{\boldsymbol{k}_2}\eta_{\boldsymbol{k}_3}\eta_{\boldsymbol{k}_4} \\ &+\frac{3\beta}{2}\eta_{\boldsymbol{k}=0}^2\sum_{\boldsymbol{k}}{}'\eta_{\boldsymbol{k}}\eta_{-\boldsymbol{k}} + \beta\eta_{\boldsymbol{k}=0}\sum_{\boldsymbol{k}_1+\boldsymbol{k}_2+\boldsymbol{k}_3=0}{}'\eta_{\boldsymbol{k}_1}\eta_{\boldsymbol{k}_2}\eta_{\boldsymbol{k}_3}\,. \end{aligned} \tag{6.19}$$

Here the primed sums do not include terms with the quantity $\eta_{k=0}$ as a cofactor; the last two sums include the products of the Fourier components of the order parameter in which $\eta_{k=0}$ occurs twice and once, respectively (note that the presence of terms with $\eta_{k=0}$ occurring three times is ruled out because of the condition $\boldsymbol{k}_1 + \boldsymbol{k}_2 + \boldsymbol{k}_3 + \boldsymbol{k}_4 = 0$. For simplicity we will use the notation $\eta_{k=0} = \eta$. Then the sought-for quantity will be given by

$$\Phi(p,T,\eta) = -k_{\mathrm{B}}T\ln\int_{-\infty}^{\infty}\exp\left[-\frac{\Phi(p,T,\eta,\ldots,\eta_{\boldsymbol{k}},\ldots)}{k_{\mathrm{B}}T}\right]\prod_{\boldsymbol{k}}{}'\mathrm{d}\eta_{\boldsymbol{k}} \tag{6.20}$$

where the integration has been carried out over all $\eta_{\boldsymbol{k}}$, except $\eta_{k=0}$. In evaluating the integral in (6.20) we will try to simplify the problem, assuming that the general trends can be ascertained without taking account of the sums which contain $\eta_{\boldsymbol{k}}$ in a power higher than the second. Then,

$$\begin{aligned}\Phi(p,T,\eta) &= \Phi_0 + A\eta^2 + \frac{1}{2}\beta\eta^4 \\ &-k_{\mathrm{B}}T\ln\int_{-\infty}^{\infty}\exp\left\{-\frac{1}{k_{\mathrm{B}}T}\sum_{\boldsymbol{k}}\frac{v}{2}(A+3\beta\eta^2+\delta k^2)\eta_{\boldsymbol{k}}\eta_{-\boldsymbol{k}}\right\}\prod_{\boldsymbol{k}}\mathrm{d}\eta_{\boldsymbol{k}}\,. \end{aligned} \tag{6.21}$$

Comparing the above expression with (6.10) and performing the integration over the real variables, we obtain

$$\Phi(p,T,\eta) = \Phi_0 + A\eta^2 + \frac{1}{4}\beta\eta^4 + \frac{k_B T}{2\pi^2}\int_0^{k_{max}} k^2 \ln(A + 3\beta\eta^2 + \delta k^2)\mathrm{d}k. \tag{6.22}$$

We see that if we take into account, even partially, fluctuations of the order parameter, this will lead to a renormalization of the coefficients in the expansion of the thermodynamic potential in even powers of the order parameter. The last term in (6.22) may be expanded into a series in η; then, retaining only the first two terms in the expansion, we obtain new renormalized coefficients:

$$\tilde{A} = A + 3\beta\frac{k_B T}{2\pi^2}\int_0^{k_{max}} \frac{k^2 \mathrm{d}k}{A + \delta k^2} \tag{6.23}$$

$$\tilde{\beta} = \beta - 9\beta^2\frac{k_B T}{2\pi^2}\int_0^{k_{max}} \frac{k^2 \mathrm{d}k}{(A + \delta k^2)^2}. \tag{6.24}$$

In the approximation under consideration the phase transition temperature T_c appears to have been shifted relative to the temperature contained in the incomplete thermodynamic potential:

$$\tilde{A} = \frac{1}{2}\alpha(T - T_c) + 3\beta\frac{k_B T}{2\pi^2\delta}k_{max} - 3\beta\frac{k_B T\alpha^{1/2}}{2\pi^2\delta^{3/2}}(T - T_c)^{1/2}$$

$$\times \arctan\frac{k_{max}}{\sqrt{A/\delta}} \approx \frac{1}{2}\alpha(T - T_c^*) - 3\beta\frac{k_B T\alpha^{1/2}}{4\pi^2\delta^{3/2}}(T - T_c)^{1/2}$$

$$T_c^* = T_c - 3\beta\frac{k_B T}{\pi^2\alpha\delta}k_{max}.$$

The coefficient β becomes temperature dependent, decreasing as $T \to T_c$; the integral in (6.24) is analogous to the one considered above (see (6.13)); so we obtain immediately

$$\tilde{\beta} = \beta - d/\sqrt{T - T_c}, \qquad d = \text{const}.$$

It should be stressed that the coefficient $\tilde{A}$ calculated with account taken of fluctuations is now a nonlinear function of temperature, which is an indication of a departure from the Curie–Weiss law near the phase transition point. The true phase transition temperature is evidently T_c^*. It can be shown that on closer inspection the temperature T_c in the formula for A is replaced by T_c^* in the second term as well.

Of course, we can only speak of tendencies in the variation of the coefficients because the calculations made are approximate. Nonetheless, the very fact of an anomalous (from the standpoint of the Landau theory) variation of the heat capacity and the coefficients in the expansion of the thermodynamic potential is evidence of the presence of a singularity at $T = T_c$, which points to the necessity of constructing a more exact theory.

6.3 Applicability of Landau Theory to Phase Transitions in Uniaxial Ferroelectrics

Let us now apply our approach to the case of ferroelectric crystals, where the spatially inhomogeneous distributions of the order parameter are inevitably associated with the appearance of a macroscopic electric field, which can have an influence upon fluctuations.

We will first consider the case of a uniaxial ferroelectric crystal, in which a spontaneous polarization arises only in one crystallographic direction and the order parameter is proportional to the component of the polarization vector along this direction; the polarization along other directions will be neglected. Equation (6.3) will correspond to

$$P_z(\boldsymbol{r}) = \sum_{\boldsymbol{k}} P_{z,\boldsymbol{k}} \mathrm{e}^{\mathrm{i}\boldsymbol{k}\cdot\boldsymbol{r}} \tag{6.25}$$

where, as before,

$$P_{z,\boldsymbol{k}} = a\eta_{\boldsymbol{k}}. \tag{6.26}$$

From (6.26) it follows that the spatially inhomogeneous distribution of polarization in the crystal volume exactly corresponds to the order parameter distribution, but the electric effects attendant on the order parameter depend on the effective charges of the sublattices being displaced and on the character of these displacements.

While considering the incomplete thermodynamic potential of a crystal and its changes associated with fluctuations of the order parameter that cause fluctuations in electric polarization, it is necessary to take account of the specific form of the long-range dipole–dipole interaction between polarization changes at various points in the crystal. Electric forces that act on a certain small part of the crystal located at a point determined by the radius-vector $\boldsymbol{r}$, and exhibiting polarization $\boldsymbol{P}(\boldsymbol{r})$, from the side of other portions of the crystal may be specified with the aid of a macroscopic electric field $\boldsymbol{E}(\boldsymbol{r})$. This field is determined by electrostatic equations:

$$\operatorname{rot} \boldsymbol{E} = 0 \tag{6.27}$$

$$\operatorname{div} \boldsymbol{D} = 0. \tag{6.28}$$

In particular, if the polarization distribution has the form of (6.25), then associated with the Fourier component of polarization $\boldsymbol{P}_{\boldsymbol{k}}$ is the spatially inhomogeneous distribution of the electric field:

$$\boldsymbol{E_k} = -4\pi(\boldsymbol{P_k}\,\boldsymbol{k}/k)\boldsymbol{k}/k \tag{6.29}$$

and the long-range dipole–dipole interaction can be taken into account in the thermodynamic potential by the term

$$-\int_v \frac{1}{2}\boldsymbol{P}(\boldsymbol{r})\,\boldsymbol{E}(\boldsymbol{r})\mathrm{d}v = -\frac{v}{2}\sum_{\boldsymbol{k}} P_{z,\boldsymbol{k}}E_{z,-\boldsymbol{k}} = 2\pi v\sum_{\boldsymbol{k}}\left(\frac{k_z}{k}\right)P_{z,\boldsymbol{k}}P_{z,-\boldsymbol{k}}. \tag{6.30}$$

Here k_z is the projection of the wave vector $\boldsymbol{k}$ on to the direction of the axis z along which the vector $\boldsymbol{P}(\boldsymbol{r})$ is oriented (Fig. 6.2). It is obvious that $k_z/k = \cos\theta$, where θ is the angle between the directions of the polar axis z and the wave vector $\boldsymbol{k}$.

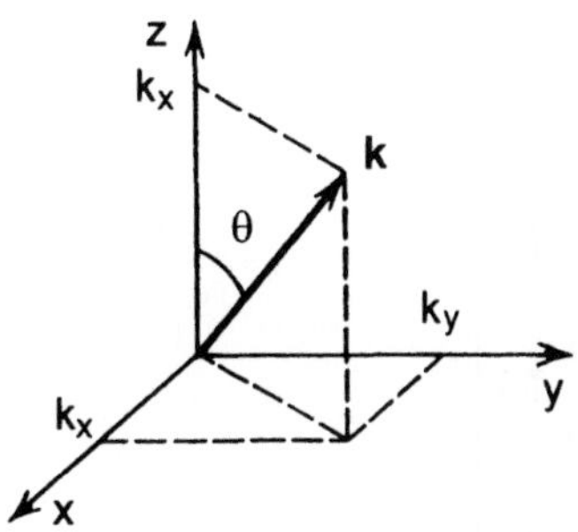

Fig. 6.2. Orientation of the wave vector of the polarization wave relative to the polar axis z.

Let us show that relation (6.29) is valid. From (6.27) we have $[\boldsymbol{k}\times\boldsymbol{E_k}] = 0$, i.e., the electric field vector in the kth polarization wave is always directed along the wave vector $\boldsymbol{k}$.

On the other hand, from (6.28) we obtain

$$\operatorname{div}\boldsymbol{E} = -4\pi\operatorname{div}\boldsymbol{P} \qquad \text{or} \qquad (\boldsymbol{k}\boldsymbol{E_k}) = -4\pi\,(\boldsymbol{P_k}\boldsymbol{k})\,. \tag{6.31}$$

Since $\boldsymbol{k}\parallel\boldsymbol{E_k}$, it follows that

$$E_{\boldsymbol{k}} = -4\pi\left(\boldsymbol{P_k}\frac{\boldsymbol{k}}{k}\right)$$

or, multiplying both sides of this equation by the unit vector $\boldsymbol{k}/k$, we obtain (6.29). From (6.29) it follows, in particular, that for the transverse wave ($\boldsymbol{P_k}\perp\boldsymbol{k}/k$), $\boldsymbol{E_k} = 0$, whereas for the longitudinal wave $\boldsymbol{E_k} = -4\pi\boldsymbol{P_k}$. Figure 6.3 clearly shows the cause of this difference: in the case of longitudinal waves in the crystal volume there arise charged regions which create a macroscopic electric field oriented in a direction opposite to the direction of the polarization vector at a given point of the crystal.[2]

[2] An electric field induced by bound charges on the surface or in the bulk of the crystal is therefore called *depolarizing*.

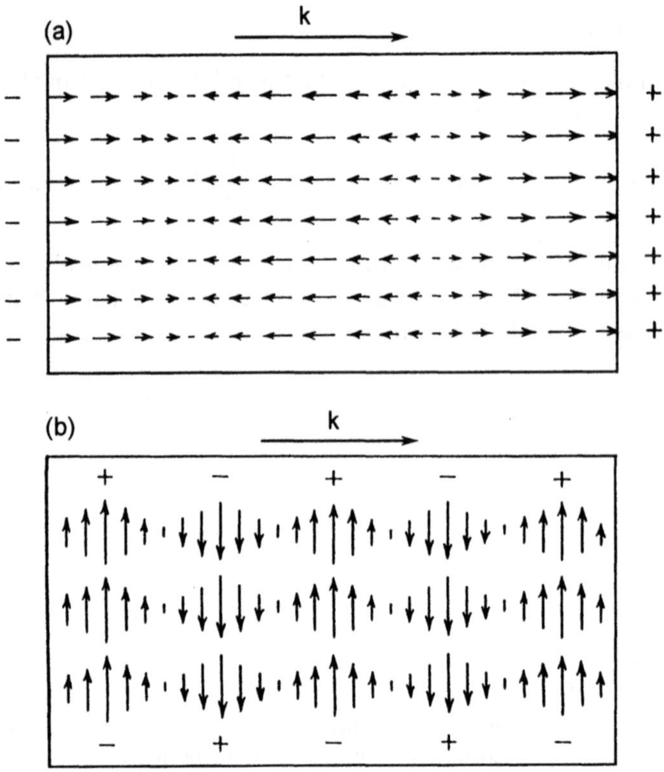

Fig. 6.3. (**a**) Longitudinal and (**b**) transverse polarization waves in a crystal.

We write the expression for the incomplete thermodynamic potential of a uniaxial ferroelectric crystal with account taken of the macroscopic electric field caused by long-range Coulomb forces.

Taking into account (6.9), (6.30) and (6.26), we get

$$\Phi(p, T, \eta_0, \ldots, \eta_{\boldsymbol{k}})$$
$$= \varphi_0 v + \frac{v}{2} \sum_{\boldsymbol{k}} \left[A + \delta k^2 + 4\pi \left(\frac{k_z}{k} \right)^2 a^2 \right] \eta_{\boldsymbol{k}} \eta_{-\boldsymbol{k}} + \ldots . \qquad (6.32)$$

One can see that if the displacements of atoms along the z axis are not associated with the occurrence of polarization ($a = 0$), we obtain the case considered above and heat capacity anomalies of the type (6.15). On the other hand, in case $a \neq 0$ we may expect a relative decrease of the contribution of critical fluctuations in order parameter to the thermodynamic potential.

Instead of (6.10) we now have

$$\Phi(p,T) \approx \Phi_0 - k_{\mathrm{B}}T\ln \int\limits_{-\infty}^{\infty} \exp\left\{-\frac{1}{k_{\mathrm{B}}T}\sum_{k}\frac{v}{2}\left[\alpha(T-T_{\mathrm{c}})\right.\right.$$

$$\left.\left. +\delta k^2 + 4\pi\left(\frac{k_z}{k}\right)^2 a^2\right]\eta_k\eta_{-k}\right\}\prod_{k}\mathrm{d}\eta_k\mathrm{d}\eta_{-k}$$

$$= \Phi_0 + k_{\mathrm{B}}T\sum_{k}\ln[\alpha(T-T_{\mathrm{c}}) + \delta k^2 + 4\pi a^2\cos^2\theta]$$

$$= \Phi_0 + k_{\mathrm{B}}T\frac{v}{(2\pi)^3}\int\limits_0^{\pi}\mathrm{d}\theta\int\limits_0^{2\pi}\mathrm{d}\varphi\,\times$$

$$\times\int\limits_0^{k_{\max}} k^2\sin\theta\ln[\alpha(T-T_{\mathrm{c}}) + \delta k^2 + 4\pi a^2\cos\theta]\mathrm{d}k. \tag{6.33}$$

Integrating over φ and differentiating (6.33) with respect to T, we obtain for the entropy of the symmetric phase

$$S = -\frac{\partial\Phi}{\partial T} = S_0 - \frac{k_{\mathrm{B}}T\alpha v}{(2\pi)^2}\int\limits_0^{k_{\max}} k^2\mathrm{d}k\int\limits_{-1}^{+1}\frac{\mathrm{d}(\cos\theta)}{\alpha(T-T_{\mathrm{c}}) + \delta k^2 + 4\pi a^2\cos^2\theta}$$

$$\approx S_0 - \frac{k_{\mathrm{B}}T\alpha v}{2(2\pi)^3a^2}\int\limits_0^{k_{\max}} k^2\mathrm{d}k\int\limits_{-1}^{+1}\frac{\mathrm{d}(\cos\theta)}{[\alpha(T-T_{\mathrm{c}}) + \delta k^2]/(4\pi a^2) + \cos^2\theta}$$

$$= S_0 - \frac{k_{\mathrm{B}}T\alpha v}{2(2\pi)^3a^2}\int k^2\mathrm{d}k\sqrt{\frac{4\pi a^2}{\alpha(T-T_{\mathrm{c}}) + \delta k^2}}$$

$$\times\arctan\sqrt{\frac{4\pi a^2}{\alpha(T-T_{\mathrm{c}}) + \delta k^2}}.$$

In the entire temperature region of interest the argument of the tangent is sufficiently large, so that

$$\arctan\sqrt{\frac{4\pi a^2}{\alpha(T-T_{\mathrm{c}}) + \delta k^2}} \approx \frac{\pi}{2}$$

and

$$S = S_0 - \frac{k_{\mathrm{B}}T\alpha v}{\sqrt{2}(2\pi)^{3/2}a}\int\limits_0^{k_{\max}}\frac{k^2\mathrm{d}k}{\sqrt{\alpha(T-T_{\mathrm{c}}) + \delta k^2}}. \tag{6.34}$$

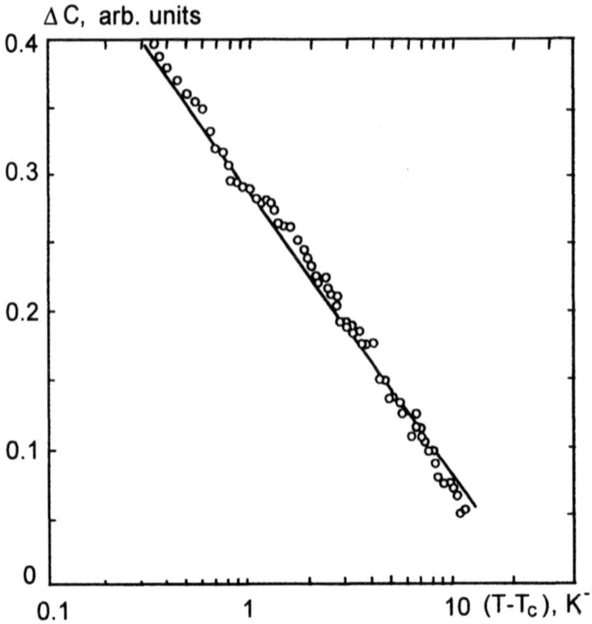

Fig. 6.4. Temperature dependence of excess heat capacity at $T > T_c$ in a TGS crystal on a semilog scale [11].

The heat capacity is given by

$$\Delta C_p = \frac{1}{v} T \frac{\partial S}{\partial T} = C_{p0} + \frac{k_B T^2 \alpha^2}{\sqrt{2}(2\pi)^{3/2} a} \int_0^{k_{max}} \frac{k^2 dk}{[\alpha(T - T_c) + \delta k^2]^{3/2}}$$

$$\approx C_{p0} + \frac{k_B T^2 \alpha^2}{\sqrt{2}(2\pi)^{3/2} \delta^{3/2} a} \ln\left(\frac{T - T_c}{T_c}\right)^{-1} \tag{6.35}$$

in the same approximation as in (6.15). We thus arrive at the conclusion that in the presence of a dipole–dipole interaction characteristic of uniaxial ferroelectrics the heat capacity associated with fluctuations in the symmetric phase exhibits not an inverse root but a logarithmic, weaker temperature dependence in a certain temperature range, which does not include the nearest vicinity of the phase-transition point. The physical factor responsible for this difference is the relative suppression of fluctuations of the order parameter by a depolarizing electric field.

As a result, since the logarithmic dependence is rather weak, the properties of unaxial ferroelectric crystals are described well enough by the Landau theory without taking account of fluctuation effects; in most cases of interest in experiments and in practice the properties of crystals can be described with the aid of the phenomenological theory outlined in the preceding chapters.

The fluctuational anomaly in the heat capacity in the paraelectric phase in uniaxial ferroelectrics can evidently be detected only by means of precision experiments. The available data confirm the inferences of the theory. Figure 6.4 shows the temperature dependence of the heat capacity of triglycine sulphate in the paraelectric phase on a semilog scale. The small heat capacity anomaly is well described by a logarithmic function.

6.4 Fluctuational Phenomena in Ferroelectric-Ferroelastics and in Phase Transitions in Multiaxial Ferroelectrics

First a few remarks should be made concerning fluctuational phenomena in two other classes of crystals – uniaxial ferroelectric-ferroelastics (exhbiting a piezoelectric effect at $T > T_c$) and multiaxial ferroelectrics, the phase transition in which is described by a multicomponent order parameter.

In uniaxial ferroelectric crystals the fluctuations in order parameter prove to have been suppressed because of the appearance of a long-range electric field; as a result, the contribution to the anomaly is made only by fluctuations with certain directions of the wave vector lying in a plane perpendicular to the polar axis. We are thus speaking of a strong decrease of the phase volume of fluctuations, which increase anomalously as $T \to T_c$. It turns out that there occurs a further decrease in the phase volume of intensive fluctuations if in the symmetric phase there is a linear relationship between order parameter and mechanical stresses. In this case, an anomalous increase in order parameter fluctuations occurs only for wave vectors coinciding with certain crystallographic directions in the crystal. As a consequence, as $T \to T_c$ the heat capacity at the phase transition point remains finite. The expansion of the thermodynamic potential into a series in powers of the order parameter holds up to the immediate vicinity of the transition point. This conclusion is also valid for ferroelastics that are not ferroelectrics.

On the other hand, in ferroelectric crystals of the $BaTiO_3$ type, which are described by a multicomponent order parameter, the situation is essentially different. In the symmetric phase the order parameter fluctuations are isotropic. For any direction of the vector $\boldsymbol{k}$, instead of fluctuations in the three components of the polarization vector P_x, P_y, P_z, one may consider fluctuations in some other three components – one longitudinal and two transverse relative to $\boldsymbol{k}$. In this case, no depolarizing electric field arises for the transverse components (see (6.29) and Fig. 6.3). Therefore, the probability of the appearance of longitudinal and transverse fluctuational polarization waves in cubic crystals is different; the mean square fluctuation in the longitudinal component proves to be equal to

$$\langle P_{k\parallel}^2 \rangle = \frac{k_B T a^2}{v/2[\alpha(T - T_c) + \delta k^2 + 4\pi a^2]} \tag{6.36}$$

whereas for the transverse components we have

$$\langle P_{k\perp}^2 \rangle = \frac{k_B T a^2}{v/2[\alpha(T - T_c) + \delta k^2]}. \tag{6.37}$$

One can see that $\langle P_{k\perp}^2 \rangle \to \infty$ as $T \to T_c$ and $k \sim 0$, while $\langle P_{k\|}^2 \rangle$ practically does not change near T_c because of the large term $4\pi a^2$, which is due to the presence of a long-range Coulomb electric field. It is clear that the incomplete thermodynamic potential of a multiaxial ferroelectric crystal will be similar to the potential of a nonferroelectric crystal (see (6.9)); the transverse components of fluctuational changes of polarization will introduce a contribution to the thermodynamic potential, which is analogous to that of the fluctuations of the order parameter not coupled at all to polarization.

Multiaxial ferroelectrics could have been convenient objects for the study of fluctuation effects in phase transitions in crystals; however, in all such ferroelectrics there occur first-order phase transitions. It has been shown that a discontunuous change of polarization at the Curie point may be a consequence of increasing fluctuations in order parameter: upon attainment of a certain level of fluctuations in the order parameter the system "runs away" into another phase through a first-order phase transition. To put it differently, the renormalization of the coefficients in the expansion of the thermodynamic potential due to fluctuations, which was mentioned above, may lead to a change of the sign of β and a corresponding change in the type of phase transition.

It should be pointed out that the experimental study of fluctuation phenomena near the points of structural phase transitions is being carried on intensively at present. For the temperature region near T_c where the Landau theory is a priori inapplicable and where we cannot, in principle, restrict ourselves to first corrections to it, use is made of the fluctuational theory of anomalies. This theory, known as the theory of critical phenomena, has made significant advances over the last decade and is widely used for systems in which the temperature region where the Landau theory is invalid is large: at the λ-point in liquid helium, at the critical liquid–gas point, in magnetics, etc. However, for structural phase transitions there are no sufficiently reliable experimental data that could point to the inapplicability of the Landau theory to these systems in almost all of the experimentally accessible region. It is for this reason that we are not going to discuss these problems.

7. Structural Phase Transitions in the Single-Ion Model

7.1 Problems of the Microscopic Theory

We have so far used an approach which is usually referred to as the phenomenological theory of phase transitions. Recall that within the framework of this theory we proceeded from the presence of a phase transition and used at the outset only the assumption of a regular, nonsingular temperature dependence of the coefficients contained in the theory (Landau coefficients of the thermodynamic potential). We also found that this assumption is invalid, unsubstantiated, but this was also done within the scope of the phenomenological theory. In this theory we were not interested, as a matter of fact, in the concrete structure, the chemical composition of a crystal, and the character of forces that are operating between particles (examples given in Chap. 1 were aimed at rendering the treatment more spectacular, but they were not required for constructing the theory). The only essential point that we took into account was the character of variation of the crystal symmetry at the Curie point. Therefore, we could say nothing of the values of the coefficients included in the theory and interpret the nature of the interactions leading to a phase transition.

In this chapter we will consider, for the first time, the microscopic theory.

What tasks are set by the microscopic theory? Its maximum program is to determine and calculate the structure and properties of a crystal provided that its chemical composition is known. Had we known how to do it, we would have been able to say what substances are to be synthesized in order to prepare crystals with required properties (in particular, ferroelectric).

In actual fact, however, this program can never be realized in practice. What is called the microscopic theory often requires that the Hamiltonian of a crystal be specified and the model itself is chosen from structural data or on the basis of crystallochemical considerations. At this stage a phase transition is commonly classified as belonging to displacive or order–disorder transitions, though no sharp boundary line exists between these types. Furthermore, for displacive-type transitions the requirements imposed on the Hamiltonian parameters are such that a phase transition actually occurs in a given crystal. As regards order–disorder transitions, they occur, in a certain sense, inevitably – a disordered lattice becomes ordered with a decrease in temperature, except for special cases when quantum-mechanical effects come into play (see Chap. 8). Once the Hamiltonian is chosen, the thermodynamic potential of the crystal is calculated, and use is commonly made of one of the approximations allowing for the applicability of the Landau theory near the Curie point, i.e., the neglect of fluctuation effects.

What information going beyond the scope of the phenomenological theory can be obtained on such a path, which may be called semiphenomenological or semimicroscopic? In final theoretical formulas we express the thermodynamic potential coefficients in terms of the constants of the initial Hamiltonian, which are, strictly speaking, unknown. However, several constants of the Hamiltonian can be used to express the coefficients of the thermodynamic potential of any order. It is also possible (provided, of course, that the starting model is sufficiently detailed) to find the temperature dependence of the coefficients in a wider temperature range than that claimed by the Landau theory. In this theory the coefficients of the thermodynamic potential are expanded in a series in the difference between the current temperature and the transition temperature and only the first term is taken into account for each of the coefficients. Rather far from the transition point the Landau theory becomes inapplicable, of course, since terms of higher order in $T - T_{\mathrm{c}}$ become important. Moreover, terms of higher order in η and other variables may also become essential.

The temperature at which this occurs depends on the character of interactions in the system and may, in principle, lie within this region, which is taken by experimentalists to be the region sufficiently close to the transition point. In this case, the region of applicability of the Landau theory appears to be too narrow to describe experimental data and the interpretation of these data must be based on the microscopic theory.

What has been said above is important for substances that undergo several phase transitions and that exhibit a complex phase diagram. For this situation to be described phenomenologically one would have to introduce, generally speaking, several transition parameters, several temperature-dependent coefficients, etc. At the same time, with the microscopic approach it is sometimes possible to describe the entire complex set of phase transitions on the basis of a single Hamiltonian containing one or two parameters.

The microscopic theory proves to be also useful for the region of applicability of the Landau theory, where it gives values of the coefficients that have been used by us above. It becomes informative when the coefficients are found to be unusual, i.e., when they differ from those which can be obtained from simple estimates based on dimensionality considerations.

Apart from such practical value, the microscopic theory is of great methodological importance since the order parameter figuring in it is not simply a certain quantity exhibiting some transformation properties but a quite concrete physical quantity, say a combination of displacements of certain atoms. The coefficients of the thermodynamic potential assume a definite physical meaning. More often than not even a small step toward a microscopic treatment appears to be efficient, and this will be demonstrated in this chapter.

One can thus speak of three levels, three stages of a theoretical description of phase transitions. The first, simplest stage (provided that fluctuational effects are ignored) is a phenomenological theory which takes into account only the crystal symmetry. The second one is a model microscopic theory in which account has already been taken of the structure of the crystal and the character of interactions that are important to the problem of phase transitions. The third level is a theory that makes it possible to calculate the constants of model Hamiltonians and to improve the form of these Hamiltonians.

7.2 The Single-Ion Model of a Diatomic Crystal

We will describe the simplest model which we call a *single-ion model.* Suppose we have a diatomic crystal consisting of atoms of species A and B (Fig. 7.1) and let us assume that in a phase transition atoms of species A are displaced along the z axis from their positions in the center of the unit cell and atoms of species B remain immovable. We are thus concerned here with the motion of only one atom in the unit cell, assuming that the role of other atoms boils down to the creation of a potential field. This is what is referred to as the single-ion model. What forces must act on atom A for a phase transition to take place in our system? We divide these forces into two types: forces that operate from the side of atoms B – to simplify the problem, we take into account only the nearest neighbors of atoms A and forces that act from the neighboring A atoms. It should be pointed out immediately that the introduction of an interaction between various A atoms is necessary since the neglect of this interaction will render a phase transition impossible in our model. Indeed, without an interaction the various A atoms will move independently of each other in a potential created by the immovable B atoms, i.e., the statistical sum of the model will be split into the product of statistical sums of individual atoms, and the thermodynamic potential of the system will be the sum of the thermodynamic potentials for individual A atoms. As for the thermodynamic potential for a single atom which is moving independently of the other particles of the system, it is well known that this potential as a function of temperature exhibits no singularities.

Thus, no phase transition is possible here, regardless of whether the treatment is within or beyond the framework of the Landau theory since, according to this theory, the thermodynamic potential possesses a mathematical singularity – a jump of the second derivative with respect to temperature (a heat capacity jump at the Curie point). At the same time, the thermodynamic potential of a system consisting of a single particle is a smooth function of temperature and there is no jump of any of its derivatives. It is this circumstance which implies that the phase transition is a cooperative phenomenon.

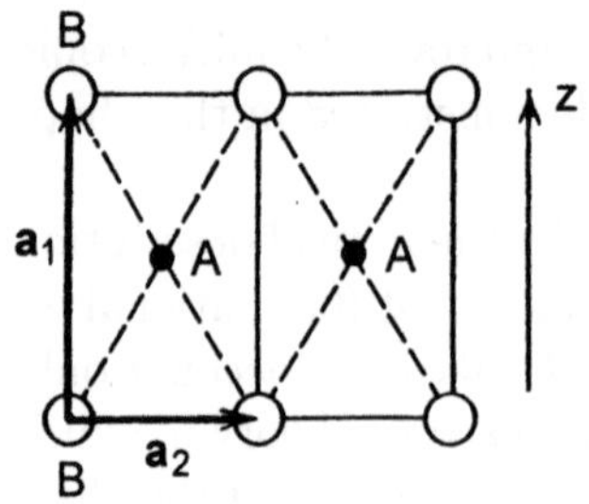

Fig. 7.1. A two-dimensional analog of the model discussed in the text.

Let us now consider the forces acting on atoms A from the side of atoms B, assuming, for the sake of simplicity, that atoms A can move only along the z axis. From Fig. 7.1 it is clear that the potential created by these forces is a symmetric (with respect to the center of the unit cell) function of z. On the other hand, these forces must displace atom A from the central position because at $T = 0$, as was assumed above, a nonsymmetric phase is realized. It is therefore clear that the potential of atom A in the field of atom B has the form of a two-well potential (Fig. 7.2).

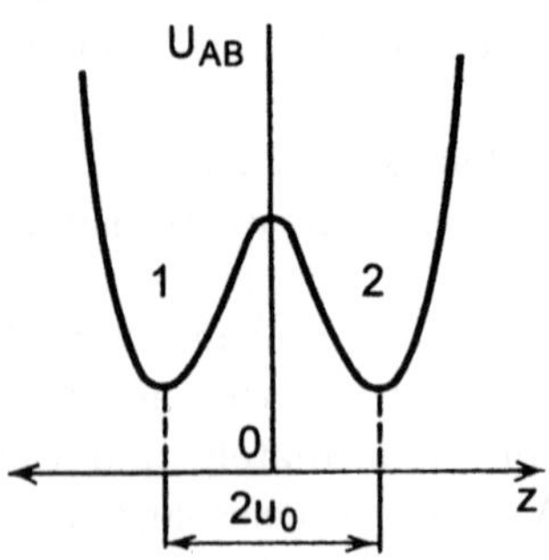

Fig. 7.2. The potential energy of atom A in the field of atom B.

It is hoped (this will be clear from what follows) that we have introduced a sufficient number of interactions in order to provide a phase transition in our system: the forces that operate from the side of atoms B displace atoms A from the central position and the forces of interaction of A atoms must provide a concerted (cooperative) displacement of these atoms.

The unidimensional analog of our system may be clearly represented as an array of particles aligned on a straight line and connected by springs, each of the particles being in its two-minimum potential (Fig. 7.3).

Now we write an expression for the Hamiltonian (potential energy) in the three-dimensional case, assuming that the interaction of A atoms is described by the potential of the elastic spring:

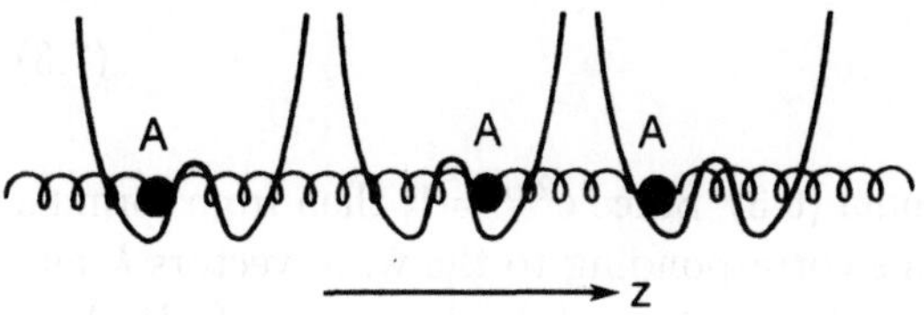

Fig. 7.3. Diagram illustrating the operation of forces for the model shown in Fig. 7.1.

$$\begin{aligned} H &= \sum_{R} U_{\mathrm{AB}}(u_{\boldsymbol{R}}) + \sum_{\boldsymbol{R},\boldsymbol{R}'} U_{\mathrm{AA}}(u_{\boldsymbol{R}}, u_{\boldsymbol{R}'}) \\ &= \sum_{\boldsymbol{R}} \left(\frac{1}{2}au_{\boldsymbol{R}}^2 + \frac{1}{4}bu_{\boldsymbol{R}}^4\right) + \frac{1}{2}c\sum_{\boldsymbol{R},\boldsymbol{R}'}(u_{\boldsymbol{R}} - u'_{\boldsymbol{R}})^2 . \end{aligned} \tag{7.1}$$

Here $u_{\boldsymbol{R}}$ is the displacement along the axis z of atom A present in the unit cell, the center of which has a coordinate given by

$$\boldsymbol{R} = n_1\boldsymbol{a}_1 + n_2\boldsymbol{a}_2 + n_3\boldsymbol{a}_3 , \tag{7.2}$$

where $\boldsymbol{a}_i$ are the vectors of unit translations, $a < 0$, $b > 0$; and c is the stiffness coefficient for springs linking the particles (it is assumed that when the particles are in the central position, the springs are not extended).

From (7.1) it can be seen that to the minimum of the potential energy H there corresponds a configuration of the system such that all particles occupy identical (say, left-hand) minima. Indeed, both the energy of the springs (which are not extended) and the energy of the particles are at a minimum. So, at $T = 0$ the particles are positioned nonsymmetrically relative to the center of the unit cell – we have a nonsymmetric phase.

With $T \neq 0$ the disposition of the particles corresponds to the minimum of the free energy (or of the thermodynamic potential; in our model where the system's volume is fixed, this is unimportant). Naturally we are speaking here of the average or most probable arrangement – the particles take part in thermal motion. If we perform the transformation $u_{\boldsymbol{R}} \to \eta(\boldsymbol{r})$[1] and compare formulas (7.1) and (6.8) (ignoring for the moment the temperature dependence of the coefficients), we will immediately find that the first is a discrete analog of the second, i.e., in (7.1) the discreteness of the crystal structure has been explicitly taken into account and in (6.8) the crystal is treated as a continuous medium (continuum).

For a further comparison of the continuum and discrete cases it is convenient to pass over from variables $u_{\boldsymbol{R}}$ to variables $u_{\boldsymbol{k}}$ in keeping with the formula

[1] In the three-dimensional case the atomic displacement $u_{\boldsymbol{R}}$ is, of course, a vector, so that the exact position of each particle is specified by the radius-vector $\boldsymbol{r} = \boldsymbol{R}+u_{\boldsymbol{R}}$. For simplicity, throughout this chapter we assume that the displacements of atoms occur along a single direction (say, z) and take into account, basically, only one displacement component $u_{\boldsymbol{R}}^z = u_{\boldsymbol{R}}$.

$$u_{\boldsymbol{R}} = \sum_{\boldsymbol{R}} u_{\boldsymbol{k}} \mathrm{e}^{\mathrm{i}\boldsymbol{k}\cdot\boldsymbol{R}} \tag{7.3}$$

which is a discrete analog of formula (6.3). Since $\mathrm{e}^{2\pi\mathrm{i}} = 1$, then from formula (7.3) it follows that the sum terms corresponding to the wave vectors $\boldsymbol{k}$ and $\boldsymbol{k} + 2\pi\boldsymbol{b}$ describe one and the same distribution of displacements if $(\boldsymbol{R} \cdot \boldsymbol{b})$ is equal to an integer. The vector $\boldsymbol{b}$, which has this property, is termed the reciprocal lattice vector; the unit vectors of the reciprocal lattice $\boldsymbol{b}_j$ are introduced by the relations

$$\boldsymbol{a}_i \boldsymbol{b}_j = \delta_{ij}. \tag{7.4}$$

It is easy to check that the vectors

$$\boldsymbol{b} = m_1\boldsymbol{b}_1 + m_2\boldsymbol{b}_2 + m_3\boldsymbol{b}_3 \tag{7.5}$$

are really the vectors of the reciprocal lattice (here m_i are whole numbers). If the vectors $\boldsymbol{k}$ differ from one another by $2\pi\boldsymbol{b}$, then to such vectors there corresponds one and the same distribution of displacements. It will therefore suffice to limit oneself to those vectors $\boldsymbol{k}$ which differ to a lesser extent than by any vector of the reciprocal lattice. The region of such vectors $\boldsymbol{k}$, which includes the point $\boldsymbol{k} = 0$, is known as the first Brillouin zone (Fig. 7.4). Thus, the summation in (7.3) is restricted to vectors $\boldsymbol{k}$ lying in the first Brillouin zone. However, even within this zone not every point corresponds to the permissible vector $\boldsymbol{k}$. If, just as in Chap. 6, we apply periodic boundary conditions for the distribution of $u_{\boldsymbol{R}}$, the values of the permissible wave vectors $\boldsymbol{k}$ will be those given by relations (6.5). So, with respect to vectors $\boldsymbol{k}$ figuring in the problem, the discreteniess of the medium manifests itself only in a natural limitation of the region of allowable values of these vectors of the first Brillouin zone; in the continuum case we artificially restricted this region by introducing $k_{\max}$ (see Chap. 6).

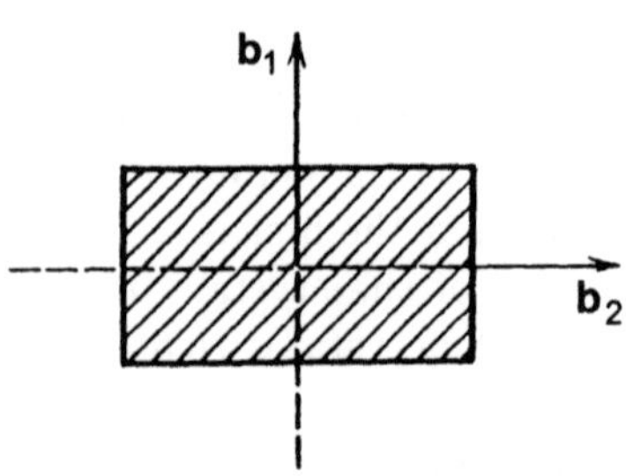

Fig. 7.4. Unit vectors of the reciprocal lattice and the first Brillouin zone (*hatched*) for the model shown in Fig. 7.1.

Before substituting (7.3) into (7.1), let us write the last term in (7.1) in the form of a single sum, i.e., the summation over the springs is replaced by the summation over A atoms. This can be done since to each atom there correspond six springs. Let us consider a certain atom A and two of its nearest neighbors located on the axis x. In other words, we are interested in A atoms positioned at points $\boldsymbol{R}$, $\boldsymbol{R}-\boldsymbol{a}_1$ and $\boldsymbol{R}+\boldsymbol{a}_1$. The sum of the energies of the springs that approach the atom A along the z axis is equal to

$$\frac{1}{2}c(u_{\boldsymbol{R}-\boldsymbol{a}_1}-u_{\boldsymbol{R}})^2+\frac{1}{2}c(u_{\boldsymbol{R}+\boldsymbol{a}_1}-u_{\boldsymbol{R}})^2. \tag{7.6}$$

Analogous expressions can be written for the energies of springs linking the atom A to its nearest neighbors along the axes y and z. If this procedure is performed for each atom A, then the energy of a given spring will appear twice, which is why the last term in (7.1) is written in the form

$$\frac{c}{4}\sum_{\boldsymbol{R}}\sum_{i=1,2,3}\left[(u_{\boldsymbol{R}-\boldsymbol{a}_i}-u_{\boldsymbol{R}})^2+(u_{\boldsymbol{R}+\boldsymbol{a}_i}-u_{\boldsymbol{R}})^2\right]$$
$$=\frac{c}{4}\sum_{\boldsymbol{R}}\sum_{i=1,2,3}(u^2_{\boldsymbol{R}-\boldsymbol{a}_i}+u^2_{\boldsymbol{R}+\boldsymbol{a}_i}+2u^2_{\boldsymbol{R}}-2u_{\boldsymbol{R}-\boldsymbol{a}_i}u_{\boldsymbol{R}}-2u_{\boldsymbol{R}+\boldsymbol{a}_i}u_{\boldsymbol{R}}). \tag{7.7}$$

We now substitute (7.3) into (7.7). As an example, let us see how the sum $\sum_{\boldsymbol{R}} u^2_{\boldsymbol{R}}$ is transformed. We have

$$\sum_{\boldsymbol{R}} u^2_{\boldsymbol{R}}=\sum_{\boldsymbol{R}}\sum_{\boldsymbol{k},\boldsymbol{k}'}u_{\boldsymbol{k}}u_{\boldsymbol{k}'}\mathrm{e}^{\mathrm{i}(\boldsymbol{k}+\boldsymbol{k}')\boldsymbol{R}}=\sum_{\boldsymbol{k},\boldsymbol{k}'}u_{\boldsymbol{k}}u_{\boldsymbol{k}'}\sum_{\boldsymbol{R}}\mathrm{e}^{\mathrm{i}(\boldsymbol{k}+\boldsymbol{k}')\boldsymbol{R}}. \tag{7.8}$$

The last sum in (7.8) is equal to the number N of atoms A in the system if $\boldsymbol{k}+\boldsymbol{k}'=\boldsymbol{0}$ or $\boldsymbol{k}+\boldsymbol{k}'=2\pi\boldsymbol{b}$, and is equal to zero if $\boldsymbol{k}+\boldsymbol{k}'\neq\boldsymbol{0}$, $\boldsymbol{k}+\boldsymbol{k}'\neq 2\pi\boldsymbol{b}$. This conclusion is arrived at as a result of the periodic boundary conditions adopted. Since (as has been pointed out above) the values considered lie within the first Brillouin zone, i.e., $\boldsymbol{k}+\boldsymbol{k}'<2\pi\boldsymbol{b}$, we will take into account only the condition $\boldsymbol{k}+\boldsymbol{k}'=\boldsymbol{0}$. So we have

$$\sum_{\boldsymbol{R}} u^2_{\boldsymbol{R}}=N\sum_{\boldsymbol{k}}u_{\boldsymbol{k}}u_{-\boldsymbol{k}}. \tag{7.9}$$

Analogous formulas also hold for $\sum_{\boldsymbol{R}} u^2_{\boldsymbol{R}-\boldsymbol{a}_i}$ and $\sum_{\boldsymbol{R}} u^2_{\boldsymbol{R}+\boldsymbol{a}_i}$. The last two terms in (7.7) yield the sum

$$\sum_{\boldsymbol{R}} u_{\boldsymbol{R}}(u_{\boldsymbol{R}-\boldsymbol{a}_i}+u_{\boldsymbol{R}+\boldsymbol{a}_i}). \tag{7.10}$$

Substituting (7.3) into (7.10), we obtain

$$\sum_{\boldsymbol{k},\boldsymbol{k}'}u_{\boldsymbol{k}}u_{\boldsymbol{k}'}\left(\mathrm{e}^{-\mathrm{i}\boldsymbol{k}\cdot\boldsymbol{a}_i}+\mathrm{e}^{\mathrm{i}\boldsymbol{k}\cdot\boldsymbol{a}_i}\right)\sum_{\boldsymbol{R}}\mathrm{e}^{\mathrm{i}(\boldsymbol{k}+\boldsymbol{k}')\cdot\boldsymbol{R}}$$
$$=N\sum_{\boldsymbol{k}}[2\cos(\boldsymbol{k}\cdot\boldsymbol{a}_i)]u_{\boldsymbol{k}}u_{-\boldsymbol{k}}. \tag{7.11}$$

Thus, expression (7.7) is transformed to the form

$$\frac{Nc}{4}\sum_{\boldsymbol{k}}\sum_{i=1,2,3}[4-4\cos(\boldsymbol{k}\cdot\boldsymbol{a}_i)]u_{\boldsymbol{k}}u_{-\boldsymbol{k}}$$
$$=2Nc\sum_{\boldsymbol{k}}\sum_{i=1,2,3}\sin^2\left(\frac{\boldsymbol{k}\cdot\boldsymbol{a}_i}{2}\right)u_{\boldsymbol{k}}u_{-\boldsymbol{k}}. \tag{7.12}$$

Using considerations analogous to those given above, we see that

$$H = \frac{N}{2}\sum_{\boldsymbol{k}}\left(a + 4c\sin^2\frac{\boldsymbol{k}\cdot\boldsymbol{a}_1}{2} + 4c\sin^2\frac{\boldsymbol{k}\cdot\boldsymbol{a}_2}{2} + 4c\sin^2\frac{\boldsymbol{k}\cdot\boldsymbol{a}_3}{2}\right)u_{\boldsymbol{k}}u_{-\boldsymbol{k}}$$

$$+\frac{Nb}{4}\sum_{\boldsymbol{k}_1+\boldsymbol{k}_2+\boldsymbol{k}_3+\boldsymbol{k}_4=0;2\pi\boldsymbol{b}} u_{\boldsymbol{k}_1}u_{\boldsymbol{k}_2}u_{\boldsymbol{k}_3}u_{\boldsymbol{k}_4}. \tag{7.13}$$

In writing the last sum it has been taken into account that the sum of more than two vectors $\boldsymbol{k}$ lying within the Brillouin zone may be equal to a nonzero vector of the reciprocal lattice multiplied by 2π. Let us consider those terms in the sums of formula (7.1) which correspond to small values of $\boldsymbol{k}$. Here we can replace the sines in (7.13) by their arguments and also take into account that the sum of four sufficiently small vectors $\boldsymbol{k}$ is always less than $2\pi\boldsymbol{b}$, where $\boldsymbol{b}$ is the nonzero vector of the reciprocal lattice. For the longwave part of the Hamiltonian (7.13) we have

$$H_{\text{lw}} = \frac{N}{2}\sum_{\boldsymbol{k}}\left(a + ca_1^2k_x^2 + ca_2^2k_y^2 + ca_3^2k_z^2\right)u_{\boldsymbol{k}}u_{-\boldsymbol{k}}$$

$$+\frac{N}{4}b\sum_{\boldsymbol{k}_1+\boldsymbol{k}_2+\boldsymbol{k}_3+\boldsymbol{k}_4=0} u_{\boldsymbol{k}_1}u_{\boldsymbol{k}_2}u_{\boldsymbol{k}_3}u_{\boldsymbol{k}_4}. \tag{7.14}$$

Comparing this formula with (6.9), we see that they coincide if we ignore the terms before the sums and the anisotropies of the $\boldsymbol{k}$-dependence of the term in parentheses in the first sum. By the way, this anisotropy could also have been taken into account in (6.9), but we ignored it instead. We neglect it here too, assuming that $a_1 = a_2 = a_3 = d$.[2] The terms before the sums in (6.9) and (7.14) can be made equal to each other by multiplying and dividing (7.14) by the volume of the unit cell $v_{\text{uc}} = d^3$. Finally we establish the correspondence between relations (6.9) and (7.14):

$$a/v_{\text{uc}} \to A, \quad cd^2/v_{\text{uc}} \to \delta, \quad b/v_{\text{uc}} \to \beta. \tag{7.15}$$

Thus, the longwave part of the Hamiltonian (7.13) coincides with (6.9). It is natural that the coefficient a in (7.14) is independent of temperature, in contrast to A; moreover, $a < 0$, as follows from the form of the potential U_{AB} (see Fig. 7.2).

Since expression (6.9) is simpler than (7.13), it would be more convenient to deal with an expression of this kind for the Hamiltonian. We will try to do so, i.e., we will assume that expression (7.14) which is equivalent to (6.9), is valid not only for $\boldsymbol{k}$ small but for all $\boldsymbol{k}$, assuming that the summation over $\boldsymbol{k}$ in (7.14) is carried out over the entire Brillouin zone. Of course, in this case

[2] Earlier in this chapter we were not accurate either in the consideration of anisotropy. In writing down the Hamiltonian (7.1) we should have assumed that the springs arranged along different axes have different stiffness coefficients, i.e., we should have introduced several quantities c.

the region of large $\boldsymbol{k}$ will not be described precisely, but, as will be clarified below, this inaccuracy is unimportant to us.

So, we will use the Hamiltonian

$$H=\frac{N}{2}\sum_{\boldsymbol{k}}\left(a+cd^2k^2\right)u_{\boldsymbol{k}}u_{-\boldsymbol{k}}+\frac{N}{4}b\sum_{\boldsymbol{k}_1+\boldsymbol{k}_2+\boldsymbol{k}_3+\boldsymbol{k}_4=0}u_{\boldsymbol{k}_1}u_{\boldsymbol{k}_2}u_{\boldsymbol{k}_3}u_{\boldsymbol{k}_4}. \quad (7.16)$$

Let us now calculate, by proceeding from the Hamiltonian (7.16), the coefficients of the thermodynamic potential in the Landau theory $\Phi(\eta)=\Phi(\eta_{k=0})$, where $\eta_{k=0}\equiv u_{k=0}$. Separating from the Hamiltonian the terms containing $\eta_{k=0}$, just as was done in formula (6.19), we find

$$\frac{1}{N}\Phi(\eta_{k=0})=\frac{a}{2}\eta_{k=0}^2+\frac{b}{4}\eta_{k=0}^4-\frac{k_{\mathrm{B}}T}{N}\ln\int\exp\left\{-\left(\frac{N}{k_{\mathrm{B}}T}\right)\right.$$
$$\times\left[\frac{1}{2}\sum_{\boldsymbol{k}}{}'\left(a+cd^2k^2\right)u_{\boldsymbol{k}}u_{-\boldsymbol{k}}+\frac{b}{4}\sum_{\boldsymbol{k}_1+\boldsymbol{k}_2+\boldsymbol{k}_3+\boldsymbol{k}_4=0}{}'u_{\boldsymbol{k}_1}u_{\boldsymbol{k}_2}u_{\boldsymbol{k}_3}u_{\boldsymbol{k}_4}\right.$$
$$\left.\left.+\frac{3b}{2}\eta_{k=0}^2\sum_{\boldsymbol{k}\neq 0}{}'u_{\boldsymbol{k}}u_{-\boldsymbol{k}}+\frac{3b}{2}\eta_{k=0}\sum_{\boldsymbol{k}_1+\boldsymbol{k}_2+\boldsymbol{k}_3=0}{}'u_{\boldsymbol{k}_1}u_{\boldsymbol{k}_2}u_{\boldsymbol{k}_3}\right]\right\}\prod_{\boldsymbol{k}}{}'\mathrm{d}u_{\boldsymbol{k}}. \quad (7.17)$$

We write some terms of the coefficient at $\eta_{k=0}^2$:

$$\frac{1}{N}\left(\frac{\partial^2\Phi}{\partial\eta_{k=0}^2}\right)_{\eta_{k=0}=0}=a+\frac{3b}{N}\sum_{\boldsymbol{k}}{}'\langle u_{\boldsymbol{k}}u_{-\boldsymbol{k}}\rangle$$
$$-\frac{b^2}{Nk_{\mathrm{B}}T}\left\langle\sum_{\boldsymbol{k}_1+\boldsymbol{k}_2+\boldsymbol{k}_3=0}u_{\boldsymbol{k}_1}u_{\boldsymbol{k}_2}u_{\boldsymbol{k}_3}\sum_{\boldsymbol{k}'_1+\boldsymbol{k}'_2+\boldsymbol{k}'_3=0}u_{\boldsymbol{k}'_1}u_{\boldsymbol{k}'_2}u_{\boldsymbol{k}'_3}\right\rangle \quad (7.18)$$

where the angular brackets $\langle\ldots\rangle$ signify an averaging with the Hamiltonian (7.17), in which we put $\eta_{k=0}=0$. For example,

$$\langle u_{\boldsymbol{k}}u_{-\boldsymbol{k}}\rangle=\int_{-\infty}^{\infty}u_{\boldsymbol{k}}u_{-\boldsymbol{k}}\exp\left\{\left[-\frac{1}{k_{\mathrm{B}}T}\left(\frac{N}{2}\sum_{\boldsymbol{k}}{}'\left(a+cd^2k^2\right)u_{\boldsymbol{k}}u_{-\boldsymbol{k}}+\right.\right.\right.$$
$$\left.\left.\left.+\frac{Nb}{4}\sum_{\boldsymbol{k}_1+\boldsymbol{k}_2+\boldsymbol{k}_3+\boldsymbol{k}_4=0}{}'u_{\boldsymbol{k}_1}u_{\boldsymbol{k}_2}u_{\boldsymbol{k}_3}u_{\boldsymbol{k}_4}\right)\right]\right\}\prod_{\boldsymbol{k}}{}'\mathrm{d}u_{\boldsymbol{k}}$$
$$\times\left(\int_{-\infty}^{\infty}\exp\left\{-\frac{1}{k_{\mathrm{B}}T}\left(\frac{N}{2}\sum_{\boldsymbol{k}}{}'\left(a+cd^2k^2\right)u_{\boldsymbol{k}}u_{-\boldsymbol{k}}\right.\right.\right.$$
$$\left.\left.\left.+\frac{N}{4}\sum_{\boldsymbol{k}_1+\boldsymbol{k}_2+\boldsymbol{k}_3+\boldsymbol{k}_4=0}{}'u_{\boldsymbol{k}_1}u_{\boldsymbol{k}_2}u_{\boldsymbol{k}_3}u_{\boldsymbol{k}_4}\right)\right\}\prod_{\boldsymbol{k}}{}'\mathrm{d}u_{\boldsymbol{k}}\right)^{-1}. \quad (7.19)$$

As pointed out in Chap. 6, such integrals cannot be evaluated. We will try to use the same approximation as in Chap. 6, i.e., we omit fourth-order terms. Then, the integrals in (7.19) are split into products of simple integrals. However, it immediately becomes obvious that here we encounter difficulties. The point is that the integrals corresponding to $u_{\boldsymbol{k}}$ with small $\boldsymbol{k}$ diverge since $a < 0$ and, hence, $a + cd^2k^2 < 0$ at small k. Nevertheless, this is not important if the volume of k space, in which $a + cd^2k^2 < 0$, is not large, i.e., $|a| \ll c$. We will limit ourselves to this case. As might be expected, the transition temperature will be low, which is why we ignore the third term in formula (7.18), which, as can be shown, is proportional to T^2. As a reasonable approximation, in calculating the second term in (7.18) we put $a = 0$. Then, the result of the integration with an accuracy of up to redesignating the quantities will coincide with (6.37), i.e., the second term in (7.18) gives a term proportional to temperature. As a result, we obtain for the coefficient at $\eta_{k=0}^2$:

$$a + \frac{3b}{8\pi}\frac{k_{\mathrm{B}}T}{c} = \frac{3bk_{\mathrm{B}}}{8\pi c}(T - T_{\mathrm{c}}) \tag{7.20}$$

that is, this coefficient has the same temperature dependence as the coefficient A in the Landau theory with the transition temperature in our approximation being equal to (as can be seen from (7.20))

$$T_{\mathrm{c}} = \frac{8\pi ac}{3bk_{\mathrm{B}}}. \tag{7.21}$$

From (7.21) it is clearly seen that the remarks made at the beginning of this chapter are valid: at $c = 0$ the transition temperature tends to zero. Moreover, we see that for a phase transition to occur the coefficient a must be negative and b positive.

In the approximation used here the coefficient at $\eta_{k=0}^4$ in the Landau potential is given by (7.15), i.e.,

$$\beta = b/v_{\mathrm{uc}}. \tag{7.22}$$

Thus, we have expressed the coefficients of the Landau thermodynamic potential in terms of the parameters of our model: a, b, c, d. But what have we gained by using our model, apart from the elucidation of the physical meaning of the coefficients of the thermodynamic potential? Is it an achievement to express some coefficients in terms of others? Does it not look like expressing the unknown through the incomprehensible, as is jokingly said? A model theory may be regarded as efficient if it gives something new to thermodynamic theory, a new relationship between experimental quantities which would have been left unsuspected without a model theory.

Here we have a similar situation. If we compare our model with some real crystal, then the atoms A and B are to be considered charged (only purely covalent crystals consist of atoms of a single species). So, what we have actually considered is a ferroelectric phase transition. We express the polarization at $T = 0$ via the Hamiltonian parameters. For this temperature we have

$$\eta_0^2 = \frac{|\alpha'|}{b}, \qquad P_0^2 = \frac{a^2|\alpha'|}{b} \tag{7.23}$$

where a is the coefficient in formula (2.35). The dependence of polarization on electric field strength E is given by the relation

$$\tilde{A}P + \tilde{B}P^3 = E \tag{7.24}$$

where

$$\tilde{A} = \frac{3bk_\mathrm{B}}{8\pi ca^2}(T - T_\mathrm{c}), \quad \tilde{B} = \frac{b}{d^3a^4}.$$

Using formulas (3.8), we find for the Curie–Weiss constant for $T > T_\mathrm{c}$:

$$C_{\mathrm{C-W}} = \frac{4\pi T_\mathrm{c}}{\tilde{B}P_0^2}. \tag{7.25}$$

A formula of this kind cannot be derived, in principle, in thermodynamic theory because there are no data on the transition temperature in it. It would be naive to think that formula (7.25) agress well with experimental data. The model used to derive it is overly simplified: it is not a single ion (as an example) that is displaced in real ferroelectrics. More realistic models will, of course, provide other fomulas, which will still be similar to the simple formula (7.25).

7.3 Phase Transitions of Displacive and Order–Disorder Types in the Single-Ion Model

In the case under consideration, i.e., when $|a| \ll c$, it is said that a displacive phase transition occurs. We will explain what this means.

Let us pose a question: What is the average potential in which the atom A is moving at $T > T_\mathrm{c}$? It is clear that the mean positions of other atoms A are in the centers of unit cells, i.e., in formula (7.1) we may set $u_{\boldsymbol{R}} = 0$ for $\boldsymbol{R} \neq \boldsymbol{R}_0$, where $\boldsymbol{R}_0$ is the coordinate of the center of a given unit cell. As a result, we obtain

$$U = \frac{1}{2}(a + 6c)u_{\boldsymbol{R}_0}^2 + \frac{1}{4}bu_{\boldsymbol{R}_0}^4. \tag{7.26}$$

(The factor 6 at c reflects the fact that the given atom A is linked to the remaining six by springs.) Thus, in our case the coefficient of the first term in (7.26) is positive. In other words, an individual atom is moving in an effective potential with a single minimum (Fig. 7.5), i.e., it has one equilibrium position. A phase transition is a cooperative phenomenon: it implies the displacement of a sublattice of atoms A from the equilibrium position in the center of the unit cell. Such a transition is known as a displacive phase transition, as has already been mentioned in Chap. 1.

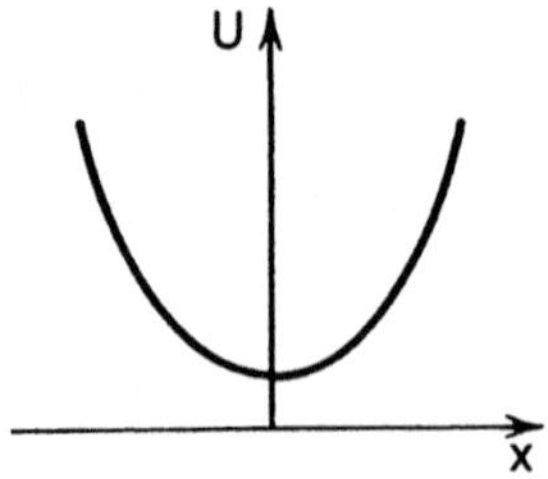

Fig. 7.5. The average potential for one atom A for the case of a crystal with a displacive-type phase transition.

In the case opposite to the one in question, i.e., $|a| \gg c$, an individual atom is moving, as follows from (7.26), in a symmetric phase in a potential of the same type as the one shown in Fig. 7.2.

The calculation that led to formula (7.20) does not work in this case since the quantity $a + cd^2k^2$ is negative now for all k. So, in formula (7.26) we may ignore the term $6c$ compared with ca in the coefficient of the first term. Simplifying the picture of thermal motion of atoms A to a certain extent, we may say that each such atom occupies by turns one of the two equilibrium positions corresponding to the potential energy minima in Fig. 7.2:

$$u_{1,2} = \pm\sqrt{|a|/b} = \pm u_0 \tag{7.27}$$

which is characteristic for order–disorder phase transitions.

What occurs in a phase transition in this case? Above the transition point each atom A is, with equal probability, in one of the two equilibrium positions. At $T = 0$ the atoms A are arranged in a strictly ordered manner, i.e., they are in minima of a single type in each unit cell (say, in left-hand ones in Fig. 7.2). It is obvious that this state corresponds to the minimum of the crystal's energy: the energy U_{AB} is a minimum and the energy U_{AA}, which cannot be negative, is equal to zero: the springs are not extended. With $T \neq 0$ the equilibrium configuration of the system is determined by the minimum of the free energy F (rather than of the energy); in our model the volume of the crystal is fixed and the thermodynamic potential coincides with the free energy. It is known that

$$F = U - TS\,, \tag{7.28}$$

where S is the entropy. The entropy increases with increasing disorder in the system, which is why, as can be seen from (7.28), at finite temperatures partially disordered configurations become favorable. In a configuration in which atom A occupies its possible equilibrium positions with equal probability, to the same energy there correspond two positions of the atom. The entropy per atom is here equal to

$$S = k_{\mathrm{B}} \ln 2. \tag{7.29}$$

The occurrence of disorder naturally increases the system's energy; with a completely chaotic position of atoms A half of the springs appear to be deformed, so that the energy per atom increases as compared with the energy at $T = 0$ by an amount

$$\frac{3}{4}c(2u_0)^2 = 3cu_0^2. \tag{7.30}$$

Here it has been taken into account that, on average, there are three springs for each atom. The phase transition temperature is naturally estimated from the condition that the increase in energy is compensated for in (7.28) by an increase in entropy. We therefore have

$$3cu_0^2 = -3ca/b = k_B T \ln 2 \tag{7.31}$$

that is,

$$k_B T_c = -\frac{3ca}{b \ln 2}. \tag{7.32}$$

Note that we have obtained for T_c a formula which differs only by a not too large numerical factor from (7.21). A more thorough analysis with the calculation of the coefficients of the thermodynamic potential will be made in Chap. 8.

At this point it should only be noted that the displacement of atom A from the center of the unit cell is in our treatment a quantity which assumes only two values with different signs. Therefore, we may write

$$u_{\boldsymbol{R}} = u_0 \sigma_{\boldsymbol{R}}, \tag{7.33}$$

where $\sigma_{\boldsymbol{R}}$ is a quantity taking on the value +1 or −1. The set of $\sigma_{\boldsymbol{R}}$ values describes now the picture of displacements of atoms A. Substituting (7.33) into (7.1) and taking into account that $\sigma_{\boldsymbol{R}}^2 = 1$, we find that

$$H = \frac{1}{2} \sum_{\boldsymbol{R},\boldsymbol{R}'} J \sigma_{\boldsymbol{R}} \sigma_{\boldsymbol{R}'} \tag{7.34}$$

where $J = 2cu_0^2$ and the terms that do not contain $\sigma_{\boldsymbol{R}}$ have been omitted. The Hamiltonian (7.34) is known as the Ising Hamiltonian and is commonly used to describe phase transitions of the order–disorder type (see Chap. 8).

7.4 Applicability of the Landau Theory to Phase Transitions of Displacive and Order–Disorder Types

Let us now compare, in the light of our model, the properties of systems with displacive and order–disorder phase transitions. It is hoped that, despite the simplified character of our model, we will succeed in obtaining correct notions of the ratio of the orders of values for the macroscopic constants of real crystals with such phase transitions.

We will first explain how the coefficients contained in the Hamiltonian of a crystal can be estimated. Suppose we have an "ordinary" crystal, i.e., a crystal with no phase transition and there occurs a strong local disturbance of the crystalline structure – a displacement of an atom to distances of the order of the lattice constant. It is clear that such a disturbance causes a significant increase of the potential energy, which in order of magnitude is equal to the bond energy of atoms in the crystal per atom. A characteristic value of this energy, which is called the atomic energy, has an order of a few electron-volts or 10^4–10^5 K. The atomic energy is the natural scale of measurement of the interaction energy between atoms in a crystal. The corresponding atomic temperature $T_{\rm at}$ is characteristic for the determination of the temperature dependence of the material constants of the crystal. As an example, one can give an estimate for the thermal linear expansion coefficient, which yields $K \approx 1/T_{\rm at} \approx 10^{-5}$, this being close to the experimentally observed values for ordinary dielectrics.

We will now consider displacive phase transitions in the single-ion model. From (7.20) and (7.21) we see that both the transition temperature and the Curie–Weiss constant are determined by the Hamiltonian constants a, b and c; for displacive transitions it is necessary that the condition $|a| \ll c$ be fulfilled. Since for "ordinary" crystals the constants that characterize the interaction of atoms have a single scale – the atomic energy then for the phase transition to occur it is necessary that at least one of these constants have an unusual, anomalous value. There are rather weighty grounds for believing that it is exactly the constant a which has an anomalously small value. In the model under discussion it is negative, which is why a phase transition is realized. However, in the same model but without a phase transition $a > 0$. This means that the constant may, in particular, be close to zero as well. As regards the constant c, it is hardly probable for it to have an anomalously large value: it would mean that the energy of interaction between two atoms in neighboring unit cells is unusually large, being much greater than the atomic energy. Based on these considerations, we can estimate for example, the value of the Curie–Weiss constant for displacive phase transitions. As can be seen from (7.20), the temperature-dependent part of the coefficient A includes only constants having a normal (atomic) order of magnitude (the constant b need not be regarded as anomalous). Therefore, the characteristic temperature scale of variation of the coefficient A is $T_{\rm at}$. In other words, it may be expected that the Curie–Weiss constant ($C_{\rm C-W}$) will be of order $T_{\rm at} \approx 10^4$–10^5 K. As has been shown in Chap. 1, this is what actually occurs, for example, in the case of ferroelectrics of the perovskite group.

At the same time, by virtue of the smallness of the constant a, $T_{\rm c} < T_{\rm at}$; this is clear from formula (7.21), where b and c have a normal (atomic) order of magnitude and the constant a is anomalously small. It is this circumstance that makes possible a structural phase transition before the crystal melts.

As will be shown in Chap. 9, the Curie–Weiss constant for order–disorder transitions is close to $T_{\rm c}$ in order of magnitude; we come to the conclusion that the Curie–Weiss constants for displacive-type transitions are considerably larger than in the case of order–disorder transitions.

For transitions of the order–disorder type the smallness of T_c as compared with T_{at} is associated with the smallness of the coefficient c, i.e., the ions being ordered must interact with one another relatively weakly.

Thus, even on the basis of a rather simplified model and approximate estimates we can get an insight into the various macroscopic properties of systems undergoing phase transitions of displacive and order–disorder types. Naturally we are speaking here only of the limiting cases of $|a| \gg c$ and $|a| \ll c$. Of course, intermediate cases are also possible, when $|a|$ and $|c|$ are of the same order of magnitude.

In conclusion, let us consider the question of the applicability of the Landau theory to displacive and order–disorder phase transitions. For the former we have expressed the coefficients of the thermodynamic potential in terms of the Hamiltonian constants. Now we will use these constants to express the condition for the applicability of the Landau theory (see (6.18)). Using (7.20)–(7.22), we find

$$(T - T_c)/T_c \gg \frac{a}{c}. \tag{7.35}$$

Since $a/c \ll 1$ for displacive transitions, the inequality (7.35) does not contradict the condition

$$(T - T_c)/T_c \ll 1 \tag{7.36}$$

which, as pointed out in Chap. 2, is also the condition for the validity of the Landau theory.

From (7.25) we see that when the Hamiltonian constants change in such a way that a displacive-type transition "moves" to an order–disorder one, the conditions for the application of the Landau theory are impaired. It may thus be expected that it cannot be applied at all for order–disorder transitions. It should, however, be emphasized that our reasoningis rather approximate. A more detailed treatment shows that for order–disorder transitions the Landau theory may be valid provided that the interaction of atoms A not only with the nearest, but also with the next nearest, neighbors becomes essential.

Moreover, it is natural that the suppression of fluctuations of the order parameter by long-range (electric or elastic) fields extends the region of applicability of the Landau theory even further. Based on the reasoning given in Chap. 6, we arrive at the conclusion that for proper uniaxial ferroelectrics and proper ferroelastics the Landau theory may be expected to be also valid for order–disorder transitions.

What has been said in this chapter actually covers qualitative results that can be obtained within the framework of the simplest microscopic theory of the thermodynamic properties of crystals with a displacive-type phase transition. The chapter that follows will therefore be entirely devoted to order–disorder phase transitions.

8. Statistical Theory of Ferroelectric Phase Transitions of the Order–Disorder Type

8.1 The Hamiltonian of a Uniaxial Ferroelectric with an Order–Disorder Phase Transition

As has been noted earlier, from the standpoint of the microscopic mechanism of a phase transition it is reasonable to consider separately a group of ferroelectric crystals with an order–disorder phase transition. In accordance with Chap. 1, we will use the term order–disorder ferroelectrics for crystals in which at least one of the sublattices consists of particles having two or more equilibrium positions. It may be said that the remaining (ordered) sublattices create, as it were, a "skeleton" of the crystal and the atoms that belong to these sublattices perform slight oscillations relative to fixed equilibrium positions. It is their configuration that creates potentials of two or more minima for sublattice atoms (see Fig. 7.2).

While considering the potential relief shown in Fig. 7.2, it may be noted that the motion of atoms of a disordered sublattice must be characterized by the following specific features. The particles perform slight oscillations relative to each of the equilibrium positions; they also pass from one equilibrium position to another. In doing so, the particle moves at once to a distance of $2u_0$, which by far exceeds the mean displacement of the particle near each of the equilibrium positions. A particle having a charge e and performing jumps 1→2 and 2→1 under the action of random thermal forces may be regarded as a reversible dipole. As a matter of fact, for an electrically neutral unit cell the dipole moment is

$$p = \pm e u_0, \tag{8.1}$$

if it is assumed that the central position of the particle (point 0) corresponds to the absence of a dipole moment. Thus, a disordered crystal is a system of dipoles aligned in a random fashion in space along one or several crystallographic directions (Fig. 8.1). In a phase transition of the order–disorder type, at the transition point the equiprobability of the distribution of particles over equilibrium positions is upset. This means that above the Curie point each disordered particle occupies position 1 during one half of a sufficiently long period of time and position 2 during the other half (see Fig. 1.7). Slightly below the Curie point the character of the motion of the particles practi-

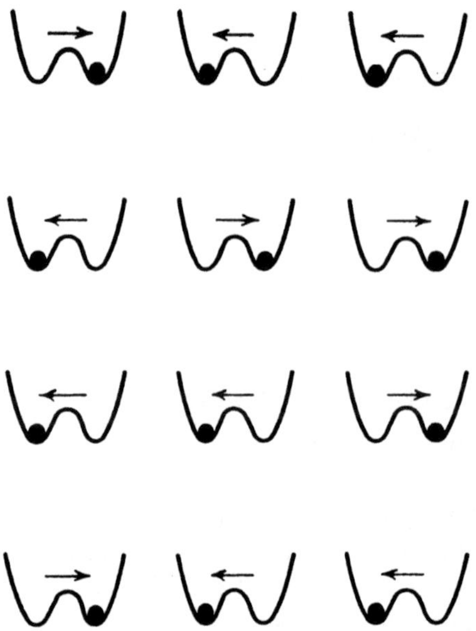

Fig. 8.1. A system of disordered reversible dipoles in a ferroelectric crystal of the order–disorder type.

cally does not change, but each of the particles begins to occupy one of the positions for a longer period of time than in the other.

This is equivalent to the appearance of a mean dipole moment in each unit cell and a spontaneous polarization in the entire crystal. It is clear that the equilibrium is upset at the Curie point because of the interaction of particles being ordered: if the energy of the configurations of particles at which they are simultaneously either in position 1 or in position 2 during a shorter period of time than in the case of the "antiparallel" arrangement, in the system there will appear a tendency towards the appearance of a "parallel" arrangement and a spontaneous polarization.

An analogous situation arises in the *Ising model*, which is used to analyze phase transitions in strongly anisotropic ferromagnetics. This model is interesting since in a two-dimensional variant it allows for a precise analytic determination of thermodynamic functions; moreover, it reflects many of the characteristic features of phase transitions in various ordering systems, including ferroelectrics of the order–disorder type.

In the three-dimensional Ising model, for each of the points of a simple cubic lattice there is a variable $\sigma_{\boldsymbol{R}}$, which may assume two values: either $+1$ or -1 (Fig. 8.2). For a magnetic substance this variable may be assigned a certain physical meaning – that of the magnetic moment of an atom present in a lattice point; then $\sigma_{\boldsymbol{R}} = +1$ if the moment is directed upward and $\sigma_{\boldsymbol{R}} = -1$ if it is directed downward. If there are no other degrees of freedom in the system, its potential energy for a certain configuration of the variables (the possible number of configurations is 2^N, where N is the total number of lattice points) has the following form:

$$U(\sigma) = -\frac{1}{2}\sum_{\boldsymbol{R},\boldsymbol{R}'} J(\boldsymbol{R}-\boldsymbol{R}')\sigma_{\boldsymbol{R}}\sigma_{\boldsymbol{R}'}. \tag{8.2}$$

The constants $J(\boldsymbol{R}-\boldsymbol{R}')$ determine the interaction of a pair of particles – *pseudospins* disposed at points $\boldsymbol{R}$ and $\boldsymbol{R}'$. An analogous expression has been derived by us earlier in the single-ion approximation and for the interaction of nearest neighbors only (see (7.34)).

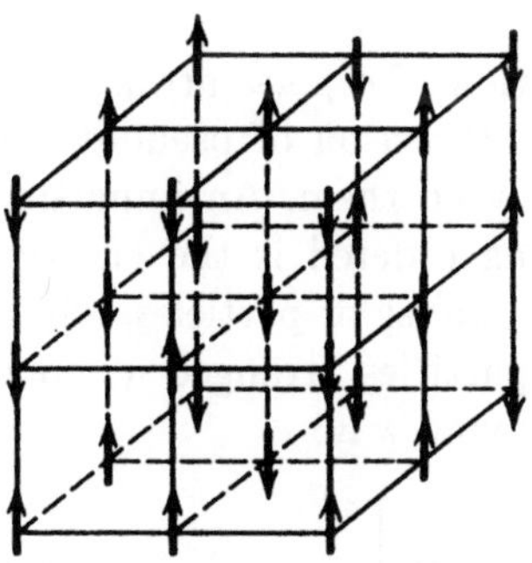

Fig. 8.2. The Ising model.

The determination of the thermodynamic functions of the Ising model with the Hamiltonian (8.2) is a problem of theoretical physics which has not yet been solved completely: it is necessary to calculate the statistical sum over all possible configurations of the variables σ:

$$Z = \sum_{\sigma_{\boldsymbol{R}_1}=\pm 1} \cdots \sum_{\sigma_{\boldsymbol{R}_N}=\pm 1} \exp\left[-\frac{U(\sigma_{\boldsymbol{R}_1}\ldots\sigma_{\boldsymbol{R}_N})}{k_{\mathrm{B}}T}\right] \tag{8.3}$$

and also the free energy

$$F = -k_{\mathrm{B}}T\ln Z. \tag{8.4}$$

The results of the calculation of (8.2)–(8.4) are of fundamental importance; in particular, a ferroelectric phase transition of the order–disorder type has many common features with the system just described, but a certain physical meaning should be assigned to the variable $\sigma_{\boldsymbol{R}}$.

We will assume that to the "right-hand" position of a disordered particle in Fig. 8.1 there corresponds the value of $\sigma = +1$ and to the "left-hand" position, the value of $\sigma = -1$, and that the crystal's energy is determined only by the method of distribution of particles over equilibrium positions.[1] Then, formally our crystal is an electric analog of the magnetic Ising model and now electric dipoles rather than magnetic moments are arranged in lattice points.

[1] Here and henceforth we assume, for simplicity, that there is only one disordered particle in each unit cell of the crystal, the disorder being characteristic for the motion along a single direction (see Fig. 8.1).

The spin variable can be determined in the same way as

$$\sigma_{\boldsymbol{R}} = p_{\boldsymbol{R}}/|p_{\boldsymbol{R}}|. \tag{8.5}$$

This quantity will evidently assume either of the values +1 and −1, depending on the position of the particle on the Rth link (see Fig. 8.1). The potential energy of the crystal takes the form

$$U(\sigma) = -\frac{1}{2}\sum_{\boldsymbol{R},\boldsymbol{R}'} J_{\boldsymbol{R},\boldsymbol{R}'} p_{\boldsymbol{R}} p_{\boldsymbol{R}'} = -\frac{1}{2}\sum_{\boldsymbol{R},\boldsymbol{R}'} J'_{\boldsymbol{R},\boldsymbol{R}'} \sigma_{\boldsymbol{R}} \sigma_{\boldsymbol{R}'}\,, \tag{8.6}$$

where the constant $J_{\boldsymbol{R},\boldsymbol{R}'}$ determines the interaction of dipoles arranged at points with the radius-vectors $\boldsymbol{R}$ and $\boldsymbol{R}'$. The interaction of pseudospins, of course, decreases with increasing distance between them. An important characteristic of the interaction of particles being ordered is the effective interaction radius which characterizes the total number of particles, whose position is still influenced by the orientation of particles arranged at point $\boldsymbol{R}$. It can be determined, for example, in the following way:

$$R_0^2 = \sum_{|\boldsymbol{R}-\boldsymbol{R}'|} |\boldsymbol{R}-\boldsymbol{R}'|^2 J(\boldsymbol{R}-\boldsymbol{R}') \left[\sum_{|\boldsymbol{R}-\boldsymbol{R}'|} J(\boldsymbol{R}-\boldsymbol{R}')\right]^{-1}. \tag{8.7}$$

It is expedient to determine R_0 in dimensionless units by referring it to a, the unit cell parameter (the mean distance between the nearest particles being ordered):

$$r_0 = R_0/a.$$

If the system is in a homogeneous electric field, its Hamiltonian may be represented in the following form:

$$H = -\frac{1}{2}\sum_{\boldsymbol{R},\boldsymbol{R}'} J_{\boldsymbol{R},\boldsymbol{R}'} \sigma_{\boldsymbol{R}} \sigma_{\boldsymbol{R}'} - E_z p_z \sum_{\boldsymbol{R}} \sigma_{\boldsymbol{R}}. \tag{8.8}$$

Here the last term takes account of the interaction of dipoles with a homogeneous macroscopic electric field.

8.2 The Free Energy of an Order–Disorder Crystal in the Self-Consistent Molecular Field Approximation

In order to discuss the thermodynamic properties of a system described by the Hamiltonian (8.8), let us consider some approximate methods of calculating the thermodynamic functions. It should be noted at the outset that in a disordered system the order–disorder phase transition implies the appearance of a long-range order, i.e., a partial ordering, at a certain temperature. The degree of long-range order may be characterized by the mean value of the quantity

$$\frac{N_1 - N_2}{N_1 + N_2} = \frac{N_1 - N_2}{N} \tag{8.9}$$

where N_1 and N_2 are the numbers of ions present in the left and right wells, respectively (see Fig. 8.1) and N is the total number of ions. In the case $N_1 = N_2$, there is in the system a state corresponding to the complete disorder and the absence of a spontaneous polarization. Long-range order arises when $N_1 \neq N_2$, i.e., in each lattice point there appears, as it were, a permament mean dipole; if we consider it to be aligned along the z axis, we obtain

$$\langle p_z \rangle = \frac{N_1 - N_2}{N} p_z = p_z \langle \sigma \rangle \tag{8.10}$$

and the spontaneous polarization of the entire crystal is given by

$$P_s = \frac{p_z}{v_{\text{uc}}} \langle \sigma \rangle . \tag{8.11}$$

Here $\langle \sigma \rangle = (N_1 - N_2)/N$ is the mean value of a pseudospin – a measure of long-range order in the system – and v_{uc} is the volume of the unit cell. Thus,

$$\langle p_z \rangle = 0 \quad \text{at} \quad T > T_{\text{c}} \tag{8.12}$$

$$\langle p_z \rangle \neq 0 \quad \text{at} \quad T < T_{\text{c}} . \tag{8.13}$$

There exist several methods for an approximate calculation of the thermodynamic functions of systems that undergo a phase transition of the order–disorder type. Use is most widely made of the mean-molecular-field (MMF) approximation which enables one to obtain a qualitative (and in some cases, also quantitative) description of the phenomenon and to express the parameters of phenomenological theory in terms of the macroscopic characteristics of the system.

In the mean-field approximation the real interaction of particles is described by the mean field acting on each particle, the field being independent of the position of the particle and its nearest environment. According to this theory, this mean field depends only on the large-scale properties of the system, say on the degree of long-range order. The mean-field approximation proves to be the more accurate, the larger the effective radius of interaction of particles being ordered: with $r_0 \gg 1$ the forces acting on each ion from the side of the others will be approximately equal, representing the averaged contributions from a large number of ions present in the interaction sphere. Thus, the essence of the mean-field aoproximation consists of the neglect of the fluctuations of the resultant force acting on each particle.

Let us carry out an identity transformation of the Hamiltonian (8.10). We write down

$$\sigma_{\boldsymbol{R}} = s + (\sigma_{\boldsymbol{R}} - s),$$

where $s = \langle \sigma_{\boldsymbol{R}} \rangle$ is the statistical mean value of the spin variable; then,

$$H = -\frac{1}{2}\sum_{\boldsymbol{R},\boldsymbol{R}'} J_{\boldsymbol{R},\boldsymbol{R}'}\left[s^2 + s(\sigma_{\boldsymbol{R}} - s) + s(\sigma_{\boldsymbol{R}'} - s) + (\sigma_{\boldsymbol{R}} - s)(\sigma_{\boldsymbol{R}'} - s)\right]$$

$$-E_z p_z \sum_{\boldsymbol{R}} \sigma_{\boldsymbol{R}} = J_0 \frac{Ns^2}{2} - \frac{1}{2}s\sum_{\boldsymbol{R},\boldsymbol{R}'} J_{\boldsymbol{R},\boldsymbol{R}'}(\sigma_{\boldsymbol{R}} + \sigma_{\boldsymbol{R}'})$$

$$-\frac{1}{2}s\sum_{\boldsymbol{R},\boldsymbol{R}'} J_{\boldsymbol{R},\boldsymbol{R}'}(\sigma_{\boldsymbol{R}} - s)(\sigma_{\boldsymbol{R}'} - s) - E_z p_z \sum_{\boldsymbol{R}} \sigma_{\boldsymbol{R}}, \tag{8.14}$$

where we have used $J_0 = \sum_{\boldsymbol{R}'} J_{\boldsymbol{R},\boldsymbol{R}'}$, to represent the constant that characterizes the interaction of a particle in an arbitrary position $\boldsymbol{R}$ with all the other particles being located within the interaction radius r_0. Expression (8.14) differs from the original Hamiltonian only by the identity transformation and is accurate; the mean-field approximation consists of the neglect of the term $\frac{1}{2}\sum_{\boldsymbol{R},\boldsymbol{R}'} J_{\boldsymbol{R},\boldsymbol{R}'}(\sigma_{\boldsymbol{R}} - s)(\sigma_{\boldsymbol{R}'} - s)$. The structure of this term shows that it describes the interaction energy of fluctuations of the spin variable associated with a deviation of its value from the mean value of s at different points $\boldsymbol{R}$ and $\boldsymbol{R}'$. From the foregoing it follows that the effect of this term in thermodynamics must be smaller, the larger the effective radius of interaction of particles being ordered (r_0).

So, in the mean-molecular-field approximation we have

$$H = J_0\frac{Ns^2}{2} - s\sum_{\boldsymbol{R},\boldsymbol{R}'} J_{\boldsymbol{R},\boldsymbol{R}'}\sigma_{\boldsymbol{R}} - E_z p_z \sum_{\boldsymbol{R}} J_{\boldsymbol{R}}\sigma_{\boldsymbol{R}}$$

$$= J_0\frac{Ns^2}{2} - (sJ_0 + E_z p_z)\sum_{\boldsymbol{R}} \sigma_{\boldsymbol{R}}. \tag{8.15}$$

Then, for the statistical sum we obtain

$$Z = \sum_{\sigma_{\boldsymbol{R}_1}} \cdots \sum_{\sigma_{\boldsymbol{R}_N}} \exp\left[-\frac{1}{k_B T}\left(\frac{J_0 N s^2}{2} - (sJ_0 + E_z)p_z \sum_{\boldsymbol{R}} \sigma_{\boldsymbol{R}}\right)\right]$$

$$= \exp\left(-\frac{J_0 N s^2}{2k_B T}\right)\left[\exp\left(\frac{sJ_0 + E_z p_z}{k_B T}\right) + \exp\left(-\frac{sJ_0 + E_z p_z}{k_B T}\right)\right]^N$$

$$= \left(2\cosh\frac{sJ_0 + E_z p_z}{k_B T}\right)^N \exp\left(-\frac{J_0 N s^2}{2k_B T}\right), \tag{8.16}$$

and the free energy of the system will be given by

$$F = -k_B T \ln Z = \frac{J_0 N s^2}{2} - N k_B T \ln\left(2\cosh\frac{sJ_0 + E_z p_z}{k_B T}\right). \tag{8.17}$$

Expression (8.17) contains the mean value of the spin variable s as a parameter. Under the conditions of thermodynamic equilibrium there is a state with the value of s corresponding to the minimum of F:

$$\frac{\partial F}{\partial s} = J_0 N s - J_0 N \left(2\sinh\frac{sJ_0 + E_z p_z}{k_B T}\right)\left(2\cosh\frac{sJ_0 + E_z p_z}{k_B T}\right)^{-1} = 0,$$

or

$$s = \tanh\frac{sJ_0 + E_z p_z}{k_B T}. \tag{8.18}$$

Equations (8.17) and (8.18) solve the problem of the analysis of the thermodynamic properties of an ordering system in the mean-field approximation. We emphasize once more that this approximation is equivalent to the assumption that the field acting on each particle is exactly equal to the mean field which reflects the most probable result of the interaction of particles with their distribution specified by the field itself. Therefore, (8.18) can also be obtained as the condition of this distribution, which is established in a self-consistent manner:

$$s = \frac{\sum_{\sigma_{\boldsymbol{R}}} \sigma_{\boldsymbol{R}} \exp\left[-H/(k_B T)\right]}{\sum_{\sigma_{\boldsymbol{R}}} \exp\left[-H/(k_B T)\right]}, \tag{8.19}$$

where the Hamiltonian H is dependent on s.

Indeed,

$$H = -\frac{1}{2}\sum_{\boldsymbol{R}} \sigma_{\boldsymbol{R}} E_{\boldsymbol{R}} - E_z p_z \sum_{\boldsymbol{R}} \sigma_{\boldsymbol{R}}, \tag{8.20}$$

where $E_{\boldsymbol{R}} = \sum_{\boldsymbol{R}'} J_{\boldsymbol{R},\boldsymbol{R}'} \sigma_{\boldsymbol{R}'}$ is the effective field acting on a particle; its mean, most probable, value can be determined by

$$\langle E_{\boldsymbol{R}} \rangle = \sum_{\boldsymbol{R}'} J_{\boldsymbol{R},\boldsymbol{R}'} \langle \sigma_{\boldsymbol{R}'} \rangle = J_0 s. \tag{8.21}$$

The condition (8.19) assumes the following form:

$$\begin{aligned} s &= \frac{\sum_{\sigma_{\boldsymbol{R}}} \sigma_{\boldsymbol{R}} \exp\left[-1/(k_B T)(sJ_0 + E_z p_z)\sigma_{\boldsymbol{R}}\right]}{\sum_{\sigma_{\boldsymbol{R}}} \exp\left[-1/(k_B T)(sJ_0 + E_z p_z)\sigma_{\boldsymbol{R}}\right]} \\ &= \frac{\exp\left[1/(k_B T)(sJ_0 + E_z p_z)\right] - \exp\left[-1/(k_B T)(sJ_0 + E_z p_z)\right]}{\exp\left[1/(k_B T)(sJ_0 + E_z p_z)\right] + \exp\left[-1/(k_B T)(sJ_0 + E_z p_z)\right]} \\ &= \tanh\frac{sJ_0 + E_z p_z}{k_B T}, \end{aligned} \tag{8.22}$$

which completely coincides with (8.18). The presence of a phase transition in the system is deduced from the fact that (8.22) at $E_z = 0$ has a nontrivial solution $s \neq 0$ in a certain temperature range $T < J_0/k_B$. Indeed, at sufficiently small s

$$\tanh\frac{sJ_0}{k_B T} = \frac{sJ_0}{k_B T} - \frac{1}{3}\left(\frac{sJ_0}{k_B T}\right)^3 + \cdots,$$

and (8.22) has the following solution:

$$s^2 = \frac{k_B^3 T^3}{J_0^3}\left(\frac{J_0}{k_B T} - 1\right),$$

that is,

$$s^2 = 3(T_c - T)/T_c, \tag{8.23}$$

where $T_c = J_0/k_B$ and we have taken into account that near T_c we may put $T = T_c$.

At the temperature T_c long-range order arises in the system. A characteristic feature of phase transitions of the order–disorder type is that, no matter how weak the interaction of particles being ordered (i.e., no matter how small the constant J_0), a phase transition will occur. In a ferroelectric phase transition of the order–disorder type we have, according to (8.10),

$$P_s^2 = -\frac{3p_z^2}{v_{uc}^2 T_c}(T - T_c). \tag{8.24}$$

We see that the temperature dependence of spontaneous polarization in a ferroelectric phase transition of the order–disorder type in the mean-field approximation coincides with the plot that follows from phenomenological theory for second-order phase transitions, which is of course what must be expected. Expression (8.17) may be reduced to a form which is ordinary in phenomenological theory by expanding the free-energy function as a series in $J_0 s$. Thus, the expansion coefficients in Landau theory can be obtained as functions of T, N, p_z and J_0.

Let us consider the dielectric constant ε (or dielectric permittivity) of the model along the direction of spontaneous polarization. Evidently,

$$\frac{\varepsilon - 1}{4\pi} = \frac{\partial P_z}{\partial E_z} = \frac{p_z}{v_{uc}}\frac{\partial s}{\partial E_z}, \qquad \varepsilon = 1 + 4\pi\frac{p_z}{v_{uc}}\frac{\partial s}{\partial E_z}. \tag{8.25}$$

From (8.18) we have

$$\frac{\partial s}{\partial E_z} = \frac{p_z}{k_B T \cosh^2[(sJ_0 + E_z p_z)/k_B T] - J_0}, \tag{8.26}$$

whence

$$\varepsilon = 1 + \frac{4\pi p_z^2}{v_{uc}}\left(k_B T \cosh^2[(sJ_0 + E_z p_z)/(k_b T)] - J_0\right)^{-1}. \tag{8.27}$$

It can be seen that in the paraelectric phase near the transition point $T_c = J_0/k_B$ with $E_z \to 0$ the function $\cosh[\ldots] \to 1$ and

$$\varepsilon = 1 + \frac{4\pi p_z^2}{J_0 v_{uc}}\frac{T_c}{T - T_c}, \tag{8.28}$$

that is, the Curie–Weiss law is fulfilled with $C_{C-W} = 4\pi p_z^2 T_c / J_0 v_{uc}$. For a quantitative estimate to be obtained it is necessary to know the value of J_0 and, hence, the character of interaction forces in the system of particles being ordered. The mean electric field that acts on each particle may be represented in the form

$$E = E_z + s\frac{J_0}{p_z} = E_z + \frac{J_0 v_{uc}}{p_z^2} P_s, \tag{8.29}$$

where E_z is the mean macroscopic field.

According to this interpretation, expression (8.29) represents a local field; for a purely electrostatic interaction the difference between the local field and the mean macroscopic field is the Lorentz field and $J_0 v_{uc}/p_z^2 = \beta$ is a quantity which is known in the electrostatics of dielectrics as the Lorentz factor. In the theory of dielectrics it is shown that for cubic crystals of the type NaCl and for a dipole located in the center of the unit cell, $\beta = 4\pi/3$. It can easily be seen that in this case

$$C_{C-W}/T_c = 4\pi/\beta = 3. \tag{8.30}$$

This result is of course very approximate since the dipole-dipole interaction is far from covering the complete effective interaction of particles being ordered. For crystals that are usually classified as belonging to the order–disorder type, relation (8.30) is fulfilled only in order of magnitude:

	T_c, K	C_{C-W}, K
TGS	322	3200
$NaNO_2$	433	5000
KH_2PO_4	123	3600

For crystals of the displacive type relation (8.30) is not fulfilled even approximately, as has been pointed out above; for example, for $BaTiO_3$ crystals $T_c = 400\,K$ but $C_{C-W} = 170{,}000\,K$.

There is still another characteristic feature of order–disorder ferroelectrics; it is a complete change of the entropy of crystals in a temperature range from 0 K (complete order) to T_c (complete disorder). Differentiating (8.17) with respect to temperature at $E_z = 0$ (a short-circuited crystal), we obtain for the crystal's entropy associated with a phase transition:

$$S = -\frac{\partial F}{\partial T} = Nk_B T \ln\left[2\cosh\frac{J_0 s}{k_B T}\right] - \frac{J_0 N s}{T}\tanh\frac{J_0 s}{k_B T}. \tag{8.31}$$

Assuming here that $s = 0$, i.e., defining the total entropy of the transition as its value at $T = T_c$, we obtain $\Delta S = Nk_B \ln 2 = R\ln 2$ per mole of substance (R is the universal gas constant); naturally, this result coincides with that obtained in Chap. 7 (see (7.29)).

The entropy of a phase transition is experimentally determined on the basis of data on the measurement of the temperature dependence of heat capacity over a sufficiently wide temperature range, including the Curie point. In this case it is possible to separate the regular part of the heat capacity $C_0(T)$ and the anomalous contribution $C(T) - C_0(T)$ associated with ordering (see Fig. 1.2), and then

$$\Delta S = \int_0^{T_c} \frac{C(T) - C_0(T)}{T} \mathrm{d}T. \tag{8.32}$$

Here $C(T)$ is the total heat capacity of the crystal being measured.

Thus, if there is one ordering particle per unit cell of the crystal, the entropy change may be expected to be equal to 5.7 J/(mole·K). Considering that we are speaking of the orders of magnitude of the quantities, we may state that the entropy change $\Delta S/R \sim 1$ is an indication of the presence in the crystal of an ordering subsystem, each ordering particle of which has two possible equilibrium positions.

Ferroelectric crystals that are commonly classified as belonging to the order–disorder class exhibit the following experimentally observable values of $\Delta S/R$, which are in good agreement with the theory expounded here:

TGS $\Delta S/R = 1.1$,

$NaNO_2$ $\Delta S/R = 0.7$,

KD_2PO_4 $\Delta S/R = 0.7$.

8.3 Tunneling Effects in Hydrogen-Containing Ferroelectrics

The temperature of an order–disorder phase transition is determined by the condition $k_B T = J_0$, i.e., $T_c \neq 0$ provided that there occurs an interaction of any kind between the ordered particles ($J_0 \neq 0$).

The situation, however, will change if the form of the potential relief and the mass of the particles being ordered are such that quantum-mechanical tunneling of the particles through the potential barrier (see Fig. 8.1) becomes appreciable. This effect competes with the processes of random jumps associated with thermal fluctuations of the energy.

Tunneling effects may become essential for hydrogen-containing ferroelectrics when light protons are disordered on hydrogen bonds. For each particle taken separately the occurrence of tunneling implie the overlap of the wavefunctions of the right- and left-hand states and a periodic change of the probability of the particle being in positions 1 and 2 (see Fig. 7.2). The state of each particle will be described by the wavefunction

$$\Psi = c_1\Psi_1 + c_2\Psi_2, \qquad c_1^2 + c_2^2 = 1, \tag{8.33}$$

where Ψ_1 and Ψ_2 are the wavefunctions which describe the states of the particles on the left and right, respectively.

We will not consider the wavefunctions of excited states, which are practically unpopulated at rather low temperatures.

If the potential barrier is impermeable, the motion of a particle in each well corresponds to the minimum energy E_0 – the ground-state energy. In the presence of a tunneling effect this level is split into two levels, one with the energy $E_1 = E_0 - Q$ and the other with $E_2 = E_0 + Q$ with the wavefunctions Ψ_{I} and Ψ_{II}, which are the symmetric and antisymmetric combinations of Ψ_1 and Ψ_2:

$$\Psi_{\mathrm{I}} = \frac{1}{\sqrt{2}}(\Psi_1 + \Psi_2), \tag{8.34}$$

$$\Psi_{\mathrm{II}} = \frac{1}{\sqrt{2}}(\Psi_1 - \Psi_2). \tag{8.35}$$

Accordingly, the total wavefunction, which describes the state of the system, is given by

$$\Psi = c_1'\Psi_{\mathrm{I}} + c_2'\Psi_{\mathrm{II}}, \tag{8.36}$$

where

$$c_1' = \frac{1}{\sqrt{2}}(c_1 - c_2), \qquad c_2' = \frac{1}{\sqrt{2}}(c_1 + c_2). \tag{8.37}$$

We are dealing with the problem of a system with two basis states which is similar to well-known problems, such as, for example, the problem of exchange interaction forces in the molecular hydrogen ion or of the ammonium molecule in connection with the problem of inversion level splitting.

The splitting of energy levels in the presence of tunneling $\Delta E = 2Q$ determines the characteristic frequency of the tunneling effect $\Omega = 2Q/\hbar$. With this frequency the probability of finding the particle on the right and left of the potential barrier is changed; in the classical limit $\Omega \to 0$ and $Q \to 0$.

It is essential that in the representation of the functions Ψ_{I} and Ψ_{II} the matrix of the particle's Hamiltonian is diagonal:

$$\hat{H}_{\boldsymbol{R}}^0 = \begin{vmatrix} E_0 + Q & 0 \\ 0 & E_0 - Q \end{vmatrix} . \tag{8.38}$$

By shifting the origin of the energy by E_0, we obtain

$$\hat{H}_{\boldsymbol{R}}^0 = \begin{vmatrix} Q & 0 \\ 0 & -Q \end{vmatrix} = Q\sigma_{\boldsymbol{R}}^z, \tag{8.39}$$

where the operator

$$\sigma_{\boldsymbol{R}}^z = \begin{vmatrix} 1 & 0 \\ 0 & -1 \end{vmatrix} . \tag{8.40}$$

Let us consider the form of the interaction operator for particles $\hat{V} = \hat{V}(\boldsymbol{r} - \boldsymbol{r}')$ in the representation of the wavefunctions Ψ_{I} and Ψ_{II}; it is clear that the potential energy of the system is

$$U = \frac{1}{2}\sum V(\boldsymbol{r} - \boldsymbol{r}'), \tag{8.41}$$

where $\boldsymbol{r} = \boldsymbol{R} + \boldsymbol{u_R}$.

We expand the function V into a series in degrees of displacements $u_{\boldsymbol{R}}$, taking into account that the mean distance between the particles is much larger than the possible maximum displacement of an individual particle:

$$V(\boldsymbol{r} - \boldsymbol{r}') = V(\boldsymbol{\varrho})(u_{\boldsymbol{R}} - u_{\boldsymbol{R}'})V'(\boldsymbol{\varrho}) + \frac{1}{2}(u_{\boldsymbol{R}} - u_{\boldsymbol{R}'})^2 V''(\boldsymbol{\varrho}) + \dots . \tag{8.42}$$

Here $\boldsymbol{\varrho} = \boldsymbol{R} - \boldsymbol{R}'$.

Since each particle has two basis states, the matrix of the particle displacement operator has the form

$$u^z_{\boldsymbol{R}} = \begin{vmatrix} u_{11} u_{12} \\ u_{21} u_{22} \end{vmatrix}, \tag{8.43}$$

where

$$u_{11} = \int \Psi_{\mathrm{I}} u_{\boldsymbol{R}} \Psi^*_{\mathrm{I}} \mathrm{d}v, \qquad u_{12} = \int \Psi_{\mathrm{I}} u_{\boldsymbol{R}} \Psi^*_{\mathrm{II}} \mathrm{d}v,$$

$$u_{21} = \int \Psi_{\mathrm{II}} u_{\boldsymbol{R}} \Psi^*_{\mathrm{I}} \mathrm{d}v, \qquad u_{22} = \int \Psi_{\mathrm{II}} u_{\boldsymbol{R}} \Psi^*_{\mathrm{II}} \mathrm{d}v.$$

It is evident that $u_{\boldsymbol{R}}$ is an odd function of the coordinates; besides, of the functions Ψ_{I} and Ψ_{II} one is even and the other is odd. The point is that in the case under consideration the potential function for each particle is an even function of the coordinates. Therefore, with a change of sign of the coordinate the Schrödinger equation does not change. This means that if $\Psi(x)$ is the solution of this equation, then $\Psi(-x)$ is also the solution, and, with accuracy of up to a constant factor,

$$\Psi(-x) = C\Psi(x).$$

By changing the sign once more, we get

$$\Psi(x) = C^2\Psi(-x).$$

that is, $C = \pm 1$: the wavefunctions of stationary states may be either even ($\Psi(-x) = \Psi(x)$) or odd ($\Psi(-x) = -\Psi(x)$). In fact, the function Ψ_{I} corresponding to the stationary state $E_0 - Q$ is even and the function Ψ_{II} corresponding to the stationary state with energy $E_0 + Q$ is odd. Therefore, $u_{11} = u_{22} = 0$ and

$$u_{\boldsymbol{R}} = \begin{vmatrix} 0 & u_{12} \\ u_{21} & 0 \end{vmatrix} = u_{12}\sigma^x_{\boldsymbol{R}} \tag{8.44}$$

($u_{12} = u_{21}$ by the symmetry of the problem), where the operator

$$\sigma^x_{\boldsymbol{R}} = \begin{vmatrix} 01 \\ 10 \end{vmatrix}, \tag{8.45}$$

and, accordingly, the second term in (8.42) has the following form:

$$V'(\varrho)u_{12}(\sigma^x_{\boldsymbol{R}} - \sigma^x_{\boldsymbol{R}'}). \tag{8.46}$$

It can be shown, in an analogous manner, that the operator corresponding to the third term in (8.42) has the form

$$\frac{1}{2}V''(\varrho)(u^2_{11} + u^2_{22}) + \frac{1}{2}V''(\varrho)\frac{u^2_{11} - u^2_{22}}{2}(\sigma^z_{\boldsymbol{R}} + \sigma^z_{\boldsymbol{R}'})$$
$$-V''(\varrho)u^2_{12}\sigma^x_{\boldsymbol{R}}\sigma^x_{\boldsymbol{R}'}, \tag{8.47}$$

where $\sigma^z_{\boldsymbol{R}}$ is equivalent to the operator introduced above, (8.40). Here it must be taken into account that the matrix elements are $u^2_{12} = u^2_{21} = 0$ ($u^2_{\boldsymbol{R}}$ is an even function of coordinates) and

$$u^2_{\boldsymbol{R}} = \frac{1 + \sigma^z_{\boldsymbol{R}}}{2}u^2_{11} + \frac{1 - \sigma^z_{\boldsymbol{R}}}{2}u^2_{22}, \tag{8.48}$$

$$u^2_{\boldsymbol{R}'} = \frac{1 + \sigma^z_{\boldsymbol{R}'}}{2}u^2_{11} + \frac{1 - \sigma^z_{\boldsymbol{R}'}}{2}u^2_{22}. \tag{8.49}$$

Indeed, the matrix, say, of the quantity $u^2_{\boldsymbol{R}}$ is

$$\left(\frac{1}{2}\begin{vmatrix}1 & 0\\ 0 & 1\end{vmatrix} + \frac{1}{2}\begin{vmatrix}1 & 0\\ 0 & -1\end{vmatrix}\right)u^2_{11} + \left(\frac{1}{2}\begin{vmatrix}1 & 0\\ 0 & 1\end{vmatrix} - \frac{1}{2}\begin{vmatrix}1 & 0\\ 0 & -1\end{vmatrix}\right)u^2_{22}$$
$$= \begin{vmatrix}u^2_{11} & 0\\ 0 & u^2_{22}\end{vmatrix}.$$

In (8.47) the last term corresponds to the operator $-2u_{\boldsymbol{R}}u_{\boldsymbol{R}'}V''(\varrho)$ since $(u_{\boldsymbol{R}}u_{\boldsymbol{R}'})_{11} = (u_{\boldsymbol{R}}u_{\boldsymbol{R}'})_{22} = 0$ because the wavefunctions of various particles do not overlap. At the same time

$$-2(u_{\boldsymbol{R}}, u_{\boldsymbol{R}'})_{12} = -\int \Psi_{\mathrm{I}}(\boldsymbol{R})\Psi_{\mathrm{I}}(\boldsymbol{R}')u_{\boldsymbol{R}}u_{\boldsymbol{R}'}\Psi^*_{\mathrm{II}}(\boldsymbol{R})\Psi^*_{\mathrm{II}}(\boldsymbol{R}')\mathrm{d}v\mathrm{d}v'$$
$$= \int \Psi_{\mathrm{I}}(\boldsymbol{R})u_{\boldsymbol{R}}\Psi^*_{\mathrm{II}}(\boldsymbol{R})\mathrm{d}v \int \Psi_{\mathrm{I}}(\boldsymbol{R}')u_{\boldsymbol{R}'}\Psi^*_{\mathrm{II}}(\boldsymbol{R}')\mathrm{d}v'$$
$$= u_{12}(\boldsymbol{R})u_{12}(\boldsymbol{R}') = u^2_{12}\sigma^x_{\boldsymbol{R}}\sigma^x_{\boldsymbol{R}'}. \tag{8.50}$$

Finally, we obtain the interaction operator in the form

$$\hat{V}(\boldsymbol{r} - \boldsymbol{r}') = V(\varrho) + u_{12}V'(\varrho)(\sigma^x_{\boldsymbol{R}} - \sigma^x_{\boldsymbol{R}'}) + \frac{1}{2}V''(\varrho)(u^2_{11} + u^2_{22})$$
$$+\frac{1}{2}V''(\varrho)\frac{u^2_{11} - u^2_{22}}{2}(\sigma^z_{\boldsymbol{R}} + \sigma^z_{\boldsymbol{R}'}) - V''(\varrho)u^2_{12}\sigma^x_{\boldsymbol{R}}\sigma^x_{\boldsymbol{R}'}. \tag{8.51}$$

The Hamiltonian of the system can be obtained by summation of one-particle Hamiltonians (8.39) and pair interactions. It will then be equal to

$$\hat{H} = \sum_{R} \hat{H}_{R}^{0} + \frac{1}{2} \sum_{R,R'} \hat{V}(r - r')$$

$$= \hat{V}_0 + Q' \sum_{R} \sigma_{R}^{z} - \frac{1}{2} \sum_{R,R'} J(R - R') \sigma_{R}^{x} \sigma_{R'}^{x}. \tag{8.52}$$

Here

$$\hat{H}_{R}^{0} = Q \sigma_{R}^{z}, \qquad \hat{V}_0 = \sum_{R,R'} V(R - R'),$$

$$Q' = Q + \frac{1}{4} \sum_{R'} V''(\varrho)(u_{11}^2 - u_{22}^2),$$

$$J(R - R') = 2V''(\varrho) u_{12}^2. \tag{8.53}$$

The constant term $\hat{V}_0$ is omitted in (8.52) because it is unimportant; moreover, we perform a transformation to the old basis of Ψ_1, Ψ_2 corresponding to the position of the particle on the right or left in the potential relief (see Fig. 7.2). Note that the transformation matrix W from the functions Ψ_{I}, Ψ_{II} to Ψ_1, Ψ_2

$$\Psi_{1,2} = W \Psi_{\mathrm{I,II}}$$

has the form

$$W = \frac{1}{\sqrt{2}} \begin{vmatrix} 1 & 1 \\ 1 & -1 \end{vmatrix},$$

and in the representation of the functions Ψ_{I} and Ψ_{II} any operator A assumes the following form:

$$B = W^{*-1} A W.$$

Here $\sigma^z \to -\sigma^x$ and $\sigma^x \to \sigma^z$. The Hamiltonian is therefore transformed to the following form:

$$\hat{H} = -Q \sum_{R} \sigma_{R}^{x} - \frac{1}{2} \sum_{R,R'} J(R - R') \sigma_{R}^{z} \sigma_{R'}^{z}. \tag{8.54}$$

From comparison of (8.54) with (8.6) it follows that consideration of the tunneling effect is equivalent to the appearance of an additional term in the Hamiltonian.

In order to represent this term, use is made of a new "component" of the spin variable σ^x; it should only be kept in mind that the indices x and z here have no relation to the real coordinates and are used to designate the matrices by means of which it is convenient to express the Hamiltonian of the system in the presence of tunneling.

Let us show how the consideration of tunneling affects the thermodynamics of the system by using once more the self-consistent molecular field approximation. In this approximation, the Hamiltonian for a single particle takes the form

$$\hat{H}_{\boldsymbol{R}} = -Q\sigma^x_{\boldsymbol{R}} - J_0\langle\sigma^x\rangle\sigma^z_{\boldsymbol{R}} \tag{8.55}$$

or in matrix notation

$$\hat{H}_{\boldsymbol{R}} = -Q\begin{vmatrix}0 & 1\\ 1 & 0\end{vmatrix} - J_0\langle\sigma^z\rangle\begin{vmatrix}1 & 0\\ 0 & -1\end{vmatrix} = \begin{vmatrix}-J_0\langle\sigma^z\rangle & -Q\\ -Q & J_0\langle\sigma^z\rangle\end{vmatrix}. \tag{8.56}$$

The matrix $\hat{H}_{\boldsymbol{R}}$ can easily be diagonalized; its eigenvalues are determined by the equation

$$\lambda^2 - J_0^2\langle\sigma^z\rangle^2 - Q^2 = 0, \tag{8.57}$$

$$\lambda_{1,2} = \pm\sqrt{Q^2 + J_0^2\langle\sigma^z\rangle^2}, \tag{8.58}$$

and the free energy per pseudospin of the system with tunneling in the mean-field approximation is given by

$$\begin{aligned}F &= \frac{1}{2}J_0\langle\sigma^z\rangle^2 - k_{\mathrm{B}}T\ln[\exp(-\lambda_1/k_{\mathrm{B}}T) + \exp(-\lambda_2/k_{\mathrm{B}}T)]\\ &= \frac{1}{2}J_0\langle\sigma^z\rangle^2 - k_{\mathrm{B}}T\ln\left(2\cosh\frac{\sqrt{Q^2 + J_0^2\langle\sigma^z\rangle^2}}{k_{\mathrm{B}}T}\right).\end{aligned} \tag{8.59}$$

Minimizing the free energy with respect to $\langle\sigma^z\rangle$, we find an equation which determines the value of the average spin:

$$\frac{\sqrt{Q^2 + J_0^2\langle\sigma^z\rangle^2}}{J_0} = \tanh\frac{\sqrt{Q^2 + J_0^2\langle\sigma^z\rangle^2}}{k_{\mathrm{B}}T}.$$

For the transition point, where $\langle\sigma^z\rangle = 0$, we have

$$Q/J_0 = \tanh Q/k_{\mathrm{B}}T. \tag{8.60}$$

With $Q \to 0$ we obtain, as before, $T_{\mathrm{c}} = J_0/k_{\mathrm{B}}$ since $\tanh Q/k_{\mathrm{B}}T \approx Q/k_{\mathrm{B}}T$. In the general case, however, with tunneling being taken into account,

$$T_{\mathrm{c}} = \frac{J_0}{k_{\mathrm{B}}}\frac{Q/J_0}{\operatorname{arctanh}(Q/J_0)}. \tag{8.61}$$

Since

$$\operatorname{arctanh}(Q/J_0) = \frac{1}{2}\ln\frac{1+Q/J_0}{1-Q/J_0}, \tag{8.62}$$

we have

$$T_{\mathrm{c}} = \frac{2J_0}{k_{\mathrm{B}}}\frac{Q/J_0}{\ln\frac{1+Q/J_0}{1-Q/J_0}}. \tag{8.63}$$

From this equation it follows that with an increase of tunneling (with an increase of the value of Q – the splitting of the particle's energy levels) T_{c}

falls off and with $Q \to J_0$ the transition temperature $T_c \to 0$. In other words, qualitatively the tunneling effect in the system is equivalent to an increase of the degree of disorder and is analogous in this sense to an increase of the temperature. It can be shown that with $Q \neq 0$ there also occurs a decrease of spontaneous polarization at each temperature as compared with the classical case and the spontaneous saturation polarization also decreases: assuming that $T \to 0$, we find the possible maximum value of polarization:

$$P^*_{\mathrm{sat}} = \frac{p_z}{v_{\mathrm{uc}}} \sqrt{1 - \frac{Q^2}{J_0^2}}. \tag{8.64}$$

In the classical case, the system is evidently completely ordered at $T = 0$ and $P_{\mathrm{sat}} = p_z/v_{\mathrm{uc}}$, so that $P_{\mathrm{sat}} > P^*_{\mathrm{sat}}$; spontaneous polarization decreases due to tunneling. The total entropy change associated with the transition also decreases. The occurrence of tunneling accounts in a natural way for the experimentally observed effect of the increase of T_c and P_{sat} for hydrogen-containing ferroelectrics when the hydrogen is replaced by deuterium, whose mass is larger and tunneling is naturally lower.

8.4 The Cluster Approximation: Crystals of the KH_2PO_4 Group

It has been pointed out above that the mean-molecular-field approximation adequately reflects the system's properties when the mean radius of interaction of particles being ordered is sufficiently large. It should, however, be kept in mind that we can not always describe the mean field acting on each particle with the aid of the constant J_0 alone. The real structure of the crystalline lattice, for example, may be such that the major contribution to the potential of the pseudospin is made by the nearest neighbors, the change of whose position can radically alter this potential. For example, a situation is possible when it is energetically favorable for the neighboring particles to be displaced at the transition point in opposite directions. It is clear that in such cases the assumption that the mutual effect of the neighboring particles is reduced to the action of the averaged molecular field alone ceases to be applicable.

There is a more realistic version of the mean-molecular-field approximation, which more completely than that considered above takes account of the specific form of the relative disposition of particles being ordered. In this version, which is known as the *cluster approximation*, the interaction within a certain assembly (cluster) of particles is described accurately, in accordance with the original Hamiltonian. The interaction with the remaining environment – pseudospins adjacent to the cluster – is described, just as in the MMF approximation, with the aid of the mean field, which is a parameter determined from the self-consistency condition. It is said that in the cluster approximation the correlations in the position of the nearest (neighboring) particles are taken into account accurately and the interaction with other particles approximately, just as in the MMF approximation.[2]

The cluster approximation in the theory of ferroelectricity has been developed basically in connection with the microscopic theory of phase transitions that occur in crystals of the KH_2PO_4 group. As has been noted in Chapter 1, in such crystals the ordering of protons on hydrogen bonds is the principal mechanism leading to spontaneous polarization of crystals. The structure of KDP crystals is such (see Fig. 1.24) that the following presumption seems to be natural: the probability of a certain position of a given proton on a hydrogen bond depends to a considerable extent on the positions occupied by protons at the bonds adjacent to this PO_4 group. This circumstance will be taken into account in our further treatment. In the MMF approximation this possibility is not considered at all since it is presumed that the mean field acting on a given particle depends only on the average pseudospin s and is thus determined by the large-scale properties of the system alone. We will consider in more detail a structural unit of the KDP crystal – the PO_4 tetrahedron and the four hydrogen bonds adjacent to it (see Fig. 1.24a). Let us assume that the values of the spin variable are negative when a hydrogen atom is near the upper oxygen atoms of the PO_4 tetrahedron and positive when it is near the lower ones, so that, for example, $\sigma_1 = \sigma_2 = \sigma_3 = \sigma_4 = -1$ in Fig. 1.24a. It can be seen that each proton is surrounded by six nearest neighbors – protons of two PO_4 groups joined by the given hydrogen atom.

We choose, as a cluster, four particles – hydrogen atoms lying near one PO_4 group. From symmetry considerations it follows that the interaction of particles lying at different levels of the PO_4 tetrahedron (1 and 2, 2 and 3, 3 and 4, 4 and 1 in Fig. 1.24a) differs from the interaction at a single level (1 and 3, 2 and 4), so that

$$J_{12} = J_{23} = J_{34} = J_{41} = V, \tag{8.65}$$

$$J_{13} = J_{24} = U, \qquad U \neq V. \tag{8.66}$$

It is obvious that there are $2^4 = 16$ different positions of protons on the hydrogen bonds of one cluster. Let us determine the values of the cluster energy – its energy levels corresponding to different configurations. With account taken of only pair interactions within the cluster,

$$\begin{aligned} H_4^1 &= -\frac{1}{2} \sum_{i,j=1\, i\neq j}^{4} J_{ij}\sigma_i\sigma_j \\ &= V(\sigma_1\sigma_2 + \sigma_2\sigma_3 + \sigma_3\sigma_4 + \sigma_4\sigma_1) - U(\sigma_1\sigma_3 + \sigma_2\sigma_4). \end{aligned} \tag{8.67}$$

Table 8.1 presents all possible proton configurations corresponding to the values of the spin variables σ_i and the energies of these configurations in

[2] It is for this reason that the long-wavelength fluctuations of the order parameter are also neglected in the cluster approximation.

accordance with (8.67). It can be seen that the electrically neutral states of the structural element – the KH_2PO_4 "molecule" – in which $\sigma_i = 1$ or $\sigma_i = -1$ (configurations 15 and 16) correspond to the minimum value of the cluster energy $H_4^1 = -4V - 2U$ and are thus doubly degenerate.

Table 8.1. The energies of different proton configurations

№	Configuration	σ_1	σ_2	σ_3	σ_4	Energy	Dipole moment	Energy level
1		1	−1	1	−1	$4V-2U$	0	$w_1 = 8V$
2		−1	1	−1	1	$4V-2U$	0	
3		1	−1	1	1	0	$2\mu_z$	$w = 4V + 2U$
4		−1	−1	1	−1	0	$-2\mu_z$	
5		1	1	1	−1	0	$2\mu_z$	
6		1	−1	−1	−1	0	$-2\mu_z$	
7		−1	1	−1	−1	0	$-2\mu_z$	
8		1	1	−1	1	0	$2\mu_z$	
9		−1	−1	−1	1	0	$-2\mu_z$	
10		−1	1	1	1	0	$2\mu_z$	
11		1	−1	−1	1	$2U$	0	$\varepsilon_0 = 4V + 4U$
12		−1	−1	1	1	$2U$	0	
13		−1	1	1	−1	$2U$	0	
14		1	1	−1	−1	$2U$	0	
15		1	1	1	1	$-4V-2U$	$4\mu_z$	0
16		−1	−1	−1	−1	$-4V-2U$	$-4\mu_z$	

Single-charged configurations ($KHPO_4$, KH_3PO_4–3-10) have $H_4^1 = 0$ and are eight-fold degenerate and doubly charged configurations (KPO_4, KH_4PO_4) have $H_4^1 = 4V - 2U$ and are two-fold degenerate. Counting the cluster energy from the lowest value $-4V - 2U$, we obtain for the energy level differences corresponding to the respective configurations: Nos. 15–16 and 11–14: $4V + 4U = \varepsilon_0$; Nos. 15–16 and 3–10: $4V + 2U = w$; Nos. 15–16 and 1–2: $8V = w_1$. It is reasonable to presume that the energy of the configuration increases significantly with increasing charge, i.e.,

$$\varepsilon \ll w \ll w_1. \tag{8.68}$$

Let us now write down the total Hamiltonian of the cluster, bearing in mind that each particle also interacts with the external neighbors that do not belong to the given cluster. This interaction is taken into account approximately by introducing the mean field φ. It is natural that the contribution from the cluster particles to the magnitude of the mean field should be ignored since this interaction has been accurately taken into account. Hence, if each hydrogen atom has n nearest neighbors ($n = 6$ in the case under consideration), the value of φ will be lower by a factor of $(n-k)/n$, where $k = 3$ is the number of neighbors included in the cluster. We have

$$H_4 = H_4^1 - \frac{1}{2}\varphi(\sigma_1 + \sigma_2 + \sigma_3 + \sigma_4) - p_z E_z(\sigma_1 + \sigma_2 + \sigma_3 + \sigma_4). \tag{8.69}$$

Here, to the Hamiltonian of an isolated cluster H_4^1 there has been added the interaction of each particle with the mean field φ ($(n-k)/n = 1/2$) and with the external electric field E directed along the c axis perpendicular to the plane in which the hydrogen bonds lie. In accordance with (8.69), depending on the disposition of particles in the cluster, the dipole moment along the same axis may have the values 0, $\pm 2p_z$, $\pm 4p_z$ (see Table 8.1).

Thus, in (8.69) it is implicitly assumed that the dipole moment of the structural unit of KPO_4 associated with the displacements of the atoms K, P and O along the c axis responds continuously to the proton configuration, instantaneously assuming the corresponding value; the mean dipole moment, which determines the spontaneous polarization of the crystal, is proportional to the mean pseudospin:

$$P_s = \frac{p_z}{v_{\mathrm{H}}} s, \tag{8.70}$$

where v_H is the part of the unit-cell volume per hydrogen bond. In the version of the MMF approximation under discussion the quantity φ is a parameter determined from the self-consistency condition; in this case this condition is equivalent to the requirement that the mean value of the pseudospin of the cluster with the Hamiltonian (8.69) coincides with the mean value of any pseudospin of the system within the field φ. The Hamiltonian of a single particle in the field and the macrofield E_z is given by

$$H_1 = -\varphi\sigma_1 - p_z E_z \sigma_1, \tag{8.71}$$

and, hence, the self-consistency condition becomes

$$\sum_{\sigma_1,\sigma_2,\sigma_3,\sigma_4} \sigma_1 \exp\left(-\frac{H_4}{k_{\mathrm{B}}T}\right)\left[\sum_{\sigma_1,\sigma_2,\sigma_3,\sigma_4} \exp\left(-\frac{H_4}{k_{\mathrm{B}}T}\right)\right]^{-1}$$
$$= \sum_{\sigma_1} \sigma_1 \exp\left(-\frac{H_1}{k_{\mathrm{B}}T}\right)\left[\sum_{\sigma_1} \exp\left(-\frac{H_1}{k_{\mathrm{B}}T}\right)\right]^{-1}. \tag{8.72}$$

We assume that the external field is $E_z = 0$. Using (8.72), we determine the quantity φ and then the equilibrium value of s and, thereby, the spontaneous polarization.

Table 8.2 presents all the terms of the numerator and denominator of the first fraction in (8.72), where the last fraction is equal to

Table 8.2. The terms of the numerator and denominator of the first fraction in (8.72).

Configuration number	$\sigma_1 \exp(-H_4/k_B T)$	$\exp(-H_4/k_B T)$
1	$\exp(-(4V-2U)/k_B T)$	$\exp(-(4V-2U)/k_B T)$
2	$-\exp(-(4V-2U)/k_B T)$	$\exp(-(4V-2U)/k_B T)$
3	$\exp(-\varphi/k_B T)$	$\exp(-\varphi/k_B T)$
4	$-\exp(\varphi/k_B T)$	$\exp(\varphi/k_B T)$
5	$\exp(-\varphi/k_B T)$	$\exp(-\varphi/k_B T)$
6	$\exp(\varphi/k_B T)$	$\exp(\varphi/k_B T)$
7	$-\exp(\varphi/k_B T)$	$\exp(\varphi/k_B T)$
8	$\exp(-\varphi/k_B T)$	$\exp(-\varphi/k_B T)$
9	$-\exp(\varphi/k_B T)$	$\exp(\varphi/k_B T)$
10	$-\exp(-\varphi/k_B T)$	$\exp(-\varphi/k_B T)$
11	$\exp(2U/k_B T)$	$\exp(-2U/k_B T)$
12	$-\exp(-2U/k_B T)$	$\exp(-2U/k_B T)$
13	$-\exp(-2U/k_B T)$	$\exp(-2U/k_B T)$
14	$\exp(-2U/k_B T)$	$\exp(-2U/k_B T)$
15	$\exp((-4V-2U)/k_B T)$ $\times \exp(-2\varphi/k_B T)$	$\exp(-(-4V-2U)/k_B T)$ $\times \exp(-2\varphi/k_B T)$
16	$-\exp(-(-4V-2U)/k_B T)$ $\times \exp(2\varphi/k_B T)$	$\exp(-(-4V-2U)/k_B T)$ $\times \exp(2\varphi/k_B T)$

$$\frac{\sinh[\varphi/(k_B T)]}{\cosh[\varphi/(k_B T)]} = \tanh\frac{\varphi}{k_B T} = s. \tag{8.73}$$

After simple transformations, taking into account that $4V - 2U = 3w - 2\varepsilon_0$, $2U = \varepsilon_0 - w$, $4V + 2U = w$, $w_1 = 4w - 2\varepsilon_0$, we obtain instead of (8.73):

$$\begin{aligned} s &= \{2\exp(-w/k_B T)\sinh(\varphi/k_B T) + \sinh(2\varphi/k_B T)\}\{\exp(-w_1/k_B T) \\ &\quad + 4\exp(-w/k_B T)\cosh(\varphi/k_B T) + 2\exp(-\varepsilon_0/k_B T) \\ &\quad + \cosh(2\varphi/k_B T)\}^{-1} \\ &= \tanh\frac{\varphi}{k_B T}. \end{aligned} \tag{8.74}$$

For further analysis, to be more spectacular, we introduce the notation

$$\frac{\varphi}{k_B T} = \alpha, \qquad \exp(-\frac{w}{k_B T}) = L,$$

$$2\exp(-\frac{\varepsilon_0}{k_B T}) + \exp(-\frac{w_1}{k_B T}) = K,$$

following which (8.74) assumes the form

$$s = \frac{\sinh(2\alpha) + 2L\sinh\alpha}{\cosh(2\alpha) + 4L\cosh\alpha + K} = \frac{\sinh\alpha}{\cosh\alpha}, \tag{8.75}$$

which is convenient for the determination of φ.

Since $\sinh(2\alpha) = 2\sinh\alpha\cosh\alpha$ and $\cosh(2\alpha) = 2\cosh^2\alpha - 1$, we have from (8.75)

$$\cosh\frac{\varphi}{k_B T} = \frac{1-K}{2L} = \frac{1-2\exp[-\varepsilon_0/(k_B T)] - \exp[-w_1/(k_B T)]}{2\exp[-w/(k_B T)]}. \quad (8.76)$$

In this relation the mean field φ has been expressed in terms of the Hamiltonian constants; together with (8.72), (8.76) determines the temperature dependence of the order parameter – the values of the mean pseudospin s and the spontaneous polarization:

$$P_s = \frac{p_z}{v_H}\frac{\sinh[\varphi/(k_B T)]}{\cosh[\varphi/(k_B T)]} = \frac{p_z}{v_H}\cdot\frac{[(1-K-2L)(1-K+2L)]^{1/2}}{1-K}. \quad (8.77)$$

In (8.77) it has been taken into account that $\cosh^2\alpha - \sinh^2\alpha = 1$.

Spontaneous polarization vanishes at a temperature corresponding to the vanishing of the numerator of (8.77), i.e., to the condition $1-K=2L$:

$$1 - 2\exp\left(-\frac{\varepsilon_0}{k_B T}\right) - 2\exp\left(-\frac{w}{k_B T}\right) - \exp\left(-\frac{w_1}{k_B T}\right) = 0. \quad (8.78)$$

It is easy to show that a phase transition, the temperature of which is determined by (8.78), is a second-order transition. We will assume, for simplicity, that doubly charged configurations do not occur (see the condition (8.68)), i.e., the term $\exp(-w_1/k_B T)$ in all equations may be dropped; then, instead of (8.77), we have

$$P_s^2 = \frac{p_z^2}{v_H^2}$$

$$\times\left\{1 - \left[\frac{\exp\left(-\frac{w}{k_B T}\right)}{\left(\frac{1}{2} - \exp\left(-\frac{\varepsilon_0}{k_B T}\right) - \exp\left(-\frac{w}{k_B T}\right)\right) + \exp\left(-\frac{w}{k_B T}\right)}\right]^2\right\}. \quad (8.79)$$

Let us compare the denominator on the right-hand side of this equation with (8.78). It can be seen that with $T \to T_c$ the expression in braces tends to zero, in which case $P_s^2 \to 0$ too, this being an indication that the spontaneous polarization does not undergo a jump at $T = T_c$ and that the phase transition is really a second-order transition (Fig 8.3).

An essential point is that if the single-charged configurations are also forbidden ($w \to \infty$), the transition point will be determined by the condition

$$1 - 2\exp\left(-\frac{\varepsilon_0}{k_B T}\right) = 0, \qquad T_c = \frac{\varepsilon_0}{k_B \ln 2}. \quad (8.80)$$

Here the spontaneous polarization will be

$$P_s = p_z/v_H \qquad \text{with} \qquad T < T_c, \quad (8.81)$$

$$P_s = 0 \qquad \text{with} \qquad T > T_c, \quad (8.82)$$

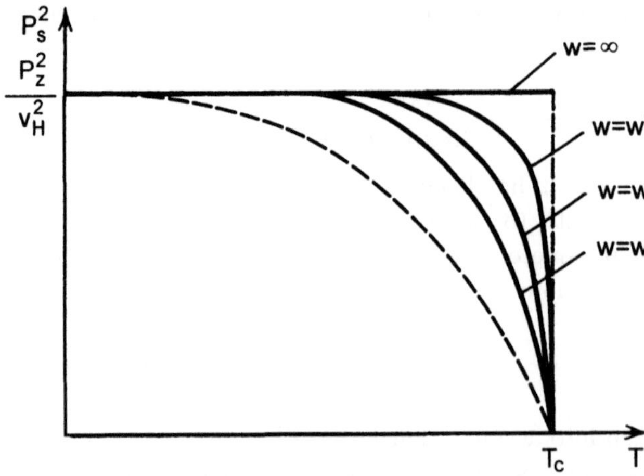

Fig. 8.3. Temperature dependence of the square of spontaneous polarization of a crystal of the KDP type in the cluster aporoximation for different values of w (the dashed line represents the mean-field approximation).

that is, it will change jumpwise at the transition point and the phase transition itself will become a first-order transition.

It is interesting to note that the first-order phase transition that results with $w, w_1 \to \infty$ is very specific. Using reasoning analogous to that applied in the derivation of (8.25)–(8.28), it can easily be shown that, in spite of the jumpwise change of polarization at $T = T_c$, the dielectric constant of the crystal, while obeying the Curie–Weiss law, attains infinitely large values at the same temperature T_c.

As follows from (8.80), the phase transition temperature determines the value of the quantity ε_0, which is the most important energy parameter of the cluster; the other parameters (w, w_1) can be chosen on the basis of a comparison with the theory of the temperature dependence of the spontaneous polarization of the crystal. The cluster theory has made it possible to describe the properties of crystals of the KDP group and their variation upon isomorphous replacements K→Pb, Cs; P→As; H→D. The working version is, however, more complicated than the version of the theory described above, which takes into account proton tunneling effects and also the interaction of protons with their non-nearest neighbors. It is also possible to account for the well-established experimental fact that the phase transitions occurring in crystals of KDP type are, as a rule, first-order transitions: the magnitude of the polarization jump at $T = T_c$ varies from 0.3 P_{sat} to 0.8 P_{sat}, depending on the composition.[3] It seems quite probable that the diversity of phase-transition types in crystals of the KDP group is associated

[3] A second-order phase transition occurs only in RbH_2PO_4 crystals. CsH_2PO_4 is not isomorphic to KH_2PO_4, though it also undergoes a ferroelectric phase transition.

with the above-discussed specific form of the correlations in the positions of the nearest particles being ordered. A slight change of the parameters of the Hamiltonian brings about a substantial change in the thermodynamic picture of the phase transition, which has been confirmed by experiment. Thus, there is microscopic evidence indicating that phase transitions in crystals of the KDP group must be close to the tricritical point. The tricritical point has been detected in the KH_2PO_4 crystal (see Fig. 3.14).

The latest experimental investigations have also shown that the temperature dependence of spontaneous polarization in crystals of the KDP group over a rather wide temperature range is described within the framework of the Landau phenomenological theory for first-order transitions similar to second-order ones with all their inherent specific features (see Chap. 3).

As can be seen, the microtheory expounded here enables one to obtain results that go beyond the scope of Landau theory, making it possible to describe specific phase transitions, in which the region of an anomalous change of the parameters is exponentially small. The problems that can be solved by means of a microtheory of this type include the elucidation of the regularities of variation of the constants of the thermodynamic potential with a change of the crystal's composition.

9. Dynamics of Displacive and Order–Disorder Phase Transitions

9.1 The Equation of Motion of the Order Parameter

In the preceding chapters we have dealt with the thermodynamic characteristics of crystals, i.e., we have considered the various equations of state relating the equilibrium values of the generalized thermodynamic forces and coordinates. An example of such an equation is (3.3) or the relation derived from it:

$$A\eta = aE \tag{9.1}$$

($A = \frac{1}{2}\alpha(T - T_c)$), which characterizes the variation of the order parameter for a proper ferroelectric under the action of a low electric field with $T > T_c$. The quantity aE may here be regarded as a "force" coupled to the order parameter and the coefficient A has the meaning of "rigidity" (corresponding to η) or of inverse susceptibility. It can be seen that with $T \to T_c$ the quantity A also tends to zero, i.e., upon approach to T_c the rigidity of the system falls off.

If the phase transition is not a proper ferroelectric (or a proper ferroelastic) transition, the experimentalist has no field at his disposal, which could be used to attain a nonzero value of the order parameter in the symmetric phase; one may however treat the coefficient A as the "rigidity" with respect to a certain (possible only in principle) field coupled to the order parameter.

In this chapter we will discuss the properties of dynamic susceptibility corresponding to the order parameter, i.e., the response of the system to an external influence variable in time. The problem thus consists of the extension of (9.1) to the case of a variable field E or, in other words, of the derivation of the equation of motion for the order parameter.

It is clear that the value of the order parameter averaged over the volume of the crystal, which appears in (9.1) (the field E is assumed to be spatially homogeneous), is a certain linear combination of the coordinates of the system's particles (see Chap. 2), i.e., it is one of the mechanical degrees of freedom of the crystal. In any real system the motion corresponding to one degree of freedom is associated with motions corresponding to all other degrees of freedom; one cannot therefore expect that in the general case the equation of motion for the order parameter will be simple – something like the equation of motion for a single particle. On the contrary, it will be natural to presume that this equation will be rather complicated. If we limit ourselves to small deviations of the order parameter from its equilibrium value, we may write the equation of motion as a linear equation containing time derivatives of any high order:

$$\ldots a_3 \dddot{\eta} + a_2 \ddot{\eta} + a_1 \dot{\eta} + A\eta = aE, \tag{9.2}$$

where the ellipsis signifies the infinite sequence of terms with higher-order derivatives.

Reasoning in the spirit of Landau theory, we may assert that there is no ground for presuming the existence of a specific temperature dependence of the coefficients a_i, in (9.2) near T_c: for a phase transition to occur, it will suffice that the sign of the coefficient A is changed.

We set $E = E_\omega \exp(i\omega t)$, in which case $\eta = \eta_\omega \exp(i\omega t)$ and, instead of (9.2), we get

$$(\ldots - ia_3\omega^3 - a_2\omega^2 + ia_1\omega + A)\eta_\omega = aE_\omega. \tag{9.3}$$

At low enough frequencies only the last two terms in parentheses may be retained, i.e., the expression for the dynamic susceptibility takes on the form

$$\chi^{-1}(\omega) = A + ia_1\omega. \tag{9.4}$$

The second term on the right-hand side of (9.4) describes the frequency dispersion of the susceptibility, which evidently becomes appreciable with $\omega \propto A/a_1$. Since $A \to 0$ with $T \to T_c$, the dispersion near T_c becomes noticeable at very low frequencies.

If we retain in (9.2) the terms corresponding to (9.4) and put $E = 0$, we will obtain an equation which describes the process of the approach of η to its equilibrium value:

$$a_1\dot{\eta} + A\eta = 0. \tag{9.5}$$

The solution of this equation is

$$\eta(t) = C\exp(-t/\tau), \tag{9.6}$$

where $\tau = a_1/A$ is the *relaxation time* of the order parameter; C is an arbitrary constant. It can be seen that $\tau \to \infty$ with $T \to T_c$. It is natural that (9.5) is valid for low frequencies and times that are sufficiently remote from the start of the process.

In order to elucidate the behavior of dynamic susceptibility in the region of high frequencies, it is necessary, in principle, to take into account the other terms in (9.1) and (9.2) too. It turns out, however, that even the simplest relaxation equation, such as (9.5), yields a sufficiently good description of the dynamics of phase transitions of the order–disorder type, and the equation of motion for an oscillator with friction provides a good description of the dynamics of displacive transitions. These equations will be substantiated at a later time, after the properties of dynamic susceptibilities, which follow from them, are considered.

9.2 Dynamic Dielectric Constant: Order–Disorder Phase Transitions

Let us write the relaxation equation (9.5) in a more general form by slightly changing the notation:

$$L\dot{\eta} = -\frac{\partial \Phi}{\partial \eta} \tag{9.7}$$

This equation shows that the rate of approach of the order parameter to the equilibrium value is faster, the farther the system is from equilibrium ($\partial\Phi/\partial\eta = 0$ at $\eta = \eta_0$; L is the kinetic coefficient). Linearizing (9.7) for the symmetric phase, we arrive at (9.5) and for the nonsymmetric phase we obtain

$$L\delta\dot{\eta} - 2\alpha(T - T_c)\delta\eta = 0, \tag{9.8}$$

where $\delta\eta = (\eta - \eta_0) \ll \eta_0$. Thus, in the nonsymmetric phase

$$\tau = -\frac{L}{2\alpha(T - T_c)}, \tag{9.9}$$

and in the symmetric phase

$$\tau = \frac{L}{\alpha(T - T_c)}. \tag{9.10}$$

The plot of the relaxation time against temperature is given in Fig. 9.1.

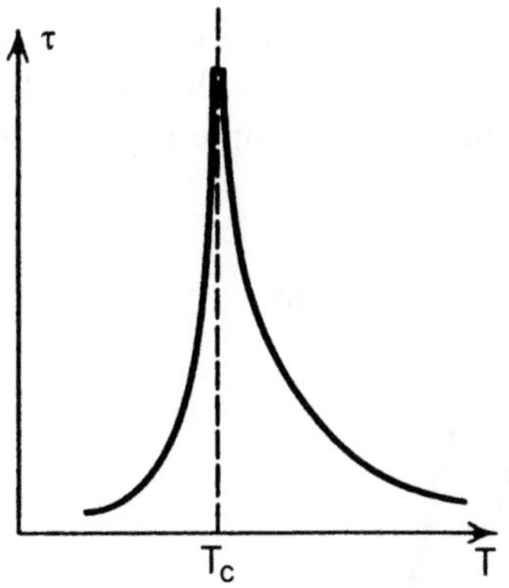

Fig. 9.1. Temperature dependence of the relaxation time of the order parameter in a ferroelectric for the case where the movement to equilibrium is relaxational.

We will now elucidate the frequency dependence of the dielectric constant at $T > T_c$. Considering that $P_\omega = a\eta_\omega$, we have

$$\varepsilon(\omega) = 1 + 4\pi\frac{P_\omega}{E_\omega} = 1 + \frac{4\pi a^2}{A + \mathrm{i}L_\omega},$$

or

$$\varepsilon(\omega) = \varepsilon'(\omega) - \mathrm{i}\varepsilon''(\omega) = \frac{\varepsilon_0}{1+\omega^2\tau^2} - \mathrm{i}\frac{\varepsilon_0\omega\tau}{1+\omega^2\tau^2}. \tag{9.11}$$

Here $\varepsilon_0 = 4\pi a^2/A$ is the static dielectric constant at $T > T_c$ (see (3.8)) and it has been taken into account that $4\pi a^2/A \gg 1$ near T_c. Note that (9.11) also holds for $T < T_c$, for which τ must be determined by formula (9.9).

One can see that the dynamic dielectric constant is a complex quantity, i.e., it is characterized by two quantities – the real (ε') and imaginary (ε'') parts, which depend on frequency. The absolute value $|\varepsilon(\omega)| = \sqrt{\varepsilon'^2 + \varepsilon''^2}$ determines the vibrational amplitude of the vector of electric displacement $D = \varepsilon(\omega)E$ and the ratio $\varepsilon''/\varepsilon' = \tan\delta$ determines the phase difference δ between D and E. The quantity δ is also known as the dielectric loss angle since the presence of a phase difference between vibrations D and E leads to the absorption of the electric field energy by the system. Figure 9.2 shows the frequency dependence of ε' and ε''. It can be seen that at a given temperature near the frequency ω_r determined by the condition $\omega_r\tau = 1$ the dispersion of ε' and also ε'' (and hence, $\tan\delta$) are at a maximum. It is clear that upon approach to the transition temperature τ increases and ω_r (frequency relaxation) decreases. At a fixed frequency ω and as T_c is approached and τ increases the dielectric constant ε' will pass through maxima at $T \neq T$ in ferroelectric and paraelectric phases where $\omega\tau = 1$. At the phase-transition point itself $\tau \to \infty$ and $\omega\tau \gg 1$, i.e., ε' will have a lower value (Fig. 9.3). The distance between the maxima depends on the frequency ω: it is the larger, the higher is this frequency. Experimental plots, such as the one shown in Fig. 9.3, are observed for some ferroelectric crystals within the range of 10^9–10^{10} Hz, i.e., in the region of superhigh frequencies (Fig. 9.4). As pointed out above, a dispersion of this type (called *relaxation* or *Debye dispersion*) is characteristic of ferroelectrics of order–disorder type; its simplified microscopic theory will be considered below.

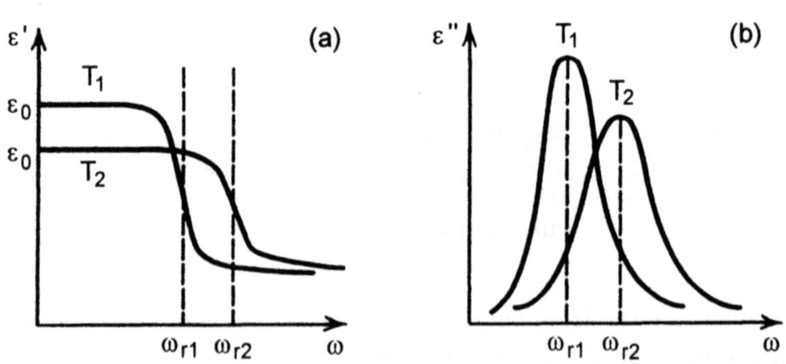

Fig. 9.2. (**a**) Frequency dependence of the real and (**b**) imaginary parts of the dielectric constant of a ferroelectric crystal for two temperatures T_1 and T_2 ($T_2 > T_1 > T_c$) in the paraelectric phase near T_c (ε_0 is the static dielectric constant).

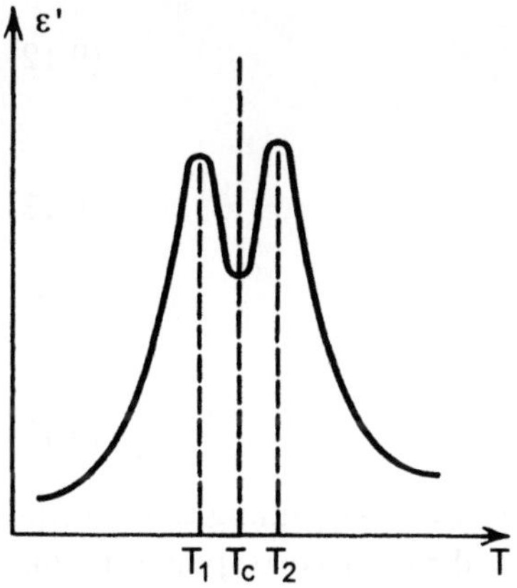

Fig. 9.3. Temperature dependence of the real part of the dielectric constant at a frequency ω_0; T_1 and T_2 are temperatures at which the condition $\omega\tau = 1$ is fulfilled.

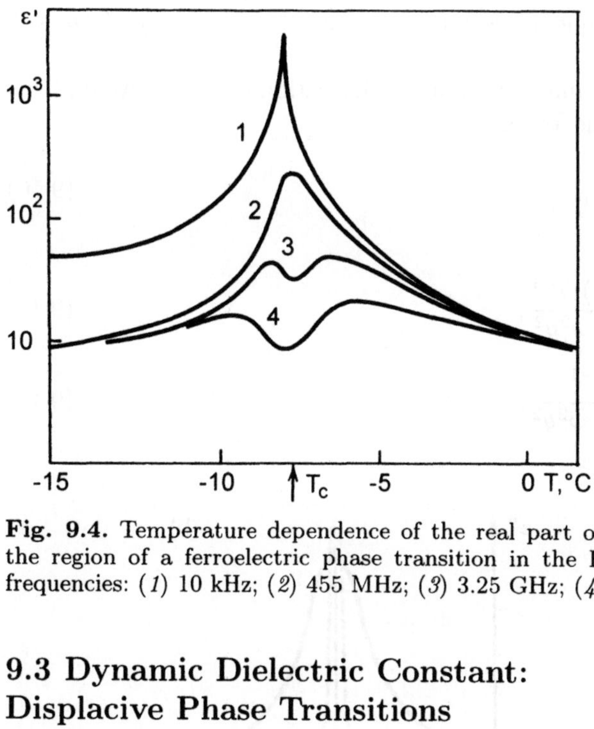

Fig. 9.4. Temperature dependence of the real part of the dielectric constant in the region of a ferroelectric phase transition in the $RbHSO_4$ crystal at various frequencies: (*1*) 10 kHz; (*2*) 455 MHz; (*3*) 3.25 GHz; (*4*) 9.50 GHz [12].

9.3 Dynamic Dielectric Constant: Displacive Phase Transitions

For displacive phase transitions, in (9.2) we retain the term with the second derivative and the coefficient at $\ddot{\eta}$ is treated as the effective mass; the equation of motion assumes a form characteristic of an oscillator with damping. This approximation relies, to a certain extent, on the fact that η is the displacement of a certain sublattice, which, as might be expected, is capable of performing, as a whole, vibrations about the equilibrium position.

We write the equation of motion for the order parameter in the following form:

$$\tilde{m}\ddot{\eta} + L\dot{\eta} = -\frac{\partial \Phi}{\partial \eta}, \tag{9.12}$$

or

$$\ddot{\eta} + g\dot{\eta} + \omega_0^2 \eta = 0, \tag{9.13}$$

where $g = L/\tilde{m}$,

$$\omega_0^2 = A/\tilde{m} = \alpha(T - T_c)/\tilde{m} \quad \text{at} \quad T > T_c, \tag{9.14}$$

$$\omega_0^2 = -2\alpha(T - T_c)/\tilde{m} \quad \text{at} \quad T < T_c. \tag{9.15}$$

We see that the natural frequency of the oscillator corresponding to the order parameter (it will be shown below that this vibration corresponds to one of the so-called normal vibrations or modes of the crystal) vanishes at the transition point. This frequency is referred to as the *frequency of the soft mode.*

Operating in the same manner as in the derivation of (9.11), we find the complex dielectric constant for this case:

$$\varepsilon(\omega) = \frac{4\pi a^2 \tilde{m}^{-1}}{\omega_0^2 - \omega^2 + \mathrm{i}\omega g}, \tag{9.16}$$

or

$$\varepsilon'(\omega) = \frac{4\pi a^2 (\omega_0^2 - \omega^2)\tilde{m}^{-1}}{(\omega_0^2 - \omega^2)^2 + \omega^2 g^2}, \tag{9.17}$$

$$\varepsilon''(\omega) = \frac{4\pi a^2 \omega g \tilde{m}^{-1}}{(\omega_0^2 - \omega^2)^2 + \omega^2 g^2}. \tag{9.18}$$

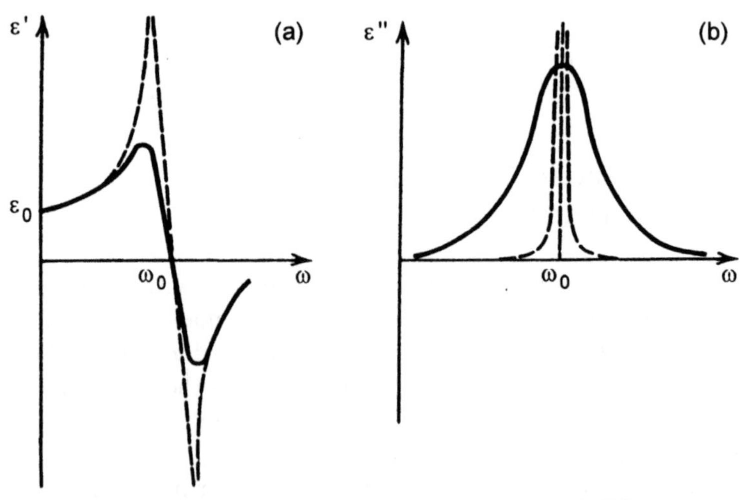

Fig. 9.5. (**a**) Frequency dependence of the real and (**b**) imaginary parts of the dielectric constant in accordance with formulas (9.17) and (9.18).

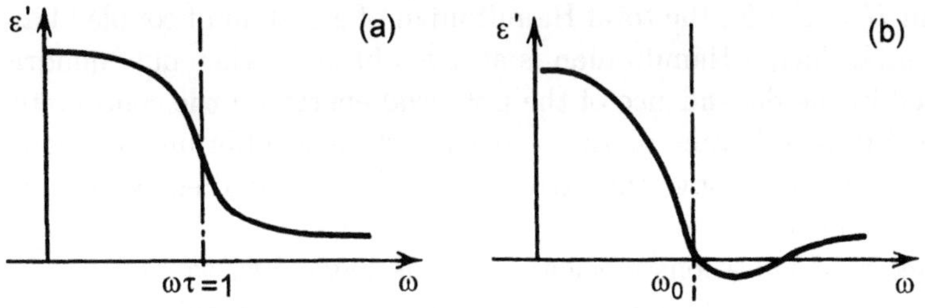

Fig. 9.6. (**a**) Frequency dependence of the real part of the dielectric constant for the cases of relaxational dispersion and (**b**) resonance dispersion with large damping.

This form of dielectric dispersion is called *resonance dispersion*. It can be seen that at the natural frequency $\omega = \omega_0$ the quantity ε' tends to zero and ε'' has a maximal value (Fig. 9.5). When the electric field frequency is much lower or much higher than ω_0, the imaginary term in the denominator of (9.16) may be omitted and then ε' and ε'' will vary in the manner shown in Fig. 9.5 by a dashed line ($g \sim 0$).

As a rule, in displacive ferroelectrics the damping constant g is large ($g \sim \omega_0$) and the difference between the two types of dispersion, Debye (relaxation) dispersion and resonance dispersion, cancels out (Fig. 9.6).

9.4 Microscopic Theory of Dynamic Processes in Displacive Phase Transitions

We will now consider the microscopic theory of the dynamic processes described above. We begin with an elementary introduction to the theory of lattice vibrations. Just as earlier in this book, we will simplify as much as possible the crystal model by retaining only the principal features required for our treatment. We will again use (see Chap. 7) the single-ion model without requiring, for the present, that a phase transition occur in it (see Fig. 7.1). The potential energy is written in the form

$$U = \sum_{\boldsymbol{R}} \frac{1}{2} a u_{\boldsymbol{R}}^2 + \frac{1}{2} c \sum_{\boldsymbol{R},\boldsymbol{R}'} (u_{\boldsymbol{R}} - u_{\boldsymbol{R}'})^2, \tag{9.19}$$

where a is now larger than zero and $u_{\boldsymbol{R}}$ represents, just as in Chap. 7, the displacement of the central mobile ion relative to the center of the unit cell. We thus assume that the potential relief created by immobile ions has one minimum and is characterized by a quadric dependence on the coordinate of a mobile ion (see Fig. 7.5). Such a dependence on the coordinates is exhibited by the potential energy of a harmonic oscillator. If we add to (9.19) the sum of the kinetic energies of mobile ions

$$T = \frac{m}{2} \sum_{\boldsymbol{R}} \dot{u}_{\boldsymbol{R}}^2, \tag{9.20}$$

we will obtain $H = U + T$, the total Hamiltonian of a system of coupled harmonic oscillators. Such a Hamiltonian is always obtained when only quadric terms are used in the dependence of the potential energy on the coordinates of ions. Here the coordinates originate from certain equilibrium positions, i.e., positions corresponding to the extremum of the potential energy (in our case it is a minimum).

In the theory of vibrations efficient methods have been worked out for the solution of problems relating to systems of coupled harmonic oscillators. The language and concepts of this theory are used whenever one is dealing with the thermal motion of atoms in a crystal including those cases when one cannot use only quadric terms in the expression for the potential energy of the system, i.e., with lattice anharmonicity being taken into account. It is for this reason that we will consider in sufficient detail the vibrations of a harmonic crystal by using the model shown in Fig. 7.1.

For an ion in the unit cell, the center of which has coordinate $\boldsymbol{R}$, we find the equation of motion:

$$m\ddot{u}_{\boldsymbol{R}} = -\frac{\partial U}{\partial u_{\boldsymbol{R}}} = -au_{\boldsymbol{R}} - c\sum_{\boldsymbol{R}'}(u_{\boldsymbol{R}} - u_{\boldsymbol{R}'}). \tag{9.21}$$

In the case of one-dimensional motion the complete system consists of N coupled equations (N is the number of mobile ions).

In order to solve the system of equations (9.21), we use the replacement of the variables of (7.3) and write the Hamiltonian of the system in terms of variables $u_{\boldsymbol{k}}$. Operating in the same manner as in Chap. 7, we find

$$H = \frac{1}{2}\sum_{\boldsymbol{k}}\left(m\dot{u}_{\boldsymbol{k}}\dot{u}_{-\boldsymbol{k}} + a(\boldsymbol{k})u_{\boldsymbol{k}}u_{-\boldsymbol{k}}\right), \tag{9.22}$$

where $a(\boldsymbol{k})$ represents the expression enclosed in parentheses in formula (7.13). We have reduced the Hamiltonian of the system to the sum of "partial" Hamiltonians, each of which depends only on the variables $u_{\boldsymbol{k}}$ and $u_{-\boldsymbol{k}}$.

For the variable $u_{\boldsymbol{k}}$ we obtain a simple equation of a harmonic oscillator:

$$m\ddot{u}_{\boldsymbol{k}} + a(\boldsymbol{k})u_{\boldsymbol{k}} = 0, \tag{9.23}$$

the solution of which has the form

$$u_{\boldsymbol{k}} = C_1\exp(\mathrm{i}\omega_{\boldsymbol{k}}t) + c_2\exp(-\mathrm{i}\omega_{\boldsymbol{k}}t), \tag{9.24}$$

where

$$\omega_{\boldsymbol{k}}^2 = a(\boldsymbol{k})/m, \tag{9.25}$$

and C_1 and C_2 are constants which are determined from the initial conditions. Expressions (9.24) and (9.25) together with (7.3) yield the general solution of the system of equations (9.21). It is important to emphasize that the variables $u_{\boldsymbol{k}}$, which are called the normal coordinates, have a rather simple (harmonic) time dependence. They represent linear combinations of the coordinates $u_{\boldsymbol{R}}$. That the transformation from ordinary to normal coordinates is convenient

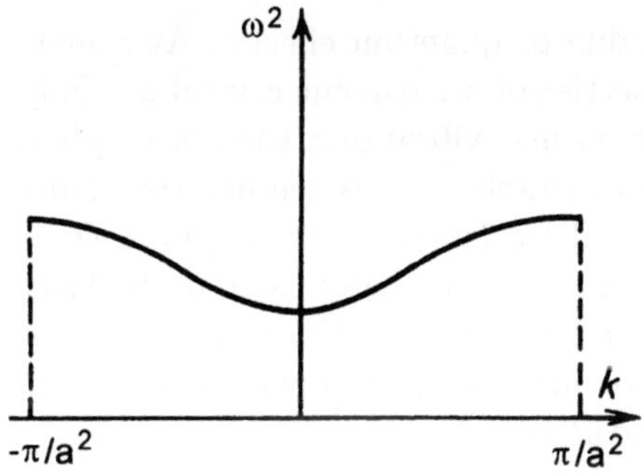

Fig. 9.7. The dependence $\omega^2(\boldsymbol{k})$ for a single-ion model of a one-dimensional crystal.

is obvious: we pass over from a system of coupled oscillators to a system of independent oscillators. Of course, each of the oscillators is no longer a concrete particle; it corresponds to a certain type of cooperative vibration of all particles – the mode of vibrations. A set of frequencies of normal vibrations is specified by the function $\omega^2(\boldsymbol{k})$ (see (9.25)), which represents the vibration dispersion curve given in Fig. 9.7 for the model under consideration. In this particular case, because of the oversimplification of the model (one-dimensional motion, the mobility of only one species of particles), to each value of the wave vector there corresponds a single frequency; in any real crystal there are observed several dispersion curves, i.e., for each $\boldsymbol{k}$ there are several frequencies. The curve shown in Fig. 9.7 corresponds to the so-called *optical normal vibrations.* The specific form of these vibrations is that with $\boldsymbol{k} \to 0$ the quantity $\omega^2(\boldsymbol{k})$ does not vanish. From lattice dynamics it is known that in any crystal there also exist *acoustic normal vibrations*, for which $\omega(\boldsymbol{k}) \to 0$ with $\boldsymbol{k} \to 0$. They correspond (at small $\boldsymbol{k}$) to elastic waves propagating in the crystal. We emphasize once more that the possibility of an analytic solution of the equations of motion for our system was associated with its harmonicity. But if the potential well, in which each ion is moving, is not considered parabolic, i.e., if, for example, the following replacement is made:

$$\frac{a}{2}u_{\boldsymbol{R}}^2 \to \frac{a}{2}u_{\boldsymbol{R}}^2 + \frac{b}{4}u_{\boldsymbol{R}}^4,$$

with the substitution of (7.3) the Hamiltonian of the system will be reduced to the sum of "partial" Hamiltonians: upon replacement of the variables of (7.3), the term $(1/4)bu_{\boldsymbol{R}}^4$ will not yield the sum that includes the quantities $u_{\boldsymbol{k}}$ with different $\boldsymbol{k}$ (cf. Chap. 6). However, the transformation to normal coordinates proves useful in this case too: although we do not pass here to a system of noninteracting oscillators, the normal oscillators will interact only due to the presence in the potential energy of anharmonic terms, whose effect can be taken into account with the aid of perturbation theory.

The most interesting phenomena in crystals are associated exactly with anharmonicity. For instance, the parameters of a harmonic crystal do not at all depend on temperature (with the exception of heat capacity in the region of low temperatures, this being due to quantum effects). As a matter of fact, as indicated above, the properties of a harmonic crystal are fully characterized by a set of frequencies of normal vibrations, each of which is expressed in terms of the Hamiltonian constants, i.e., is temperature independent. Within the framework of classical statistics, which is applicable at rather high temperatures, the mean energy per oscillator is equal to $k_{\mathrm{B}}T$. The internal energy of the crystal is therefore equal to $Nk_{\mathrm{B}}T$ and the heat capacity is equal to Nk_{B}; in application to our case of unidimensional motion, this is the well-known law of Dulong and Petit.

If the role of anharmonicity is relatively low (see below for factors that may be responsible for this), to a first approximation, the vibrations of each normal coordinate may still be regarded as nearly harmonic, but the frequency of these vibrations no longer has its original meaning: instead, it becomes a certain averaged effective characteristic of the motion of the oscillator, which depends on the motion of other oscillators, i.e., on temperature. Further, the vibrations of each normal oscillator cease to be undamped; if a sufficiently large energy is imparted to a certain oscillator, this energy will be transferred with time to other normal oscillators, to which the given oscillator is coupled by virtue of anharmonicity. It is natural that the internal energy of the crystal too will no longer be egual to $Nk_{\mathrm{B}}T$, not to mention that for an isolated oscillator as well its energy will be equal to $k_{\mathrm{B}}T$ only for the harmonic case. The heat capacity of an anharmonic crystal depends on temperature even in the classical limit.

For the majority of crystals under normal conditions the anharmonic contribution to the potential energy is much smaller than the harmonic contribution. A natural measure of the quantities $u_{\boldsymbol{R}}$ is the lattice constant d, which is why we may assume that the values of $\frac{1}{2}au_{\boldsymbol{R}}^2$ and $\frac{1}{4}bu_{\boldsymbol{R}}^4$ are of the same order for $u_{\boldsymbol{R}} \sim d$. In other words, the anharmonic contribution may be compared in magnitude to the harmonic contribution only at rather high temperatures, when the vibration amplitude becomes comparable to the lattice constant. As a rule, this situation does not occur: the crystal melts much earlier. For the majority of crystals at not too high temperatures, anharmonicity may be considered to be small, which greatly facilitates the theoretical treatment of anharmonic effects.

Let us now turn to the derivation of an equation of motion for the normal coordinate $u_{\boldsymbol{k}}$ in an anharmonic crystal. The starting point is the system of equations for all normal coordinates since, by virtue of anharmonicity, the normal coordinates are coupled and the motion of a given normal coordinate affects the motion of all the others. Each of the equations has the following form:

$$m\ddot{u}_{\boldsymbol{k}} + \frac{\partial U(\ldots, u_{\boldsymbol{k}}, u_{-\boldsymbol{k}}, \ldots)}{\partial u_{-\boldsymbol{k}}} = 0. \tag{9.26}$$

In principle, the procedure of deriving the equation for the given normal coordinate $u_{\boldsymbol{k}_0}$ consists of excluding from the equation of motion for this coordinate all the other coordinates by expressing them through $u_{\boldsymbol{k}_0}$ and $u_{-\boldsymbol{k}_0}$ with the aid of the other equations. It is then necessary to perform an averaging over the initial conditions for the coordinates excluded, assuming naturally that the initial values of the coordinates themselves and of their velocities are distributed in accordance with the Gibbs formula. It is clear that the thus derived equation will be very complicated, nonlinear and integro-differential, but if among its terms containing time derivatives we choose only the linear terms, the equation will assume the form of an infinite series:

$$\ldots + \tilde{m}\ddot{u}_{\boldsymbol{k}_0} + \gamma\dot{u}_{\boldsymbol{k}_0} + \frac{\partial \tilde{U}}{\partial u_{-\boldsymbol{k}_0}} = f_{\mathrm{r}}(t), \tag{9.27}$$

where the ellipsis represents terms with higher-order derivatives (up to infinite order) and on the right-hand side are given terms containing no $u_{\boldsymbol{k}_0}$; in fact, it is a so-called *random force* $f_{\mathrm{r}}(t)$, which describes "thrusts" from the other normal oscillators and which makes the given oscillator perform thermal motion. On the left-hand side there are explicitly present only terms that correspond to the equation of motion of an oscillator with friction; the sign "$\sim$" indicates that the effective mass and potential energy are meant. We will consider the latter in more detail.

We have

$$\frac{\partial \tilde{U}}{\partial u_{-\boldsymbol{k}_0}} = \int \frac{\partial U}{\partial u_{-\boldsymbol{k}_0}} w(\ldots, u_{\boldsymbol{k}}, u_{-\boldsymbol{k}}, \ldots) \prod_{\boldsymbol{k}}{}' \mathrm{d}u_{\boldsymbol{k}}, \tag{9.28}$$

where w is the probability distribution for different initial values of the normal coordinates and the prime for the product indicates that the differentials $u_{\boldsymbol{k}_0}$ and $u_{-\boldsymbol{k}_0}$ have been omitted. The Gibbs function for the probability distribution of the system's coordinates was given in Chap. 2. Using this function, we find

$$\begin{aligned}
\frac{\partial \tilde{U}}{\partial u_{-\boldsymbol{k}_0}} &= \int \frac{\partial U}{\partial u_{-\boldsymbol{k}_0}} \exp\left(-\frac{U(\ldots, u_{\boldsymbol{k}}, u_{-\boldsymbol{k}}, \ldots)}{k_{\mathrm{B}}T}\right) \prod_{\boldsymbol{k}}{}' \mathrm{d}u_{\boldsymbol{k}} \\
&\quad \times \left\{ \int \exp\left(-\frac{U(\ldots, u_{\boldsymbol{k}}, u_{-\boldsymbol{k}}, \ldots)}{k_{\mathrm{B}}T}\right) \prod_{\boldsymbol{k}}{}' \mathrm{d}u_{\boldsymbol{k}} \right\}^{-1} \\
&= \frac{\partial}{\partial u_{-\boldsymbol{k}_0}} (-k_{\mathrm{B}}T) \ln \int \exp\left(-\frac{U(\ldots, u_{\boldsymbol{k}}, u_{-\boldsymbol{k}}, \ldots)}{k_{\mathrm{B}}T}\right) \prod_{\boldsymbol{k}}{}' \mathrm{d}u_{\boldsymbol{k}} \\
&= \frac{\partial}{\partial u_{-\boldsymbol{k}_0}} \Phi(u_{\boldsymbol{k}_0}, u_{-\boldsymbol{k}_0}),
\end{aligned} \tag{9.29}$$

where $\Phi(u_{\boldsymbol{k}_0}, u_{\boldsymbol{k}_0})$ is a quantity that has been defined earlier as the incomplete thermodynamic potential corresponding to the variables $u_{\boldsymbol{k}_0}$ and $u_{-\boldsymbol{k}_0}$. Thus, the role of the potential energy in the equation of motion of the normal coordinate in an anharmonic crystal is played by the incomplete thermodynamic potential. It is evident that in the case of anharmonicity the effective mass $\tilde{m}$ is close to the true mass of the oscillator and depends slightly on temperature, the viscosity coefficient γ is low (the oscillator is slightly attenuating, not completely damped) and the role of terms with higher derivatives is insignificant. We thus arrive at (9.12), in which, as pointed out in Chapter 7, $\eta = u_{k=0}$. As can be seen from the derivation, we have not used model representations at all and it may be presumed that the equation derived is also valid for systems that undergo a phase transition.

So, according to (9.14) and (9.15), the frequency of one of the normal vibrations of the crystalline lattice vanishes at the transition point. It is sometimes said that this fact (i.e., the presence of a *soft mode*) is responsible for the occurrence of a displacive phase transition. As can be seen from the above treatment, this statement is tantamount in meaning to the assertion that the phase transition is caused by the loss of stability by the original structure, i.e., in essence, by a change of the form of the function $\Phi(\eta)$, which is associated with a change of the sign of the coefficient A at η^2. Naturally, in this statement nothing is said of the microscopic causes of the phase transition, not to mention that the relation of the coefficient A to the frequency has been established with the aid of an approximate equation, which holds only for a weakly anharmonic crystal, and one can hardly rely on its universality. In fact, only the conclusion that the relaxation time of the order parameter goes to infinity is universal. As regards the actual microscopic cause of a phase transition, it should be sought for in the interaction that determines the temperature dependence of the coefficient A (see Chaps. 7 and 8).

Note that the terminology used in lattice dynamics is widely employed for describing displacive phase transitions. In such cases the order parameter is represented by the normal coordinate corresponding to the soft mode; the stability condition for a crystal corresponding to the minimum of the thermodynamic potential is expressed as the condition of positiveness of the normal vibration frequencies and the loss of stability is expressed as the vanishing of one of these frequencies. The appearance of an equilibrium nonzero order parameter in a nonsymmetric phase corresponds in this terminology to the "freezing" of atomic displacements corresponding to a soft mode. In the example under discussion the soft mode occurs at $k = 0$, which corresponds to a homogeneous (throughout the crystal) displacement of atoms upon transition to a nonsymmetric phase. If the dispersion of the soft branch of vibrations is such that the lowest frequency corresponds to the boundary of the Brillouin zone or, in general, to $k \neq 0$, then in the crystal upon transition to a nonsymmetric phase inhomogeneous displacements will be "frozen" and the translational symmetry will be changed. An example of this kind was considered in Chap. 1. It should be noted that the approximation that was used in Chap. 7 for displacive phase transitions is just the weak-anharmonicity approximation; the temperature-dependent contribution to the coefficient A

was calculated by an approximation linear in b since the corrections associated with b were ignored in expressions for $\langle u_{k_0}, u_{-k_0}\rangle$. This is just what explains why the effective mass in (9.12)–(9.18) may be considered to be temperature-independent and the coefficient g small as compared to ω_0 (see (9.13)). Of course, sufficiently close to the transition point the soft mode will necessarily become "overdamped" (a strongly attenuating oscillator) and the low-frequency dynamics of the order parameter will become relaxational. It is hoped, however, that this temperature region will be smaller than the region where Landau theory is valid. Recall in this connection that a "pure" phase transition of the displacive type as well as a "pure" order–disorder phase transition, is a physical abstraction, a certain limiting case, and that there are many transitions which should be classified as intermediate. The dynamics of such "intermediate" transitions is a very complicated topic, which goes beyond the scope of this book.

9.5 Microscopic Theory of Dynamic Processes in Order–Disorder Phase Transitions

We will now consider the microscopic approach to the description of the dynamics of the order parameter for phase transitions of order–disorder type. As has been shown in Chap. 7, in such transitions the displacements of atoms relative to symmetric positions are large and the weak-anharmonicity approximation does not apply here at all. For this reason the transformation to normal coordinates is not efficient. In Chap. 8, while describing the thermodynamic (i.e., static) properties of crystals with order–disorder phase transitions, we were dealing with the state of each ion in a two-minimum potential well, with account taken of the interaction between the ions by means of the mean-field method. This method can also be used to describe the dynamic properties of such crystals.

We will first analyze the motion of an ion in an isolated potential well (see Fig. 7.2). Of interest is the case where the barrier height ΔU is considerably larger than $k_B T$. It is evident that the thermal motion of an ion is in this case the vibrations near one of the minima interrupted by relatively rare jumps to another well.

We will further be concerned with such jumps since the vibrations near each of the minima usually have a high frequency and are not of interest from the viewpoint of the elucidation of the specific features of the dynamics of the order parameter near the transition point.

Let us consider an ensemble of n particles present in two-minima potentials and suppose that at the initial moment of time N_1 ions are in the left minimum and N_2 ions are in the right one. How will the distribution change with time? Let ν denote the vibrational frequency of ions near each of the minima; then the flux of ions impinging on the barrier on the left can be estimated as νN_1. The barrier is overcome only by high-energy particles, whose kinetic energy appears to be higher than the barrier height ΔU. Thus, the number of ions that pass from minimum 1 to minimum 2 in unit time will be $\nu N_1 \exp(-\Delta U/k_B T) \equiv w_{12} N_1$. Reasoning along the same lines for the right minimum, we find that the number of particles that pass from right to left in unit time is $w_{21} N_2$, where $w_{21} = \nu \exp(-\Delta U/k_B T) = w_{12}$.

Therefore, we have

$$\frac{dN_1}{dt} = -w_{12}N_1 + w_{21}N_2, \qquad \frac{dN_2}{dt} = -w_{21}N_2 + w_{12}N_1. \tag{9.30}$$

If we introduce the notation $N_1 - N_2 = \Delta N$ and $w_{12} = w_{21} = w$, we will get

$$\frac{d\Delta N}{dt} = -2w\Delta N, \tag{9.31}$$

that is, the difference in the occupation of the minima will decrease by an exponential law with a characteristic time equal to $1/(2w)$.

We will now take into account the interaction between ions in the mean-field approximation, i.e., we will assume that each ion is acted on by a field (see Chap. 8) proportional to the order parameter, so that one energy minimum ascends and the other descends in each of the two-minimum potentials. Now $w_{12} \neq w_{21}$:

$$w_{12} = \nu \exp\left(-\frac{\Delta U + J_0 s}{k_B T}\right), \tag{9.32}$$

$$w_{21} = \nu \exp\left(-\frac{\Delta U - J_0 s}{k_B T}\right). \tag{9.33}$$

From (9.30), (9.32) and (9.33) it follows that

$$\frac{d\Delta N}{dt} = -(w_{12} + w_{21})\Delta N + N(w_{21} - w_{12}), \tag{9.34}$$

or

$$\frac{ds}{dt} = -(w_{12} + w_{21})s + (w_{21} - w_{12}), \tag{9.35}$$

where it has been taken into account that $s = (N_1 - N_2)/N$.

Retaining the terms proportional to the parameter s, we may assume that the first expression in parentheses is approximately equal to $2w$; expanding the difference $w_{21} - w_{12}$ into a series in s and retaining only the first term, we obtain

$$\frac{ds}{dt} = -2ws + \frac{2wJ_0 s}{k_B T} = 2w\left(-1 + \frac{J_0}{k_B T}\right)s = 2w\frac{T - T_c}{T}s. \tag{9.36}$$

Here we have used the relation $T_c = J_0/k_B$ from Chapter 8.

Thus, the equation of motion for the order parameter in the case of order–disorder phase transitions is relaxational in character, the relaxation time going to infinity as $T \to T_c$ in accordance with the result obtained at the beginning of this chapter. We have thus substantiated the nonobvious assumption that the coefficient a_1 in formulas (9.2) and (9.3) may be considered to be temperature independent near T_c: in accordance with (9.36), $a_1 \propto T$, i.e., it may be considered to be constant in the region where Landau theory is applicable.

It should be emphasized that our treatment is very approximate and the model used is idealized, but the main conclusion concerning the relaxational character of the low-frequency dynamics of the order parameter for order–disorder phase transitions does not cause any doubt.

9.6 Dielectric Constant and Soft Mode: The Lyddane–Sachs–Teller Relation

As pointed out in Chap. 6, the spatially inhomogeneous fluctuations of the order parameter for proper ferroelectrics are accompanied by the appearance of an electric field. From the foregoing it is clear that the spatially inhomogeneous fluctuations of the order parameter in the terminology of lattice dynamics are the normal vibrations of one of the optical vibrational branches corresponding to $k \neq 0$. Let us see how this circumstance affects the dynamics of the order parameter.

We will consider the case of a cubic ferroelectric in the paraelectric phase where the order parameter has three components: η_1, η_2 and η_3. For a displacive phase transition (9.12) may be written for each of the components η_i; by virtue of the equivalence of all the components the coefficients $\tilde{m}$ and L will be identical in all three equations. Equation (9.12) may also be generalized for the case of $k \neq 0$:

$$\tilde{m}\ddot{\eta}_{i,\boldsymbol{k}} + L\dot{\eta}_{i,\boldsymbol{k}} + \frac{\partial \Phi}{\partial \eta_{i,-\boldsymbol{k}}} = 0, \tag{9.37}$$

with the implication that the values of $\boldsymbol{k}$ are nonzero, though small, so that the quantities $\tilde{m}$ and L may be regarded as independent of $\boldsymbol{k}$. For the paraelectric phase we have

$$\Phi = \sum_{\boldsymbol{k},i} \frac{1}{2}(A + \delta k^2)\eta_{i,\boldsymbol{k}}\eta_{i,-\boldsymbol{k}} - \frac{1}{2}\sum_{\boldsymbol{k},i} aE_{i,\boldsymbol{k}}\eta_{i,-\boldsymbol{k}}, \tag{9.38}$$

where the second term is coupled to a dipole–dipole interaction induced by a macroscopic field (see Chap. 6) and where it has been taken into account that near the transition point $P_i = a\eta_i$. Hence,

$$\Phi = \frac{1}{2}\sum_{\boldsymbol{k},i,j} \left[(A + \delta k^2)\eta_{i,\boldsymbol{k}}\eta_{i,-\boldsymbol{k}} + 4\pi a^2 \frac{k_i k_j}{k^2}\eta_{i,\boldsymbol{k}}\eta_{i,-\boldsymbol{k}} \right]. \tag{9.39}$$

We see that the expression for Φ is no longer a quadratic form of $\eta_{i,\boldsymbol{k}}$ (like the first term in formula (9.38)). It is easy to see that the nondiagonal terms in (9.39) cancel out when the wave vector is directed along the coordinate axes, i.e.,

$$\boldsymbol{k} = (k, 0, 0), \tag{9.40}$$

or

$$\boldsymbol{k} = (0, k, 0), \tag{9.41}$$

or else

$$\boldsymbol{k} = (0, 0, k). \tag{9.42}$$

For the first case ($\boldsymbol{k}(k, 0, 0)$), while considering the contribution to the thermodynamic potential corresponding to $\boldsymbol{k}$, we obtain:

$$\begin{aligned}\Delta\Phi_k = {} & (A + \delta k^2 + 4\pi a^2)\eta_{1,\boldsymbol{k}}\eta_{1,-\boldsymbol{k}} + (A + \delta k^2)\eta_{2,\boldsymbol{k}}\eta_{2,-\boldsymbol{k}} + \\ & + (A + \delta k^2)\eta_{3,\boldsymbol{k}}\eta_{3,-\boldsymbol{k}}.\end{aligned} \tag{9.43}$$

It is evident that, in spite of the equivalence of the components of the order parameter, the vibrations of $\eta_{1,\boldsymbol{k}}$ correspond to a different frequency than those of the components $\eta_{2,\boldsymbol{k}}$ and $\eta_{3,\boldsymbol{k}}$. In other words, the vibrations that are accompanied by the occurrence of polarization directed along the wave vector (longitudinal vibrations) have a frequency ω_{LO} different from the frequency of vibrations ω_{TO} at which the polarization occurs perpendicular to the wave vector (transverse vibrations). Of course, the corresponding frequencies will differ with $k \to 0$ as well, i.e.,

$$\omega_{\mathrm{LO}}^2 = (A + 4\pi a^2)/\tilde{m}, \tag{9.44}$$

$$\omega_{\mathrm{TO}}^2 = A/\tilde{m}. \tag{9.45}$$

Dividing (9.44) by (9.45), we obtain

$$\omega_{\mathrm{LO}}^2/\omega_{\mathrm{TO}}^2 \approx 4\pi a^2/A = \varepsilon_0, \tag{9.46}$$

where ε_0 is the static dielectric constant. It is natural that, as follows from (9.46), if $\varepsilon_0 \to \infty$ as $T \to T_{\mathrm{c}}$, then $\omega_{\mathrm{TO}}^2 \to 0$. Formula (9.46), which relates the frequencies of transverse and longitudinal vibrations η to the static dielectric constant, allows for certain refinement. As a matter of fact, the macroscopic electric field appearing in (9.38) and being due to polarization associated with η, acts in a medium where there are other, high-frequency polarization mechanisms, which are not associated with the occurrence of η. We denote the dielectric constant of such a medium with $\eta = \text{const}$ by $\varepsilon_{ij}^{\infty}$, in which case the electric polarization is given by

$$P_{i,\boldsymbol{k}} = a\eta_{\boldsymbol{k}} + \frac{1}{4\pi}(\varepsilon_{ij}^{\infty} - 1)E_{j,\boldsymbol{k}}, \tag{9.47}$$

and the macroscopic electric field is

$$E_{\boldsymbol{k}} = -4\pi a \frac{\boldsymbol{k}(\boldsymbol{k}\eta_{\boldsymbol{k}})}{k_i \varepsilon_{ij}^{\infty} k_j}. \tag{9.48}$$

Considering that we are dealing with the case of (9.40) and omitting the indices in $\varepsilon_{ij}^{\infty}$, we obtain

$$E_{1,\boldsymbol{k}} = -4\pi a \frac{\eta_{1,\boldsymbol{k}}}{\varepsilon_{ij}^{\infty}}, \qquad E_{2,\boldsymbol{k}} = E_{3,\boldsymbol{k}} = 0. \tag{9.49}$$

Therefore, instead of (9.44), we have

$$\omega_{\mathrm{LO}}^2 = \frac{1}{\tilde{m}} \left(A + \frac{4\pi a^2}{\varepsilon^{\infty}} \right), \tag{9.50}$$

and, hence,

$$\frac{\omega_{\mathrm{LO}}^2}{\omega_{\mathrm{TO}}^2} = 1 + \frac{4\pi a^2}{A\varepsilon^{\infty}} \approx \frac{\varepsilon_0}{\varepsilon^{\infty}}. \tag{9.51}$$

The formula obtained is known as the Lyddane–Sachs–Teller (LST) relation.

Note that in the examples considered in the preceding chapters we invariably assumed that $\varepsilon^{\infty} = 1$, without making any appreciable error in the analysis of the dielectric properties of a crystal near T_{c}. Indeed, for proper ferroelectrics near T_{c}, $\varepsilon_0 = \varepsilon^{\infty} + 4\pi a^2/A$ and the first term is negligibly small since $A \to 0$ as $T \to T_{\mathrm{c}}$.

10. Domain Structure and Defects

10.1 Nucleation of Domains in a Structural Phase Transition

In Chaps. 2–9 it was assumed that the nonsymmetric phase is spatially homogeneous, i.e., that a phase transition occurs into a monodomain state. Domains were mentioned only in connection with the fact that the system of equations for an equilibrium order parameter has several solutions; it was pointed out that these solutions correspond to different domains.

It will be useful to note that the solutions of the system of equations resulting from the condition of a minimum of the thermodynamic potential are interconverted by symmetry transformations of the symmetric phase that disappear upon phase transition. This can be seen most simply by considering a one-component order parameter. Different domains in this case are regions with different signs of η. Symmetry transformations of the symmetric phase either do not change η at all or change only its sign. In a nonsymmetric phase only those symmetry elements are retained which do not change η. Thus, it is precisely the symmetry elements disappearing in a phase transition that convert the structure of one domain into the structure of another.

Extending these reasonings to the case of a many-component order parameter, we emphasize once again that the thermodynamic potential and, hence, the system of equations resulting from its minimization, do not change under all symmetry transformations of the symmetric phase. Suppose that we have found one of the solutions of this system of equations. We apply to it a certain symmetry transformation of the symmetric phase. It will alter the solution only if the corresponding symmetry element disappears upon transition into a phase to which this solution corresponds. Even such a transformation does not change the system of equations proper. We will therefore obtain, with its aid, another solution of this system equivalent to the initial one, i.e., corresponding to the same value of the thermodynamic potential. Thus, the symmetry elements that disappear in a phase transition will also "work" in a nonsymmetric phase – they interconvert the structures of different domains. In this sense it is sometimes said that if the entire set of equivalent domains (solutions) is considered instead of a single domain (solution), the symmetry lost will be restored upon transition.

Note that domains (often called twins) are also observed in many crystals that do not undergo a phase transition. However, the structure of such crystals may be represented as the result of a distortion of a certain more symmetric hypothetical structure. In other words, this could be the case had there occurred a phase transition from a certain more symmetric phase (which may not be experimentally realized) into an actually observed phase. All the above symmetry considerations are, of course, applicable to this case as well.

We will first discuss the problem of domains for the simplest case of a nonferroelectric and a nonferroelastic phase transition with a one-component order parameter. It should be pointed out at the outset that the nucleation of domains is energetically favorable in this case. The point is that the boundary between domains (which is also known as the domain wall) is a spatial inhomogeneity of the order parameter and its presence therefore increases the energy of the system as compared to a homogeneous state. Indeed, in the expression for the density of the thermodynamic potential (6.7) the coefficient δ in the term $\frac{1}{2}\delta(\text{grad}\eta)^2$ is positive. The positive nature of this coefficient implies that the spatially homogeneous distribution of η is energetically the most favorable.

Thus, if a second-order transition takes place in an ideal crystal with an infinitely slow decrease of temperature occurring homogeneously throughout the crystal (which is required in order to maintain a thermodynamic equilibrium), then a single-domain state is preferable as a result of the transition. Here both signs of the equilibrium order parameter are equally probable.

In reality, as has been shown by experiment, domains arise almost always in phase transitions. The simplest explanation would be as follows: at different portions of the crystal, which are sufficiently remote from one another a nonsymmetric phase is formed practically independently (due to the finiteness of the rate of cooling and to the inhomogeneity of the temperature field), so that the signs of the order parameter may prove different in different regions, as a result of which domains are formed. In fact, however, the situation with the nucleation of domains is much more complicated and is unclear in many respects; this is the problem of the real crystal since the major role in the nucleation of domains is played by defects and inhomogeneities of the sample.

10.2 Domain Wall Structure: One-Component Order Parameter

Here we will be concerned with the simplest problem associated with domains: suppose that an infinite crystal consists of two domains. Let us assume that in one half-space ($x > 0$) $\eta > 0$ and in the other ($x < 0$) $\eta < 0$, i.e., the domain wall coincides with the plane yz. What will be the equilibrium spatial distribution of the order parameter? It may seem that it will assume the form of a step (the dashed line in Fig. 10.1) since with this distribution at each point of the crystal the value of the order parameter is equal either to $+|\eta_0|$ or to $-|\eta_0|$, and it is only to these two values of η that there will correspond the minimum of the thermodynamic potential. Therefore, the presence of regions with other values of η would imply an energy loss – an increase of the thermodynamic potential as compared to its value at $\eta = \eta_0$. In actuality, however, having generated a step, we have not avoided an energy loss: the presence of a step implies an infinite value of $(\partial\eta/\partial x)$ and, hence, the contribution of the term $\frac{1}{2}\delta(\partial\eta/\partial x)^2$ to the density of the thermodynamic potential is infintely large in the region of the step. In order to reduce it, we have to "smear" the step, in which case there will appear regions where $|\eta| \neq |\eta_0|$. As a result, the distribution of η in which the increase of the thermodynamic potential caused by the appearance of a domain wall will be at a minimum has the form of a "smeared" step (the solid line in Fig. 10.1).

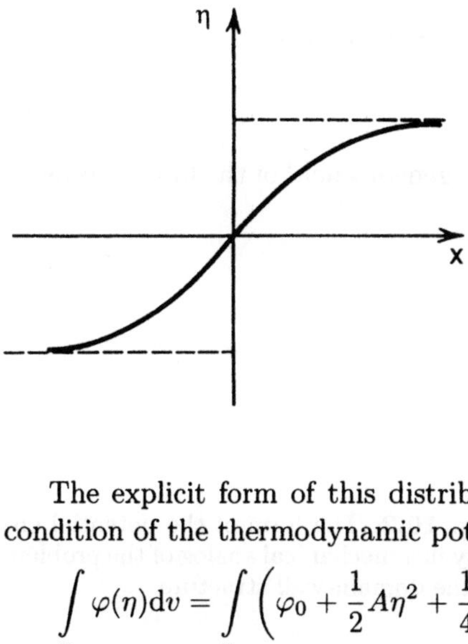

Fig. 10.1. Distribution of the order parameter in the domain boundary: the dashed line represents the distribution in the form of a step and the solid line is the actually realizable distribution.

The explicit form of this distribution will be found from the minimum condition of the thermodynamic potential of the crystal:

$$\int_v \varphi(\eta)\mathrm{d}v = \int_v \left(\varphi_0 + \frac{1}{2}A\eta^2 + \frac{1}{4}\beta\eta^4 + \frac{1}{2}\delta(\nabla\eta)^2\right)\mathrm{d}v \tag{10.1}$$

without assuming, of course, just as was done in the preceding chapters, that the equilibrium value of η is the same at all points; here η is taken to be equal to $\eta(x)$. To find the function $\eta = \eta(x)$, which minimizes the integral (10.1), is a typical problem of variational computation; it is known that such a function must satisfy the Euler equation

$$\frac{\partial}{\partial x}\left(\frac{\partial\varphi}{\partial(\partial\eta/\partial x)}\right) - \frac{\partial\varphi}{\partial\eta} = 0.$$

Thus, we obtain

$$A\eta + \beta\eta^3 - \delta\frac{\partial^2\eta}{\partial x^2} = 0. \tag{10.2}$$

This is a nonlinear second-order differential equation. There is no routine procedure for the solution of such equations. However, the required solution ($\eta \to +|\eta_0|$ for $x \to \infty$, $\eta \to -|\eta_0|$ for $x \to -\infty$) can be found by using the method of physical analogies.

As a matter of fact, (10.2) coincides with the equation of motion for a material point with mass δ and coordinate η if x is taken to mean time instead of the coordinate. We introduce the following notation: $\eta \to y$ and $x \to t$ and write down (10.2) in the form

$$\delta \frac{\mathrm{d}^2 y}{\mathrm{d}t^2} - Ay - \beta y^3 = 0. \tag{10.3}$$

Comparing (10.3) with the equation of motion for a material point

$$\delta \frac{\mathrm{d}^2 y}{\mathrm{d}t^2} = F_y = -\frac{\partial U}{\partial y} \tag{10.4}$$

where $U(y)$ is the potential energy, we see that in our case

$$U(y) = -\frac{A}{2}y^2 - \frac{\beta}{4}y^4. \tag{10.5}$$

Figure 10.2 shows a hard profile in a homogeneous field of the force of gravity corresponding to such an energy.

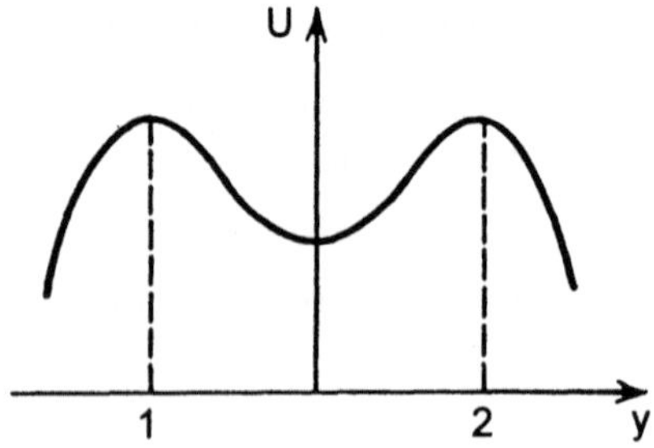

Fig. 10.2. The form of the potential energy in a mechanical analog of the problem of the domain-wall structure.

It is easy to visualize the motion of a material point (a ball) corresponding to the sought-for solution. The ball resides on hill 1 for a relatively long time and then, rolling down from it, passes the valley between the hills and comes to a standstill on hill 2. The relatively fast passage of the ball through the valley corresponds to the relatively small thickness of the domain wall (the states of near rest are represented by the value of η at $x = \pm\infty$). With an indefinite time interval (which corresponds to the infinite size of the crystal) we may assume that in positions 1 and 2 the ball is at rest.

Let us now use the law of conservation of energy. The total energy (the sum of kinetic and potential energies) is given by

$$\frac{\delta}{2}\left(\frac{\mathrm{d}y}{\mathrm{d}t}\right)^2 - \frac{A}{2}y^2 - \frac{\beta}{4}y^4 = -\frac{A}{2}y_0^2 - \frac{\beta}{4}y_0^4. \tag{10.6}$$

Here y_0 represents the coordinate of the maximum of the potential energy and it has been taken into account that with the motion of interest the kinetic energy is equal to zero at $y = y_0$. Returning now to the old variables and introducing the dimensionless quantities

$$\tilde{\eta} = \sqrt{\beta}\eta/\sqrt{-A}, \qquad \tilde{x} = x/r_{\mathrm{c}}$$

where $r_{\mathrm{c}} = \sqrt{\delta/(-2A)}$, we obtain from (10.6)

$$\frac{\mathrm{d}\tilde{\eta}}{\mathrm{d}\tilde{x}} = \pm\frac{1}{2}\left(\tilde{\eta}^2 - 1\right) \tag{10.7}$$

or

$$\frac{\mathrm{d}\tilde{\eta}}{\tilde{\eta}^2 - 1} = \pm\frac{1}{2}\mathrm{d}\tilde{x}. \tag{10.8}$$

Integrating, we find

$$\operatorname{arctanh}\tilde{\eta} = \pm\frac{1}{2}\tilde{x} + C \tag{10.9}$$

where C is an arbitrary constant, i.e.,

$$\tilde{\eta} = \pm\tanh\left(\frac{\tilde{x}}{2} + C\right) \tag{10.10}$$

To our choice of the domain wall in the plane yz there corresponds the condition $C = 0$ (it is exactly with this condition that $\eta(x)$ passes through zero at $x = 0$). Finally, we get

$$\eta = \pm\sqrt{-\frac{A}{\beta}}\tanh\frac{x}{2r_{\mathrm{c}}}. \tag{10.11}$$

The different signs in front of the root correspond to two possible signs of η in the given half-space, for example, for $x > 0$. The distribution of $\eta(x)$ corresponding to the positive sign of η in this half-space is shown in Fig. 10.1 by a solid line.

As can be seen from (10.11) and Fig. 10.1, the quantity r_{c} plays the role of the effective thickness of the domain wall. It should be emphasized that this quantity has a much broader meaning. This is the so-called *correlation radius of the order parameter*, which is an important quantity in the theory of phase transitions and which has the meaning of the characteristic scale of spatial changes of the order parameter. Suppose that in a certain plane $x = x_0$ in a single-domain crystal or inside one of the domains we recorded a certain value of order parameter $\eta = \eta_{\mathrm{loc}}$, which differs from the equilibrium value of η in a homogeneous ideal crystal $|\eta_0| = \sqrt{-A/\beta}$. How this was achieved is of no importance. Suppose, for example, that we have at our disposal a force coupled to η and different from zero in the $x = x_0$ plane. It is clear that near the plane the order parameter will also be different from η_0 since for the same reasons that were discussed above the order parameter cannot change in a jumpwise manner. The characteristic length of fall-off of $\eta - \eta_0$ to zero appears to be equal to r_c. We will assume, for simplicity, that η_{loc} does not strongly differ from η_0, so that $\eta - \eta_0 \ll \eta_0$ throughout the region of interest. Equation (10.2) can then be linearized by substituting $\eta = \eta_0 + \eta'$ and omitting the powers of η' higher than the first. As a result, we obtain

$$\delta \frac{d^2\eta'}{dx^2} + 2A\eta' = 0. \tag{10.12}$$

The solution of this equation that satisfies the required boundary conditions $\eta' = \eta_{loc} - \eta_0$ at $x = x_0$ and $\eta' \to 0$ for $x \to \infty$ has the form

$$\eta' = (\eta_{loc} - \eta_0)\exp(-|x - x_0|/r_c). \tag{10.13}$$

This is the so-called *Ornstein–Zernike correlation function* for the one-dimensional case (we took account of the dependence of η only on one coordinate). If we consider the same problem for a symmetric phase, the correlation radius will appear to be equal to $(\delta/A)^{1/2}$.

The correlation radius r_c also plays the role of the characteristic size of thermal fluctuations η: if at a certain point, as a result of a thermal fluctuation, there will arise a value of η appreciably different from the equilibrium value, it will change the values of η at neighboring points in just the same way as in the example considered above. Thus, thermal fluctuations of η do not occur independently at various points of the crystal; instead, they are intercorrelated at distances of the order of r_c. Hence, the term "the correlation radius". It is also easy to find the correlation radius in the formulas given in Chap. 6 for fluctuations of the order parameter. Formula (6.37) may be rewritten in the form

$$\langle \eta_k^2 \rangle = \frac{k_B T}{\frac{v}{2}(A + \delta k^2)} = \frac{k_B T}{\frac{v}{2}A(1 + k^2 r_c^2)}. \tag{10.14}$$

Hence, the characteristic value of the wave vector on which the fluctuations of η decline with increasing k is r_c^{-1}; it is also said provisionally that only those fluctuations are large which have wavelengths greater than the correlation radius.

With $T \to T_c$ the correlation radius and, hence, the domain-wall thickness tend toward infinity. This statement is, of course, only formal: it is meaningless to speak of a domain-wall thickness larger than the crystal's size, but the temperature region in which the correlation radius becomes comparable to the size of ordinary crystals (thin films are excluded) is negligibly small and experimentally unattainable. Therefore, we may ignore the formal character of the statement made.

Let us now calculate the energy associated with the formation of a unit surface of the domain wall (this quantity is also known as the surface tension or *specific surface energy of the wall*). This energy is evidently equal to the difference between the energy of a crystal with a domain wall and the energy of a monodomain crystal:

$$\varepsilon_{\text{sur}} = \int_{-\infty}^{\infty} [\varphi(\eta(x)) - \varphi(\eta_0)]\, \mathrm{d}x\,. \tag{10.15}$$

Using the formulas given above and tables of integrals, it is not difficult to calculate ε_{sur}. The answer can also be found without integration, though with an accuracy within a numerical coefficient. It may be expected that the energy loss per unit volume in the domain-wall thickness will be approximately equal to $A^2/(4\beta)$ (the difference in values of the density of the thermodynamic potential at $\eta = 0$ and $\eta = \eta_0$) and the wall thickness will be of the order of r_c, as pointed out above. So, we have

$$\varepsilon_{\text{sur}} \approx \frac{A^2}{4\beta} r_c \approx \frac{\delta^{1/2}}{4\beta} |A|^{3/2}. \tag{10.16}$$

The exact answer will be as follows:

$$\varepsilon_{\text{sur}} = \frac{2\sqrt{2}\delta^{1/2}}{3\beta} |A|^{3/2}. \tag{10.17}$$

Thus, the surface energy of the wall tends to zero as $T \to T_c$ according to the law $\varepsilon_{\text{sur}} \propto (T_c - T)^{3/2}$.

10.3 Domain Wall Structure: Two-Component Order Parameter

We will now make the problem more complicated and consider the case of a two-component order parameter (η_1, η_2). Formula (2.39) gives an example of the dependence of the thermodynamic potential on spatially homogeneous quantities η_1 and η_2 (as was assumed in Chap. 2). This is what we will use in our treatment.

In order to elucidate the structure of domain walls, we have to take account, in the thermodynamic potential density, of terms dependent on the derivatives of η_1 and η_2 with respect to the coordinates. The number and form of these terms are determined by the symmetry group of a symmetric phase. We will not be concerned here with any particular case – for our purposes it will suffice to consider the term $\frac{1}{2}\delta[(\nabla\eta_1)^2 + (\nabla\eta_2)^2]$ with an a priori allowable symmetry. Thus, we have

$$\begin{aligned}\varphi(\eta_1, \eta_2) &= \frac{1}{2}\alpha(T - T_c)(\eta_1^2 + \eta_2^2) + \frac{1}{4}\beta_1(\eta_1^2 + \eta_2^2)^2 \\ &\quad + \frac{\beta_2}{2}\eta_1^2\eta_2^2 + \frac{1}{2}\delta[(\nabla\eta_1)^2 + (\nabla\eta_2)^2].\end{aligned} \tag{10.18}$$

To find the functions $\eta_1(x)$ and $\eta_2(x)$, we have now to solve the system of Euler equations:

$$\alpha(T-T_c)\eta_1+\beta_1(\eta_1^2+\eta_2^2)\eta_1+\beta_2\eta_2^2\eta_1-\delta\frac{\partial^2\eta_1}{\partial x^2}=0 \tag{10.19}$$

$$\alpha(T-T_c)\eta_2+\beta_1(\eta_1^2+\eta_2^2)\eta_2+\beta_2\eta_1^2\eta_2-\delta\frac{\partial^2\eta_2}{\partial x^2}=0. \tag{10.20}$$

This is a system of coupled nonlinear second-order equations and to find its analytic solution is a very difficult task. The mechanical analog used above is no longer useful: our equations describe the motion of two particles residing in a potential well and involved in a complex interaction.

It is possible, however, to formulate conditions for the coefficients of the thermodynamic potential such that the problem will be reduced again to the consideration of a system having one degree of freedom, in which case it will become quite solvable. Let us pass to polar coordinates in the plane $\eta_1\eta_2$:

$$\eta_1=\rho\cos\psi,\qquad \eta_2=\rho\sin\psi. \tag{10.21}$$

The quantity ρ is naturally called the *modulus of the order parameter* and ψ its *phase*. The expression for the thermodynamic potential will now take on the following form:

$$\varphi(\rho,\psi)=\frac{\alpha(T-T_c)}{2}\rho^2+\frac{1}{2}\beta_1\rho^4+\frac{\beta_2}{8}\rho^4\sin^2(2\psi)$$
$$+\frac{\delta}{2}(\nabla\rho)^2+\frac{\delta}{2}\rho^2(\nabla\psi)^2. \tag{10.22}$$

Let β_2 be larger than zero, in which case for equilibrium (i.e., corresponding to the minimum of the thermodynamic potential) homogeneous states the following condition will be fulfilled:

$$\sin(4\psi_0)=0\quad\text{and}\quad\cos(4\psi_0)>0,\quad\text{i.e., }\psi_0=0,\frac{\pi}{2},\pi,\frac{3\pi}{2},$$

$$\rho_0^2=-\frac{\alpha(T-T_c)}{\beta_1}. \tag{10.23}$$

Let us now consider two domains corresponding, for example, to $\psi=0$ and $\psi=\pi/2$. It is interesting to pose the following question: when will changes in η_1 and η_2 in the domain wall occur in such a way that ρ will remain practically constant ("rotation of the vector" (η_1,η_2) with the modulus remaining unchanged)? This will evidently take place when an "incorrect" value of ψ results in a relatively small energy loss, i.e., when $|\beta_2|$ is much lower than β_1. In this case, while analyzing the domain-wall structure, we may ignore the dependence of ρ on the coordinate and then we can derive from (10.22) the Euler equation in the following form:

$$-\delta\rho^2\frac{\mathrm{d}^2\psi}{\mathrm{d}x^2}+\frac{\beta_2}{4}\rho^4\sin(4\psi)=0. \tag{10.24}$$

Introducing the variable $\zeta = 4\psi$, we obtain

$$\frac{\mathrm{d}^2\zeta}{\mathrm{d}x^2} - \frac{\beta_2\rho^2}{\delta}\sin\zeta = 0. \tag{10.25}$$

This is the equation of motion for a pendulum (if the time t is made to correspond to the coordinate x), in which the origin of the angle ζ corresponds to the "upside-down" position of the pendulum. We are interested in such a solution when the pendulum starts moving from this position, picks up speed and passes the lowest point and then again takes the "upside-down" position. The dependence $\zeta(x)$ for such a solution is well known and is given by the Jacobi function. There is no need, however, to give the form of this solution in order to determine the basic characteristics of the domain wall. It is clear that the width of the domain wall (the characteristic time of the pendulum's motion along the circumference) is as follows in order of magnitude

$$\left(\frac{\delta}{\beta_2\rho^2}\right)^{1/2} = \left(\frac{\beta_1}{\beta_2}\right)^{1/2}\left(\frac{\delta}{\alpha|T-T_\mathrm{c}|}\right)^{1/2}. \tag{10.26}$$

The second term on the right-hand side of (10.26) is the familiar correlation radius, which in this particular case is the correlation radius for the modulus ρ. The domain-wall width is much larger than this quantity (since $\beta_2/\beta_1 \ll 1$): it corresponds to the correlation radius of the phase ψ. Thus, in the case under consideration, because of the "small anisotropy" the domain-wall width may prove to be large. An analogous situation occurs for the so-called Bloch wall in a ferromagnetic. The specific energy of the wall can be estimated in the same way as was done in Sect. 10.2, by multiplying $\beta_2\rho^4/8$ by the wall width. It can be seen that in this case too $\varepsilon_\mathrm{sur} \propto (T_\mathrm{c} - T)^{3/2}$.

10.4 Motion of the Domain Wall in an Ideal and a Real Crystal

In the treatment we have been concerned with so far the wall coordinates did not appear at all, i.e., we assumed that the wall energy is independent of its position. We came to this conclusion because we treated the crystal as a continuous medium, which is quite permissible when r_c is substantially larger than the lattice constant. This condition is fulfilled only near the Curie point; it will be abandoned at a later time. At this point we will still consider the crystal to be a continuous medium.

Let us consider the motion of the domain wall for the case where the domains separated by a wall are energetically nonequivalent. The wall will naturally move in such a way that the volume of an energetically favorable domain increases at the expense of an energetically unfavorable domain. The simplest way to create an energetical nonequivalence of domains is to apply a homogeneous field h coupled to the order parameter. In the case of a ferroelectric with a one-component order parameter (one axis of spontaneous polarization) such a field is an electric field parallel to the polar axis. For nonferroelectrics (and nonferroelastics) such a field is not realized, as a rule, in experimental conditions; it is convenient, however, to introduce it, though on purely formal grounds, by inserting into the expression for $\varphi(\eta)$ in formula (10.1) the term $-h\eta$, which is quite analogous to the term $-a\eta E$ in formula (2.35). Let $h > 0$. Then the specific thermodynamic potential for a domain with $\eta_0 > 0$ will be lower than that for a domain with $\eta < 0$ by an amount $\Delta\Phi = 2h\eta_0$ (we assume that the field is low and it is possible to neglect the change in the equilibrium value of η upon application of the field). Below, for the sake of definiteness, we will deal with a ferroelectric, but it should be pointed out that all our inferences refer to nonferroelectrics as well: we have only to take into account that the difference in specific thermodynamic potentials for domains is caused not by an external electric field but by something else, say by elastic stresses.

So, suppose that a crystal composed of two domains with opposite signs of η_0 is acted on by a homogeneous field $h \equiv aE$. Our task is to describe variations of the order-parameter distribution with time, i.e., to find the function $\eta(x,t)$, which is what will describe the motion of the domain wall. The regularities of variation of η with time have already been considered in this book (see Chap. 9). At this point we are going to extend (9.5) or (9.12) to the spatially inhomogeneous case. Since the static equations for the inhomogeneous case are known (10.2), we can immediately write

$$\tilde{m}\ddot{\eta} + L\dot{\eta} + A\eta + \beta\eta^3 - \delta\frac{\mathrm{d}^2\eta}{\mathrm{d}x^2} = h. \tag{10.27}$$

We are not going to obtain the solution of this equation (which is possible, in principle); we will only determine the parameters that characterize the motion of the domain wall on the basis of the following simple considerations.

It is clear that in a certain period of time after the field is switched on the wall will move uniformly with a certain velocity v. Let us discuss the motion of the wall at this stage. If the field is low, it may be expected that the shape of the moving wall (the distribution of the order parameter in it) will be the same as that of the immovable wall. Suppose that the wall was displaced to a distance equal to its effective width $4r_\mathrm{c}$ (Fig. 10.3). For a wall of unit area the decrease of the system's energy (the work of the external field h) is equal to $2\eta_0 h \cdot 4r_\mathrm{c}$. This work must be equal to the work of viscosity forces (since the motion is uniform). From (10.27) one can easily see that the role of the viscosity force is played by the term $L\dot{\eta}$. Simplifying the real situation, we may assume that upon displacement of the wall the variation of η ("displacement") occurred only at points x lying in the interval $4r_\mathrm{c}$ marked in Fig. 10.3. Here $\dot{\eta} = (2\eta_0/\Delta t)$ and $\Delta t = 4r_\mathrm{c}/v$ and the work of viscosity forces (remember that we are speaking here of a wall of unit area) is equal to $L\dot{\eta}\Delta\eta \cdot 4r_\mathrm{c} = L \cdot 4\eta_0^2 v$. Equating the work of viscosity forces to that of the external field, we find

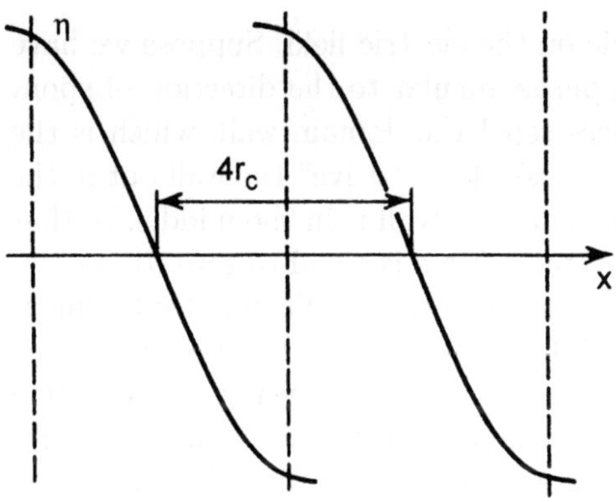

Fig. 10.3. Calculation of the velocity of motion of a wide domain wall. Shown are two successive positions of the wall when its displacement is equal to the wall width.

$$v = \frac{2hr_c}{L\eta_0}. \tag{10.28}$$

In a strict treatment there is obtained an analogous expression with an accuracy of up to the replacement of the numerical coefficient.

Thus, the velocity of the wall is directly proportional to the "external force" and inversely proportional to the viscosity coefficient, in exactly the same way as in the case of the motion of a body in a viscous liquid. This analogy may be extended by calculating the "mass" of this "body", i.e., the effective mass of the unit surface of the wall. In the derivation of (10.27) it was assumed that the density of kinetic energy associated with the change of η is equal to $\tilde{m}\dot{\eta}^2/2$. Considering that during the displacement of the wall (Fig. 10.3) the atoms lying in the interval $4r_c$ are involved in the motion, we find the kinetic energy per unit area of the wall:

$$\frac{1}{2}\tilde{m}\left(\frac{2\eta_0 v}{4r_c}\right)^2 \cdot 4r_c = \frac{1}{2}\frac{\tilde{m}\eta_0^2}{r_c}v^2$$

that is, the effective mass of the unit area of the wall is equal to

$$M = \tilde{m}\eta_0^2/r_c. \tag{10.29}$$

Thus, the wall may be looked upon as a body of mass M, the equation of motion for which is

$$M\frac{dv}{dt} + Nv = F \equiv 2\eta_0 h \tag{10.30}$$

where $N = \eta_0^2/r_c$. Equation (10.30) is convenient since it can also be used in the case of a variable field $h(E)$ in order to calculate the dielectric constant of a ferroelectric with domain walls. Recall that this equation is valid at not too high velocities, when the change of the wall shape may be neglected.

Of course, a ferroelectric with a single domain wall wide enough for us to be distracted from the discreteness of the crystal lattice is an object which is strongly different from those actually encountered in experiments. Nevertheless, it will be useful to discuss its properties, namely the dependence of the polarization of such a ferroelectric on the electric field. Suppose we have a crystal plate the plane of which is perpendicular to the direction of spontaneous polarization and the end faces repel the domain wall, which is the only one present inside, so that it is impossible to "drive" the wall out of the crystal. Suppose that at the initial moment the wall is in the middle, so that the polarization averaged over the crystal volume is equal to zero. It is obvious that the application of even a very low field will result in a displacement of the wall, which will approach one of the end faces to a distance allowed by the repulsive forces mentioned above. As a result, the average polarization of the plate will increase jumpwise to a value close to the polarization of a monodomain plate and with further increase of the field the polarization will increase due to both its increase in the domain polarized along the field and a decrease in the volume left at the domain end of opposite polarization. When a field as low as desired is applied in the opposite direction, the wall will be shifted towards the opposite end.

In other words, the dependence of the polarization on the electric field for our ferroelectric will look like that shown in Fig. 10.4.

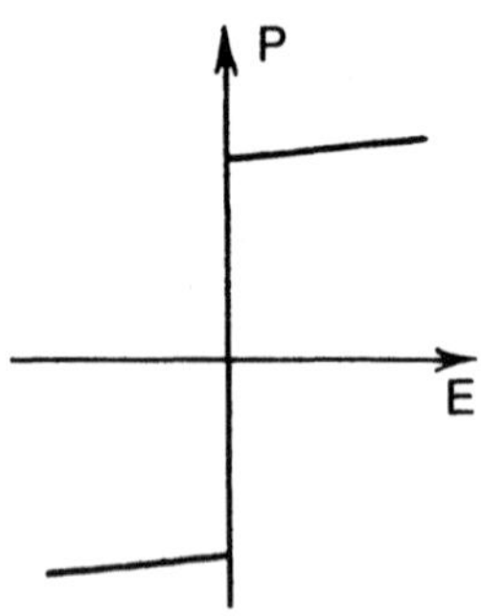

Fig. 10.4. The dependence $P(E)$ for an ideal ferroelectric with special boundary conditions.

In a real crystal the domain wall appears to be clamped by defects, crystal inhomogeneities, and sometimes by its boundaries. At a later time we will discuss certain interactions that cause the clamping of the wall; at this point we will consider the properties of an idealized crystal, in which the clamping centers (or, as is also said, the *pinning centers*) are arranged only in a single plane coinciding with the wall plane before the electric field is turned on. These centers provide a restoring force and now the wall is no longer a free body in a viscous medium; instead, it is something like a ball attached to a spring. Instead of (10.30), we now have

$$M\frac{\mathrm{d}v}{\mathrm{d}t} + Nv + Kx = 2\eta_0 h \qquad (10.31)$$

where K is the stiffness coefficient for coupling the wall to the pinning centers. Bringing our model closer to reality, we will assume that our "spring" cannot "expand" infinitely and "is disrupted" by the action of a large force, i.e., a rather strong electric field tears the wall off the pinning centers, and (10.31) holds only for sufficiently weak fields. If we assume that the wall is torn off as a whole, the dependence $P(E)$ will have the form shown in Fig. 10.5a. The coercive field E_c is determined in our case by the nature and concentration of pinning centers rather than by values of the coefficients of the thermodynamic potential, as was in the case corresponding to Fig. 3.7, where we were dealing with a homogeneously polarized ideal crystal. If the detachment of the wall from the pinning centers occurs by portions, i.e., the wall is bent in a field lower than E_c, the $P(E)$ plot will assume the form depicted in Fig. 10.5b. Naturally, in a real crystal the domain wall detached from some pinning centers may appear to be fixed on others, so that jumps will arise on the $P(E)$ plot; in short, the switching of the polarization of a ferroelectric is much more complicated.

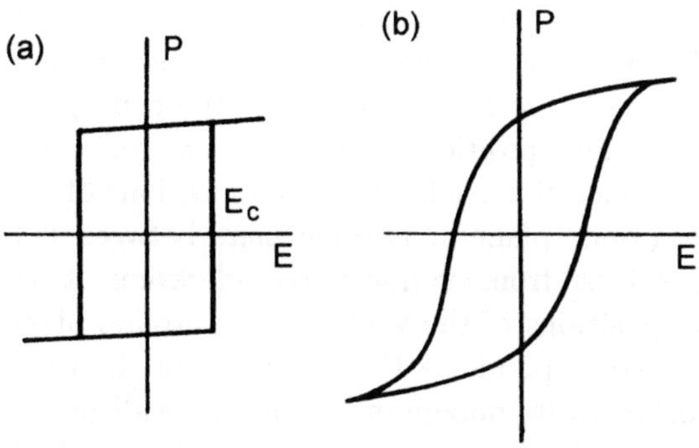

Fig. 10.5a,b. The $P(E)$ plot for a ferroelectric with pinning centers of the domain wall: (**a**) the domain wall is detached as a whole from pinning centers; (**b**) various portions of the wall are detached from pinning centers at different fields.

It should be noted that by representing the pinning center as a certain spring, which is disrupted upon application of a rather large force, we a priori simplified the situation. It would be more realistic to describe the pinning center as a certain potential well with barriers at its edges (Fig. 10.6). The spring disruption corresponds to the over-barrier motion of the wall, but in a field as low as desired (in the case of a crystal as long as desired, it is energetically favorable for the domain wall to cross over the barrier, which can be done as a result of thermal fluctuation). In brief, if it is given a rather long period of time, the domain wall will detach from the pinning centers upon application of as low a field as possible, i.e., for a strictly static field the dependence $P(E)$ will still have the form depicted in Fig. 10.4 rather than that given in Fig. 10.5. This is evidence that the shape of the hysteresis loop may substantially depend in a nontrivial way on the time during which the crystal is subjected to an electric field.

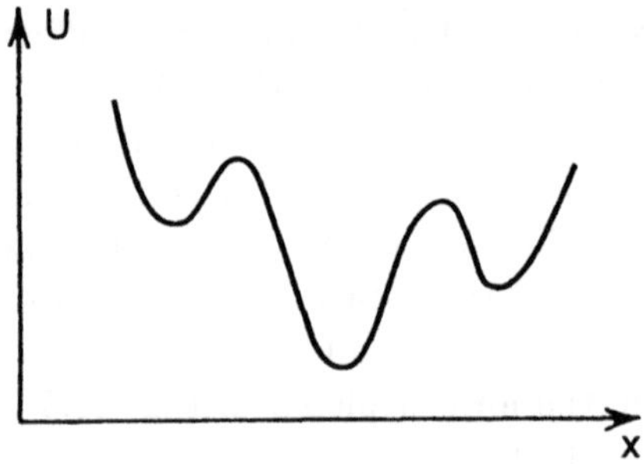

Fig. 10.6. Schematic representation of the potential relief for a domain wall created by pinning centers.

10.5 Motion of the Domain Wall: Account of the Discreteness of the Crystal

We will again consider an ideal crystal, but abandoning the continuous-medium approximation, we will take into account the discreteness ("atomic form") of its structure. To begin with, we have to abandon the concept of the energetic equivalence of the various positions of the domain wall. As a matter of fact, as is clear without special consideration, the positions of the wall when its center lies in the atomic plane or is somewhere between two planes will have different energies. Even from symmetry considerations it becomes clear that one of the two positions of the wall – in the atomic plane or midway between the planes – corresponds to the minimum of the thermodynamic potential and the other to its maximum. Thus, the wall has to overcome potential barriers by moving between the successive positions which are separated from one another at an interplane distance. This situation is quite analogous to that which occurs during the motion of a dislocation in an ideal crystal.[1]

It is obvious that the wider the domain wall, the smaller the difference in energy between the positions when the wall center lies in atomic planes and when it passes between them. Therefore, for wide walls (near T_c) or, for example, in the case of weak anisotropy for a two-component order parameter (see Sect. 10.3), the continuous-medium approximation is quite applicable. However, when the width of the boundary is comparable to the lattice constant, the discreteness of the crystal must be taken into account.

Before discussing the motion of the wall in a discrete crystalline medium, let us consider the equilibrium state of the wall. At $T = 0$, when there is no thermal motion (we stay within the scope of classical concepts; quantum effects are, as a rule, unimportant in real situations), the wall is in one of the "potential valleys" formed by the atomic relief and is absolutely planar. With $T \neq 0$ it remains planar only on average, undergoing thermal fluctuations. Apart from small fluctuations, under which the wall resides within one "valley", there occur fluctuations which transfer some portions into neighboring

[1] The analog of the barrier mentioned here is known in the theory of dislocations as the Peierls barrier.

"valleys". A snapshot would show that the wall is covered by ridges separated by steps from the remaining wall surface. The minimal height of the steps is equal to the distance between the valleys. On the whole, it may be said that the wall tends to "run away" from the valley. Note that the "escape" can be recorded without viewing the wall so as to see the interatomic distances. Even with a "small magnification" it is possible to see that the wall deflects at some sites from the crystallographic plane, i.e., the wall is no longer planar; it is partially irregular, rough.

Let us clarify what is meant by roughness. Let us pick out in the wall a limited portion – a circle of radius R and clamp it along the circumference. The wall center will naturally be displaced because of thermal fluctuations. The larger the radius R, the more freely it will move. If with a value of R as large as desired the amplitude of its movement remains finite, the wall practically remains planar, on average (strictly speaking, it is partly rough). But if the movement amplitude tends to infinity with an unlimited increase of R, i.e., if the degree of freedom is limited only by pinning along the perimeter of the portion, the wall is called rough (more exactly, completely rough). Thus, at low temperatures the wall is planar and at higher temperatures it is rough, i.e., there occurs a kind of phase transition, which is called a roughening transition (the planar wall is quite analogous to the crystal face). However, we will not discuss this rather interesting transition since no experimental data on it (concerning domain walls) are available at present; moreover this kind of transition does not occur on domain walls in ferroelectrics, which separate regions with different polarization directions.

The point is that the warping of such walls for practically all types of twists results in the appearance of bound charges (Fig. 10.7) and, hence, of an electric field. This is energetically unfavorable and it means that the domain walls in ferroelectrics (and in ferroelastics) display increased rigidity with respect to twists. As a result, these domain walls remain planar (in the sense indicated above) over the entire temperature range of existence of a nonsymmetric phase and taking account of the discreteness of the crystal structure is very important.

We will now consider the motion of a planar domain wall, assuming that neither electric nor elastic fields arise when the wall is twisted. Qualitatively, the results of our treatment will also be applicable to walls in ferroelectrics and ferroelastics. So, suppose that there is an infinite wall which separates

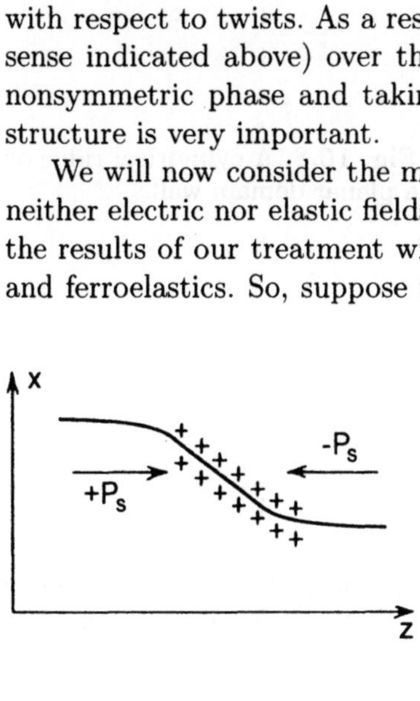

Fig. 10.7. Formation of a bound charge when the domain wall is bent in a ferroelectric.

two domains with opposite signs of η_0, and a low homogeneous field h is applied to the crystal, so that a direct transfer of the wall from one valley into another is impossible. We will call a fluctuation that corresponds to an increase (decrease) of the volume of a favorable domain a protuberance (a cavity). At $h = 0$ the probability of formation of a protuberance and a cavity is the same. However, with $h \neq 0$ a protuberance becomes energetically more favorable than a cavity. Moreover, a rather extended protuberance (in contrast to a cavity), once it is formed, will grow rather than diminish (as in the case of equilibrium conditions).

Let us consider a round protuberance of radius r and height equal to the distance between the potential valleys (Fig. 10.8). The energy gain upon formation of a protuberance is equal to $2h\eta_0 a\pi r^2$ ($a\pi r^2$ is the volume of the protuberance), and the energy loss is caused by an increase in the surface energy of the wall by an amount of $\varepsilon_{\text{sur}} 2\pi r a$ ($2\pi r a$ is the lateral surface of the protuberance). Thus, the energy of the protuberance is given by

$$U(r) = -2h\eta_0 a\pi r^2 + \varepsilon_{\text{sur}} 2\pi a r. \tag{10.32}$$

The graph of this function is given in Fig. 10.9. It has a maximum at $r = r^* = \varepsilon_{\text{sur}}/(2h\eta_0)$. From Fig. 10.9 it can be seen that it is favorable for the protuberance with $r > r^*$ to grow in order to reduce its energy. The quantity $U(r) = \pi a\varepsilon_{\text{sur}}^2/(2h\eta_0)$ plays the role of the potential barrier to the formation of growing protuberances. (We repeat here the reasoning that one uses in the discussion, say, of the formation of liquid drops in an supercooled vapor or of crystal nuclei on its growing face.) The extension of the protuberance occurs due to the movement of the step in the plane of the domain wall.

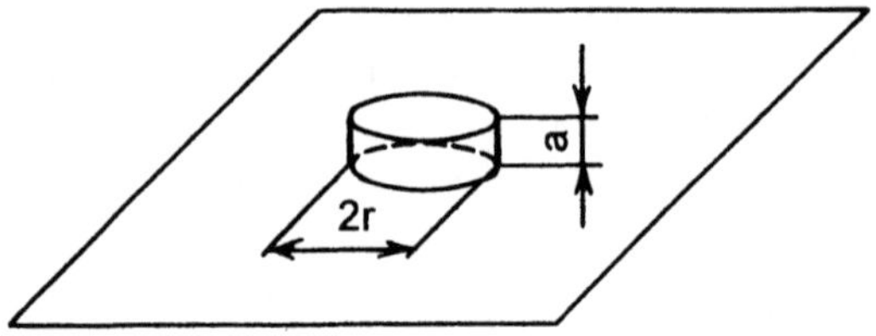

Fig. 10.8. A cylindrical ridge on a planar domain wall.

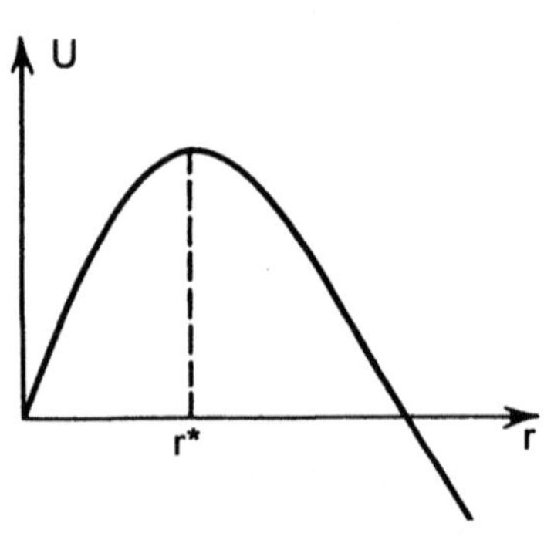

Fig. 10.9. Dependence of the energy of the ridge on the radius of its base.

Of course, during such movement the step "feels" the discreteness of the crystal structure in the same way as a dislocation "feels" the Peierls barrier. However, without going into the discussion of this problem, we note that for unidimensional objects (such as a step) thermal fluctuations are more important than for two-dimensional ones (such as a wall), which is why a step is usually "rough". Besides, the corresponding barriers are usually lower than in the case of domain walls, so that the expansion of the protuberance occurs rather easily. We will thus assume that the domain wall surface "seems" to be continuous to the step. The motion of the step can then be described by an equation analogous to (10.28). The only difference is that in the derivation of (10.28) we considered an infinite planar wall and in this case we deal with a strip with a width equal to the interatomic ("intervalley") distance a. Thus, for the rate of increase of the protuberance radius we have

$$v_{\mathrm{st}} = \mu_{\mathrm{st}} h \qquad (10.33)$$

where μ_{st} is the mobility of the step. In order to estimate the number of protuberances with $r > r^*$ that arise during time t on an initially planar wall, we split the wall surface into portions of size r^*. Recalling our reasoning used in the derivation of formula (9.30), we take into account that the rate of variation of the number of critical protuberances is proportional to $\exp[-U(r^*)/(k_{\mathrm{B}}T)]$, i.e., during time t on the unit surface of the wall there are formed

$$N(t) = \frac{b}{r^{*2}} t \exp[-U(r^*)/(k_{\mathrm{B}}T)] \qquad (10.34)$$

protuberances (the coefficient b has the same dimensions as ν in formulas (9.32) and (9.33)). The mean distance between protuberances is $N^{-1/2}$. The protuberances coalesce when this distance becomes equal to the protuberance radius, i.e., at $t = t_1$, which is determined according to the formula

$$v_c t_1 = N^{-1/2}(t_1) \qquad (10.35)$$

$$t_1 = \frac{r^{*2/3} \exp[-U(r^*)/(3k_{\mathrm{B}}T)]}{v_{\mathrm{st}}^{2/3} b^{1/3}} . \qquad (10.36)$$

During this time the domain wall moves as a whole to a distance a. So, the wall velocity is given by

$$v = \frac{a}{t_1} = a \left(\mu_{\mathrm{st}} \eta_0\right)^{2/3} \varepsilon_{\mathrm{sur}}^{-2/3} h^{4/3} b^{1/3} \exp\left(\frac{\pi a \varepsilon_{\mathrm{sur}}^2}{2h\eta_0}\right) . \qquad (10.37)$$

From formula (10.37) we see that the wall velocity in the case under consideration (at small h) depends on h very slightly in the sense that the derivatives $\mathrm{d}^n v/\mathrm{d}h^n$ for any n vanish as $h \to 0$. The $v(h)$ curve thus "lies" on the h axis near the origin. This sharply contrasts with formula (10.28) for the case where the crystal is treated as a continuous medium. Recall that formula (10.37) was derived for small h. At large h our picture of the layer-by-layer growth of a "favorable" domain proves invalid: new protuberances will arise on the existing ones before the latter fill the layer. At still larger h the domain walls will begin to pass from valley to valley, i.e., the character of motion of the wall will be the same as in a continuous medium. Recall again that near T_c , where the domain walls are wide, the barriers separating the valleys are low and formula (10.37) is no longer applicable for low fields h.

10.6 Domain Walls and Defects

We will now consider certain interactions that cause the pinning of domain walls. It should first be noted that we have not taken into account the deformation of crystals so far in this chapter. To do this, it is necessary (just as was done in Chap. 5) that the thermodynamic potential contain terms with the components of the stress (or strain) tensor and (10.2) be supplemented with the conditions of elastic equilibrium:

$$\frac{\partial \sigma_{ik}}{\partial x_k} = 0. \tag{10.38}$$

The differential equations that make up the system (10.38) will contain the space derivatives u_{ik} and η since σ_{ik} depend not only on u_{ik} but also on η (see Chap. 5). We will neither give here nor seek for the solution of the systems of four equations (10.2) and (10.38). We note only that, as follows from this solution, in the wall region (in a layer of thickness r_c) there exist elastic stresses.

These stresses are comparable to stresses that would arise in a clamped crystal undergoing a phase transition. They may be tensile or compressive, depending on the signs and ratios of the coefficients for terms of the type $\eta^2\sigma_{ik}$ in the thermodynamic potential and elastic moduli. Suppose that in the wall region, where there exist tensile stresses, there is introduced a defect which "expands" the crystal (for example, one of the crystal ions is replaced by an ion of a larger radius); such a defect is called a *dilatation center.* Tensile stresses perform here positive work: the crystal energy is reduced. Thus, it is favorable for the wall to be located so that the "expanding" defect is within its thickness, which is what implies that the wall is pinned to the defect.

There is possible another, bulk mechanism of pinning of the domain wall. Suppose that the crystal incorporates defects that are said to cause a local ordering, i.e., defects that create in their neighborhood a nonzero value of the order parameter even in the symmetric phase (Fig. 10.10). As can be seen from Fig. 10.10, such a defect may be in two states corresponding to different signs of the local value of the order parameter. In the symmetric phase the two states possess the same energy; therefore, if the crystal was grown in the symmetric phase, the orientation of defects is chaotic, i.e., in each portion of the crystal there is approximately the same number of defects with both signs of η_{loc}. We assume that the defects are "frozen", i.e., they cannot be

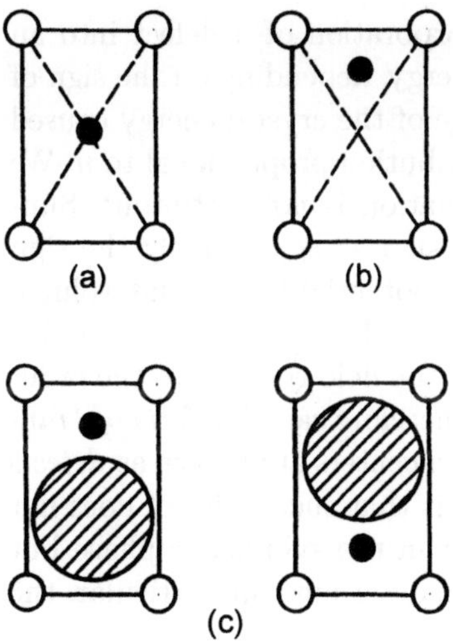

Fig. 10.10. (**a**) A defect which causes a local ordering: the unit cell of an ideal crystal in the symmetric ($T > T_c$) and (**b**) in the nonsymmetric ($T < T_c$) phase; (**c**) a unit cell containing a defect ($T > T_c$).

reoriented (an example of such a defect is the molecule of α–L–alanine in the TGS crystal). The energetically most favorable state in the nonsymmetric phase is a state when the sign of η_{loc} coincides with the sign of η_0 in a defect-containing domain. Therefore, if there is a domain wall in the crystal, it tends to be arranged in such a way that each defect is in a "respective" domain. Of course, this is not realized in full measure. Nevertheless, while being warped (it is not favorable for the wall to be strongly warped because its area and, hence, its surface energy will increase), the wall finds a certain, relatively favorable (corresponding to the minimal energy) position. In other words, it appears to be pinned. In order to discuss the effect of the pinning of the wall by defects on the character of the domain structure of a crystal, we should first answer the following question: Will the single-domain state correspond to the equilibrium (the minimum of the thermodynamic potential) after defects are introduced into the crystal as well? For defects of the type just considered the answer to this question is far from being obvious. Indeed, each defect "would like" to be in a certain domain, and since these defects are distributed and oriented in a completely random manner (let us assume so), in the system the tendency towards the formation of a multidomain state is clearly observed. The factor that prevents the separation into domains is an increase of the crystal energy due to the surface energy of the walls. A rigorous treatment of what is to be regarded as the energetically most favorable state in a crystal with defects is too complicated and goes far beyond the scope of this book. But a qualitatively correct answer is provided even by simple considerations.

Suppose we have an ideal single-domain crystal (which is naturally in a nonsymmetric phase). We introduce defects into it and arrange them randomly, ignoring the fact that there are two types of defects. A defect with the same sign of η_{loc} as that of the order parameter lowers the energy of the system and a defect with the opposite sign increases the energy. The same may be expressed in a different way: the incorporation of a defect into an ideal crystal increases or reduces the crystal energy, depending on the sign of η in the given crystal. It means that the change of the crystal energy caused by the introduction of a defect contains a contribution proportional to η. We denote this contribution by $-h_1\eta$. This designation is not fortuitous. Suppose that we introduce defects with the same sign as $\eta_{\rm loc}$, namely the sign coinciding with that of η. The thermodynamic potential in the unit volume will change by an amount $-h_1 n\eta$, where n is the defect concentration. We thus see that the quantity $h_1 n$ plays the role of the field coupled to the order parameter. Defects of the type in question are therefore called *defects of "random field" type.* (Defects of the type of dilatation centers are known as *defects of "random temperature" type*: the implication is that such a defect does not lead to the appearance of the order parameter in the symmetric phase but locally changes the constants of the crystal that determine, in particular, the transition temperature.)

Let us return to the crystal into which defects were introduced. Remember that initially the crystal was in the nonsymmetric phase. Suppose that the defects introduced by us are "weak", i.e., the defect with the sign of h_1 opposite to that of η placed at a given point lowers to a certain extent the value of the order parameter but does not cause a local change of the signs of η. However, if defects with the same sign of η gather together by chance, they can, in principle, create a field h high enough for the nucleation of a new domain.

Two situations are possible: (a) the crystal as a whole remains monodomain, i.e., though there are some domains of opposite sign, their volume is much less than that of the "main" domain; (b) the crystal is split into domains in such a way that the volume-averaged value of η is equal to zero, i.e., defects will eliminate the qualitative difference between the nonsymmetric and the symmetric phase; they eliminate ("smear") the phase transition.

In order to find out which of these situations is actually realizable, suppose that the crystal is split into domains of approximately the same size, which is assumed to be equal to L. Let us estimate the energy of such a crystal. As pointed out above, though the total number of defects with different signs of h_1 is the same, the field h associated with these defects does not vanish at all because of the fluctuations in the number of particles. It may be expected that at those points where the field h is positive there will be located domains with the positive sign of η. The characteristic magnitude of fluctuations of the field in a L^3 volume can be estimated using the formula for fluctuations of the number of particles in an ideal gas $\langle(\Delta N)^2\rangle = \langle N\rangle$; we find $h_{\rm ch} \approx (nL^3)^{1/2}h_1$. Thus, since the sign of h coincides with that of η the energy gain will be equal (per unit volume) to $(nL^3)^{1/2}h_1/L^3 = n^{1/2}h_1L^{3/2}$. The surface energy of domain walls per unit volume is calculated in a simple way: $\varepsilon_{\rm sur}L^2/L^3 = \varepsilon_{\rm sur}/L$. So, after defects are incorporated, the crystal will be split into domains with an average size L and the energy of the unit volume of the crystal will change by an amount

$$U(L) = -\frac{h_1 n^{1/2}\eta_0}{L^{3/2}} + \frac{\varepsilon_{\rm sur}}{L}. \tag{10.39}$$

The graph of this function is shown in Fig. 10.11. The function has a maximum at the point $L = L^*$. With an accuracy of up to a numerical coefficient (it may be neglected in our approximate treatment)

$$L^* = \frac{n h_1^2 \eta_0^2}{\varepsilon_{\rm sur}}. \tag{10.40}$$

It can be seen from this formula that at a low concentration and sufficient "weakness" of defects (small value of h_1) the size L^* may have any small value. But the domain size L cannot be smaller than the width of the domain wall r_c (as a matter of fact, it is $4r_c$ but we agreed to ignore the numerical coefficients). In other words, it is reasonable to consider the $U(L)$ graph only for $L > r_c$. If L^* does not enter into this ("physical") region, it will mean that over the entire range of L values the function $U(L)$ decreases with increasing L, i.e., the larger the average domain size, the lower the crystal energy. Thus, in any case with a rather low concentration of defects n (with their low "strength" h_1) the energetically most favorable state is the monodomain state of the crystal.

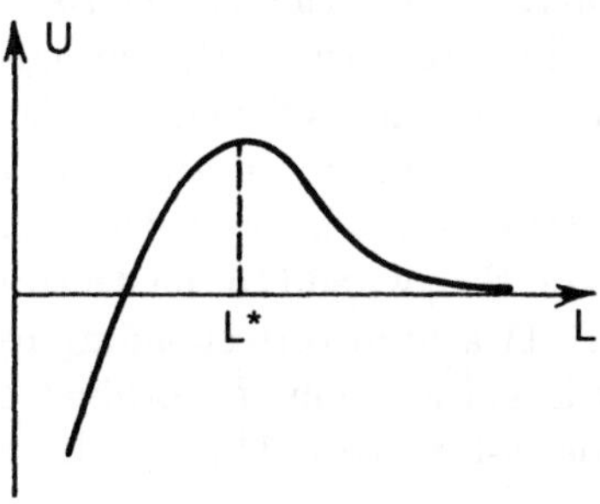

Fig. 10.11. A change in the energy of a defective crystal associated with its separation into domains with a characteristic size L.

Knowing the temperature dependence of η_0^2 and $\varepsilon_{\rm sur}$, we find that $L^* \propto (T_c - T)^{-1}$. At the same time $r_c \propto (T_c - T)^{-1/2}$. Thus, at a certain temperature $T < T_c$ the size L^* appears to be in the physical region. At a somewhat higher temperature that part of the $U(L)$ curve to which there corresponds $U(L) < 0$ will also be found in the physical region. We come to the conclusion that before T_c is reached the crystal is split into domains with a size of the order of the domain wall width. As a matter of fact, this is a symmetric phase. Hence, chaotic defects of the type of the random field lead to a decrease of the transition temperature, though they do not eliminate the phase transition into the nonsymmetric phase.

We have so far mentally introduced defects into a crystal in a nonsymmetric phase. However, in reality the crystallization temperature is usually significantly higher than the transition temperature, i.e., actually defects are commonly incorporated into a crystal in the symmetric phase. We thus have a crystal with randomly oriented defects of the type of the random field after this crystal has been cooled to a temperature corresponding to the nonsymmetric phase. Suppose that this cooling occurred in a nonequilibrium way, i.e., very rapidly (quenching). In the crystal immediately after cooling there are a large number of small domains since each defect "has chosen" the sign of the order parameter in the symmetric phase. But for the crystal, as we have already seen, it is energetically more favorable to be in the monodomain state and therefore the domain will start to enlarge. However, for this to be done the domain walls have to move througout the crystal with defects and during such movement the wall has to overcome barriers, i.e., the enlargement of domains more often than not occurs very slowly. We see that the crystal with defects of the "random field" type represents in the nonsymmetric phase a system which can relax to equilibrium for a very long time.

10.7 Defects in the Symmetrical Phase

Defects of the "random field" type exert an influence on the properties of the symmetrical phase as well. Indeed, each defect of this type is a sort of nucleus of the nonsymmetric phase inside the symmetric phase. The radius of this nucleus is approximately equal to the correlation radius r_c – the characteristic length of the falloff of the order parameter. Thus, the size of the nucleus rapidly increases upon approach to the Curie point. As long as the nucleus size is much smaller than the spacing between the defects, their interaction may be neglected. Let us find, for example, the contribution of a single defect of this type to the heat capacity of the crystal. To do this, we add to the density of the thermodynamic potential in formula (10.1) a term corresponding to the field: $-h_1\delta(r)\eta$ (we assume that the defect is at the origin of coordinates and that the field density is described by the delta-function). Thus, we have

$$\varphi(\eta) = \frac{\alpha(T-T_c)}{2}\eta^2 + \frac{\beta}{4}\eta^4 + \frac{\delta}{2}(\nabla\eta)^2 - h_1\delta(r)\eta. \tag{10.41}$$

The Euler equation now becomes:

$$\alpha(T-T_c)\eta + \beta\eta^3 - \delta\nabla^2\eta = h_1\delta(r). \tag{10.42}$$

Since we are dealing with the symmetric phase, we ignore the term $\beta\eta^3$ in this equation. The solution of the remaining equation for the one-dimensional case is already known:

$$\eta = C\exp(-x/r_c). \tag{10.43}$$

At $T = T_c$, i.e., as $r_c \to \infty$, $\eta = C$. In the three-dimensional case the solution of the linearized equation (10.42) at $T = T_c$, i.e., the equation

$$\nabla^2\eta = h_1\delta(r)/\delta \tag{10.44}$$

can be found easily if we recognize that it is in fact Poisson's equation for the electrostatic potential φ of a point charge e:

$$\nabla^2\varphi = -4\pi\rho(r) \equiv -4\pi e\delta(r). \tag{10.45}$$

The potential of a point charge is known to be e/r, i.e., what is a constant in the one-dimensional case is a function in the three-dimensional case:

$$\frac{h_1}{4\pi\delta}\frac{1}{r}. \tag{10.46}$$

The solution of the linearized equation (10.42) therefore has the following form:

$$\eta(r) = \frac{h_1}{4\pi\delta}\frac{1}{r}\exp\left(-\frac{r}{r_c}\right). \tag{10.47}$$

This solution is, of course, approximate since we have neglected the term $\beta\eta^3$. However, we can find a correction for our solution; to this end, we are to transpose the term $\beta\eta^3$ to the right-hand side of (10.42), where η is given by formula (10.47), and to solve the resultant equation. Of course, this solution will not be as simple as (10.47), but if a transformation is made to the Fourier components of the corresponding function (see Chap. 6), it can be done. We leave this as an exercise to the reader.

Let us now calculate the contribution of a defect to the thermodynamic potential of the crystal. To do this, we have to substitute the solution (10.47) into expression (10.41) and to integrate over the crystal volume. Neglecting the term $\beta\eta^4/4$ for the moment, we find for the temperature-dependent contribution of a single defect to the potential:

$$\Delta\Phi_1 = h_1^2/(8\pi\delta r_c). \tag{10.48}$$

The contribution of defects to the thermodynamic potential of the unit volume is therefore given by

$$\Delta\Phi = nh_1^2/(8\pi\delta r_c). \tag{10.49}$$

Using (10.49), we find the contribution of defects to the heat capacity:

$$\Delta C = \frac{nh_1^2\alpha^{1/2}}{32\pi\delta^{3/2}(T - T_c)^{3/2}}. \tag{10.50}$$

We see that the contribution of defects to heat capacity increases as $T \to T_c$. Recall that the contribution of fluctuations of the order parameter increases as well but by a different law, less steeply, than in the case of defects. Thus, if in an experiment the heat capacity of the symmetric phase rapidly increases upon approach to the Curie point (not predicted by the Landau theory), it may imply either the significant role played by critical fluctuations or the presence of defects in the crystal.

It becomes clear from the derivation of formula (10.50) that it is approximate since formula (10.48) for an isolated defect is approximate; moreover, the procedure of addition of the contributions of various defects (transformation from (10.48) to (10.49)) is also approximate since (10.42) is nonlinear, i.e., $\eta(r)$, the distribution of the order parameter over the crystal, cannot be, strictly speaking, represented as the sum of expressions of the type (10.47). It is clear that far from the Curie point, where the "regions of influence" of various defects practically do not overlap, this circumstance is not too important, but near T_c, when the value of r_c is high, it becomes a crucial factor and formula (10.50) is no longer valid. Nonetheless, the qualitative inference drawn from it that the contribution of defects to anomalies of thermodynamic quantities is reminiscent of the contribution of fluctuations of the order parameter is beyond any doubt, at least for a temperature region not too close to T_c. The history of the experimental study of anomalies of various quantities near the phase transition points features many cases when anomalies that do not conform to the Landau theory were initially accounted for by the effect of fluctuations in the order parameter, but at a later time it turned out that these anomalies were appreciably less strong in a more perfect crystal. One should be very careful with the interpretation of experimental data of this kind.

10.8 Domains in Proper Ferroelectrics

We will now be concerned with the formation of domains in a proper ferroelectric. At first we will limit ourselves to the case of a maximally anisotropic uniaxial ferroelectric for which $P = (0, 0, P_z)$. The process of formation of a domain structure in a ferroelectric is different from the nonferroelectric case considered above. Indeed, if the conductivity of a crystal (which is very low more often than not) is neglected, one may conclude that the formation of domains in a ferroelectric under certain conditions is inevitable even for an ideal crystal which passes over into a ferroelectrie phase by way of a second-order transition and as a result of uniform infinitely slow cooling.

Suppose a crystal has the form of a plate, whose plane is perpendicular to the direction of spontaneous polarization. We will be interested in an electrodeless crystal, i.e., we are to take special account of the field created by spontaneous polarization. If the polarization is homogeneous throughout the crystal, the density of the thermodynamic potential has the form

$$\varphi(\eta) = \frac{A}{2}\eta^2 + \frac{\beta}{4}\eta^4 - \frac{1}{2}E_z p_z\,, \tag{10.51}$$

where the last term takes account of the effect of a depolarizing field produced by spontaneous polarization (see (6.29)). For a homogeneously polarized plate $E_z = -4\pi p_z$. Substitution of this expression for the field into (10.51) yields

$$\varphi(\eta) = \frac{1}{2}(A + 4\pi a)\eta^2 + \frac{\beta}{4}\eta^4. \tag{10.52}$$

Thus, the coefficient of η^2 does not vanish at $T = T_c$; the principal condition for a phase transition is not fulfilled. In other words, in an unshorted (and nonconducting) crystal of the given form the phase transition into a homogeneously polarized state is impossible (a reservation concerning the form of the crystal is made because for a crystal having the form of a long needle with the polarization directed along the needle axis the depolarizing field is close to zero; here we are dealing with a crystal in the form of a plate). What has been said does not mean, however, that the phase transition is not possible at all; now we are to probe the possibility of a transition into an inhomogeneous, multidomain state.

Our purpose is to find such an inhomogeneous distribution of polarization when the electric field associated with it is at a minimum. In this case the contribution to the thermodynamic potential connected with this field will also be minimal. If we were dealing with an infinite crystal, the answer would be quite clear. Indeed, as was shown in Chap. 6, the polarization distribution of the form

$$P_z(\boldsymbol{r}) = P_z(\boldsymbol{k})\mathrm{e}^{\mathrm{i}\boldsymbol{k}\cdot\boldsymbol{r}} + P_z(-\boldsymbol{k})\mathrm{e}^{-\mathrm{i}\boldsymbol{k}\cdot\boldsymbol{r}}, \quad P_x = P_y = 0 \tag{10.53}$$

does not produce an electric field if the vector $\boldsymbol{k}$ is perpendicular to the z axis. And since the continuous function specified at a finite interval may be represented as the sum of sines and cosines (a Fourier series), we come to the conclusion that any polarization distribution (which does not contain a homogeneous component) is such that $P_z = P_z(x, y)$ does not create an electric field in an infinite crystal.

However, for a finite crystal the situation is quite different. Consider a distribution $P_z = P_z(x, y)$, i.e., let us assume that the domain walls are parallel to the z axis. The domains themselves will be assumed to be columns having at the base a square with side l (the column height is naturally equal to the plate thickness, which is assumed here to be much larger than l (Fig. 10.12)). It is clear that such a polarization distribution does produce an electric field since there is now a bound surface charge (a charged net) on the plate surfaces. The character of the field created by such a net can be understood from simple considerations. For points located near the net at a distance much smaller than the size of the unit cell l the corresponding unit cell is a practically infinite charged plane, i.e., the field at these points is close

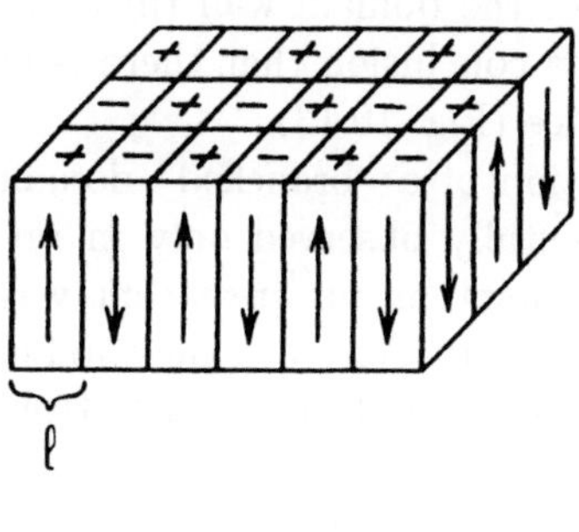

Fig. 10.12. A hypothetical version of the domain structure.

to $2\pi P$. On the other hand, at a distance from the net much larger than the unit-cell size the net represents a practically neutral object, i.e., the field is very low. We may thus assume approximately that near each surface of the slab in a layer having a thickness of the order of l there will exist an electric field of the order of $2\pi P$.

An essential point is that in each domain the field direction is opposite to the polarization direction. How will this field affect the polarization? As can be seen from Fig. 3.7, the field, whose direction is opposite to that of the polarization, will not change the direction of polarization but will only reduce it if the magnitude of this field does not exceed the value corresponding to the point B in Fig. 3.7. It is not difficult to calculate: the latter is proportional to AP, where P is the polarization corresponding to the point B. But this field is a priori lower than $2\pi P$ (since A vanishes at $T = T_c$). We have thus encountered a contradiction and have to conclude that our tentative polarization distribution cannot be realized. In order to understand how this distribution can be corrected, we apply the following reasoning. Suppose we have "imposed" upon a crystal the distribution given in Fig. 10.12, forcing the ions to shift to those positions which correspond to this distribution and fixing them in these positions. We will now gradually set the ions free and watch the crystal choose a polarization distribution. We begin this operation in the bulk of the slab at a distance L from the surface and then move to the surface. Since in the bulk of the crystal with $L \gg l$ the field is close to zero, the "release" of the ions will not lead to a change in their positions, i.e., to a change in polarization. But upon approach to the surface the electric field builds up and, finally, it attains a value corresponding to the point B in Fig. 3.7. Naturally, it occurs at first at a certain point lying at the domain axis, i.e., perpendicular to the slab surface through the center of the unit cell.

Then, there occurs a switching of polarization, i.e., there appears a domain with an opposite polarization direction (Fig. 10.13). The field will decrease to a certain extent since the effective size of the unit cell is reduced. Upon further approach to the surface the field attains a critical value and again the polarization is switched or, as is sometimes said, there appears a new branching of the domain structure, which does not completely solve the problem with the field either, and so on. To put it differently, upon approach to the surface the domain structure has to be expected to experience an infinite branching. As a matter of fact, we are speaking, of course, of a reduction of the size of domains to values of the order of the domain wall thickness, but one way or another we have arrived at the conclusion that there is an extremely tiny domain structure near the surface (Fig. 10.13).

However, the tendency towards the formation of a "branched" domain structure near the crystal surface is experimentally observed only in rare cases. This is quite natural, in fact. The key point in our treatment was, indeed, the assumption of the absence of conductivity. Actually, the conductivity is always different from zero; even a low conductivity leads to a strong

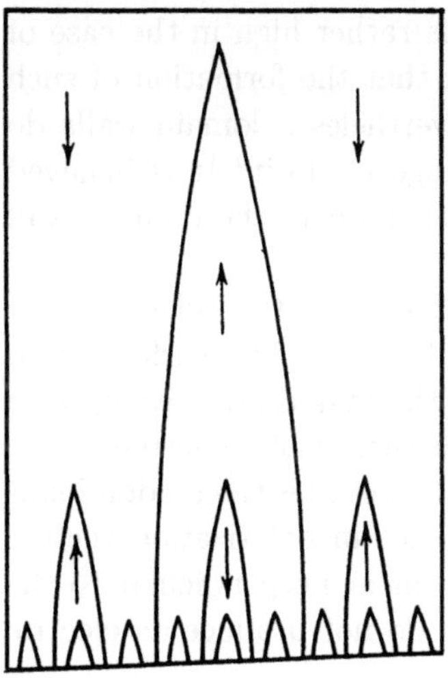

Fig. 10.13. Branching out of a domain structure (the arrows indicate the directions of polarization in domains).

shielding of the electric field due to the recharging of the electron centers coupled to impurity ions and intrinsic defects of the crystal. As a result, the picture of the domain structure becomes significantly changed as compared with the hypothetical case of a nonconducting crystal. If there exist shielding charges near the crystal surface, then, in particular, a structure of the type depicted in Fig. 10.12 (cf. Fig. 10.13) becomes a possibility. If the electric field were shielded completely, the problem of domains in a ferroelectric would have reduced entirely to an analogous problem in a nonferroelectric, i.e., we would have to recognize that, because of the positive energy of the domain wall, there are no domains in an ideal crystal and that their appearance is associated with the real structure and prehistory of the specimen. This is true in many respects, but not completely, since the region of the shielding charge has a finite width (usually, of the order of fractions of a micrometer) and there still exists a field inside this region. Moreover, in order to set up an equilibrium shielding, a long period of time will be needed. So, on the whole, the effects associated with the electric field that arise because of the spatial inhomogeneity of polarization are very important in ferroelectrics. It has been pointed out, in particular, that, because of these effects, the domain wall in a ferroelectric remains smooth over the entire temperature range of existence of the ferroelectric phase.

Consider, as another example, the mechanism of the motion of the domain wall in an ideal uniaxial ferroelectric. It is clear that, in contrast to the situation discussed in Sect. 10.5, the appearance of a circular ridge is no longer favorable, since bound charges are formed on the lateral surface. To reduce the field created by these charges, the base of the ridge should be extended along the axis of spontaneous polarization. However, with such an extended form the specific contribution of the surface energy of the lateral surface of the ridge will be much greater than in the case of a nonferroelectric.

In other words, the energy of a crystal nucleus is rather high in the case of a ferroelectric; estimates show that it is so high that the formation of such nuclei practically becomes hardly probable. Nevertheless, domain walls do move in a ferroelectric. Bot how can they manage to do it? It is believed that the appearance of ridges occurs at those sites where the domain wall intersects the sample surface.

Indeed, near the surface the electric field created by the ridge may be compensated due to the redistribution of the charge on the surface itself (whose conductivity is usually much higher than the bulk conductivity) or in electrodes which are usually applied directly to the surface of the ferroelectric.

Still another effect of an electric field of bound charges is the predominant orientation of domain walls such that the difference in polarization vectors between the adjacent domains contains no component perpendicular to the wall. On such walls there is no charge, i.e., they do not produce an electric field (Fig. 10.14).

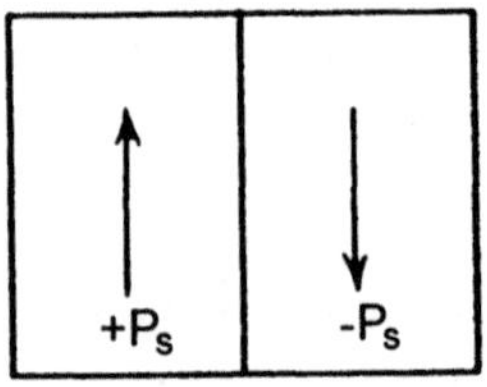

Fig. 10.14. An uncharged domain wall in a ferroelectric.

At first sight it may be said that the domain wall energy strongly depends on the angle between the normal to the wall and the polarization difference vector in domains. This statement, however, would be incorrect. Indeed, let us visualize a domain wall for which the angle mentioned is different from 90°. This wall is charged and, like any charged plane, it creates an electric field, which is practically independent of the distance to the wall, up to distances comparable with the wall size. The energy of this field is proportional to $S^{3/2}$ where S is the wall area and, hence, if the energy of this field is assumed to be a contribution to the surface energy, which is defined as the total wall energy divided by S, then we come to the conclusion that this energy itself is proportional to $S^{1/2}$ and, hence, is very high for large S. To speak of the surface energy in the ordinary sense of this term is meaningful if we are dealing with the directions of the normal to the wall, which are perpendicular to the vector of the domain polarization difference. The energy of the domain wall will be, generally speaking, different for different directions of the normal. In order to find this dependence, it is necessary to take into account in the density of the thermodynamic potential derivatives with respect to the coordinates in the corresponding plane and also the coupling of η to elastic strains. For example, for a uniaxial proper ferroelectric with the axis of spontaneous polarization directed along z ($\eta \propto P_z$), the gradient terms that have to be taken into account have the form $\sum_{i,k} \delta_{zzik}(\partial\eta/\partial x_i)(\partial\eta/\partial x_k)$, where i, $k = 1, 2$ (x and y). The coefficients δ_{zzik} constitute a fourth-rank tensor: a set of its nonzero components is determined by a point symmetry group of the symmetric phase.

What has been said above is sufficient to determine the equilibrium form of domains if the spontaneous deformation in adjacent domains is the same (otherwise, it will also affect the form of the domain; see below). We limit ourselves to the case of a slab of a uniaxial ferroelectric crystal with the polarization axis perpendicular to the slab plane. Suppose the crystal as a whole is polar, i.e., the greater part of its volume is occupied by regions with the same polarization and the lesser part is occupied by regions with polarization of opposite sign (reversed domains). Taking account of what has been said above, it is natural to expect that reversed domains will have the form of cylinders piercing the slab. Determination of the equilibrium form of the cylinder base requires a knowledge of the dependence of the domain wall energy on the direction of the normal to the plane perpendicular to spontaneous polarization. In experiment the situation is different, as a rule, from the ideal situation. For example, there may be encountered factors that will prevent the growth of domains through the slab. This tendency, however, manifests itself in the cigar-shaped form of domains extended along the polar axis (Fig. 10.15).

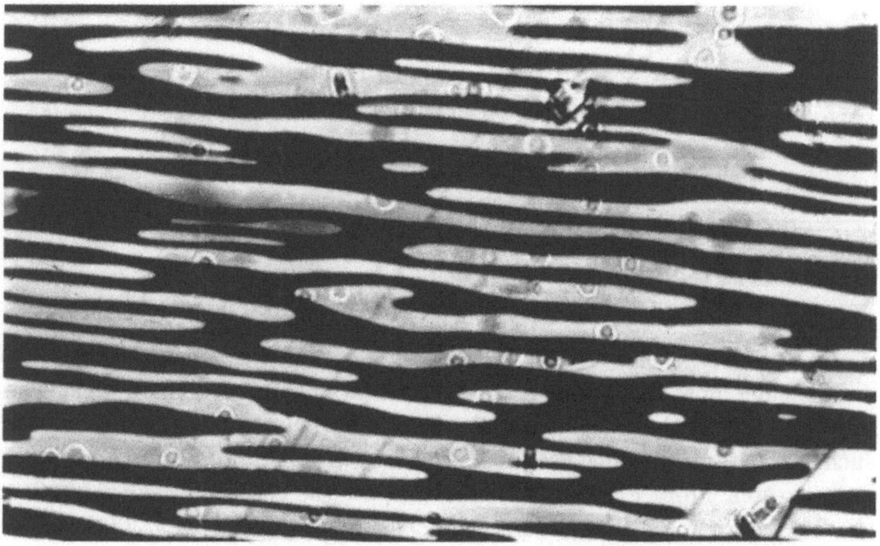

Fig. 10.15. A picture of the domain structure of the TGS crystal (the cross-section is perpendicular to the direction of P_s) [13].

10.9 Domains in Ferroelastics

As pointed out in Chap. 5, the phase transition in a clamped proper ferroelastic may occur only into a multidomain state. The factor responsible for the separation of a crystal into domains is here quite analogous to that observed for an unshorted ferroelectric; in the latter case it was necessary to diminish the electric field created by spontaneous polarization and here the elastic stresses caused by spontaneous polarization have to be reduced.

Let us explain the foregoing by an illustration. Suppose $\eta \propto u_{12}$, just as in Fig. 5.5. Upon phase transition in a free crystal its form undergoes a change; one of the possibilities is illustrated in Fig. 10.16. When the crystal is clamped, such a change is impossible; instead, the structure shown in Fig. 10.17 may be formed. Of course, stresses arise in such a structure too, but only near the surface. Just as in the case of a ferroelectric, the problem of the character of the structure near the surface is complicated; here again the conclusion as to its branching is inevitable. We will not discuss this problem in detail, especially as experiments with clamped crystals are the exception than the rule.

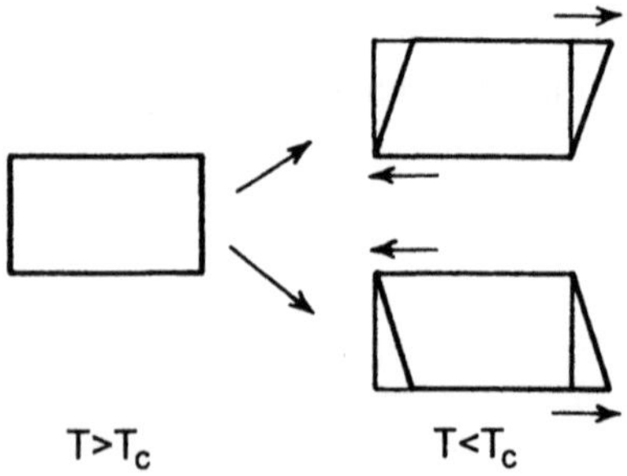

Fig. 10.16. A change of the form of a crystal in a ferroelectric phase transition.

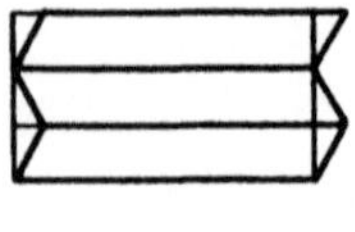

Fig. 10.17. The domain structure in a clamped ferroelectric.

Consider now the problem of the orientation of the domain wall. The approach here is the same as in the case of a ferroelectric: we have to choose such orientations at which the existence of a wall would not lead to additional bulk stresses. From Fig. 5.5 it follows that for the case under consideration this occurs for walls that are perpendicular to the axis x or y. The wall in such cases "sews" two domains together without "pulling the material being sewn". This does not refer, however, to walls located at other angles to these axes. Figure 10.18 shows a plane (dashed line), which we would like to treat as a boundary between domains. Let us mentally cut the crystal throughout

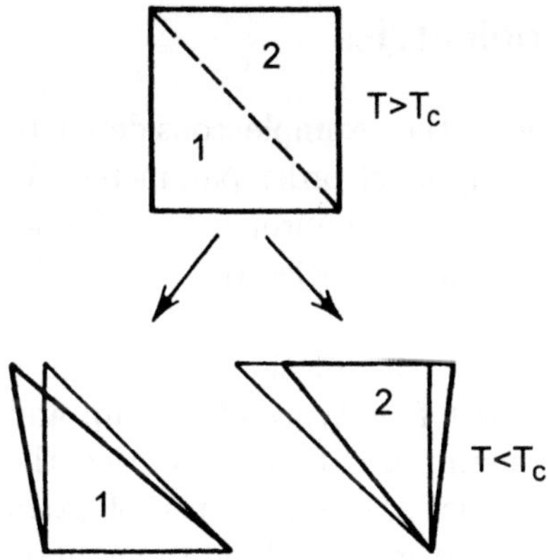

Fig. 10.18. For the determination of the energetic favorableness of an arbitrarily oriented domain wall in a ferroelectric.

this plane and transfer it into a nonsymmetric phase so that the signs of η are different for different parts of the crystal. It can be seen that for part 1 of the crystal the boundary area is reduced and for part 2 it increases and we can no longer "sew" these two parts together without deforming them. The same conclusion can easily be made with respect to a boundary perpendicular to the x axis.

Thus, in the case under consideration the wall may be perpendicular only to axis x or y. So, we should expect the existence of planar walls and layered domains. This conclusion is confirmed by experiment (Fig. 10.19).

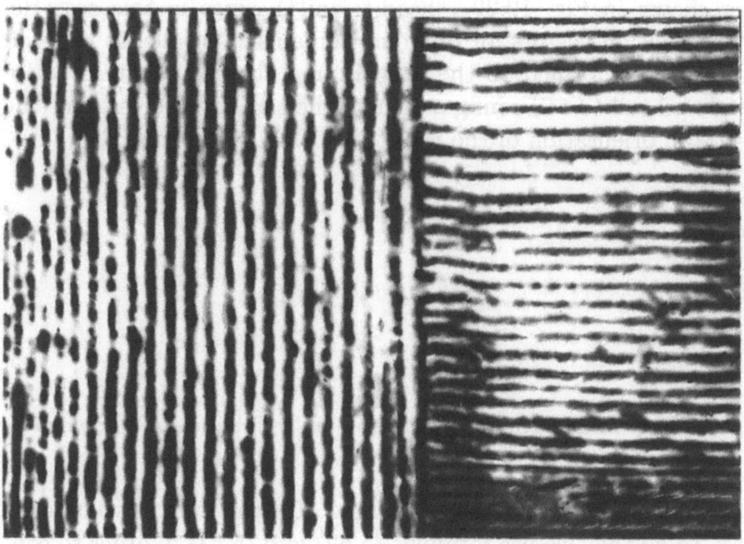

Fig. 10.19. A photograph of the domain structure of the KH_2PO_4 crystal (the section is perpendicular to the direction of P_s) [14].

10.10 Domains in Polyaxial Ferroelectrics

We begin the treatment of this problem with the example considered in Chap. 5: a proper ferroelectric with a two-component order parameter. As pointed out in Chap. 5, such a ferroelectric is also an improper ferroelastic. So, it is not accidental that the section on polyaxial ferroelectrics follows the section on ferroelastics.

Let us restrict ourselves to the phase defined by the solutions (5.52). Since the number of such solutions is four, there exist four types of domain and it would seem that there are six types of domain boundaries. However, the requirement that the boundary should not cause the appearance of extra electric or elastic fields reduces their number and imposes limitations upon their orientation. We take into account that the equilibrium values of the components of the order parameter are proportional to the components of the polarization vector and so in what follows we will use P_x and P_y instead of η_1 and η_2. The angle between the polarization vector in adjacent domains may be equal to 180° or 90°. Accordingly, we have 180°- and 90°-domain walls. It is evident that the properties of the 180°-wall are the same as those of the domain wall of the uniaxial ferroelectric considered in Section 10.8, which is not a ferroelastic. In particular, elastic fields do not arise with any orientation of the wall since the strains in both domains are the same and proportional to the square of the order parameter; electric fields do not appear either if the wall is parallel to the spontaneous polarization axis. In the case of 90°-domain walls no electric field appears when the domain wall is arranged in the manner shown in Fig. 10.20 ("head-to-tail"), since in this case the difference in polarization vectors for different domains is parallel to the wall and no bound charge is formed on the wall. Bound charges inevitably arise in head-to-head and tail-to-tail configurations. The same is also valid for the general head-to-tail orientation of the wall (Fig. 10.21).

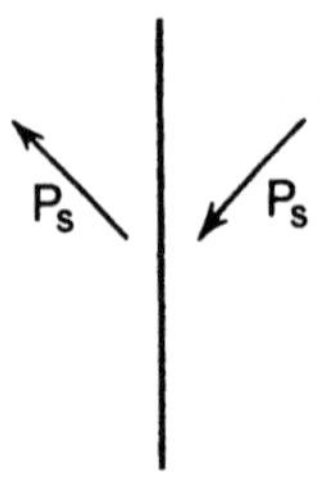

Fig. 10.20. An uncharged domain wall between 90°-domains.

We are now to check whether the uncharged 90°-wall creates volumetric elastic stresses. To do this, we use formulas (5.52) and (5.53) in order to consider strains in two neighboring 90°-domains. It is easy to see that in these domains the spontaneous dilatation $u_{11}^0 + u_{22}^0$ is the same (a two-dimensional analog of volume strain) and the strains $u_{11}^0 + u_{22}^0$ have opposite signs

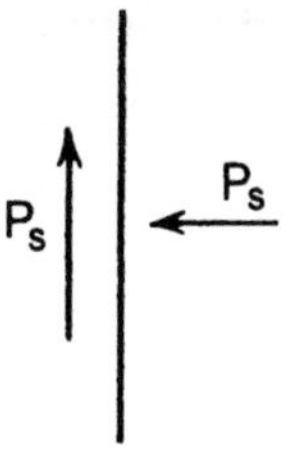

Fig. 10.21. A charged domain wall between 90°-domains.

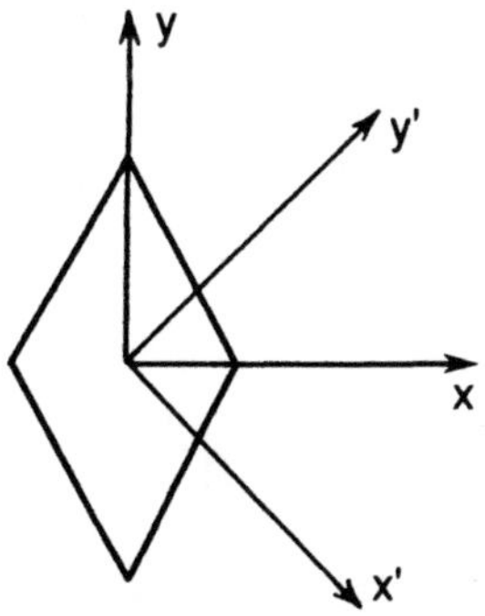

Fig. 10.22. Explanation of the equivalence of the strains $u_{xx} - u_{yy}$ (at $u_{xx} + u_{yy} = 0$) to shear strain.

(Fig. 10.22); the strain $u_{11} + u_{22}$ is the shear strain $u_{1'2'}$, in the system of coordinates x', y' rotated by 45° relative to the original one. Thus, with respect to strains the 90° boundary is quite analogous to the boundary between ferroelastic domains considered in Sect. 10.9. We see that with the "uncharged orientation" of the 90° boundary no additional elastic fields arise.

So, we have established the type and orientations of those domain boundaries in a ferroelectric with a two-component order parameter which do not create additional electric and elastic fields.

It will now be easy to consider the case of a ferroelectric with a three-component order parameter. It is obvious that the 90° boundaries here will be planar because the restrictions imposed upon ferroelectric bounadries are valid for them and the 90° boundaries may be nonplanar. Figure 10.23 shows, as an illustration, a photograph of the domain structure of the $BaTiO_3$ crystal and the orientation of spontaneous polarization at various points of the specimen.

The presence of the domain structure is responsible for most of the important properties of ferroelectrics. As a rule, these properties are tied up in one way or another with the specific dependence of polarization on the electric field. In cases where the polarization is homogeneous throughout the specimen, the dependence $P(E)$ is given by (3.1) and (3.3); it is shown as a curve in Fig. 3.7.

As pointed out in Chap. 3, this dependence leads to the presence of dielectric hysteresis. At first glance it would seem that the coercive field of the corresponding hysteresis loops must be close to the field corresponding to the point B in Fig. 3.7. However, as has been shown by experiment, the actually

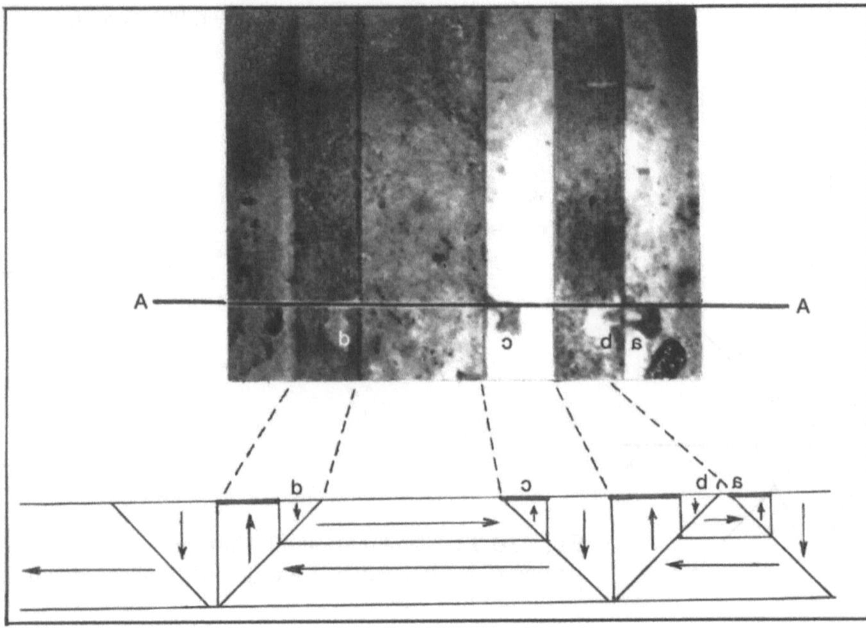

Fig. 10.23. The domain structure of the $BaTiO_3$ crystal revealed by the polarization-optic technique [15].

observed coercive fields are lower by several orders of magnitude. This is due to the fact that the switching of polarization occurs not as a change of polarization homogeneous throughout the specimen but as a result of a change in the volume of existing domains because of the motion of domain walls and nucleation of new domains.

The changes in the domain structure caused by heating and by the action of external stresses make more often than not a certain contribution to the pyro- and piezoelectric coefficients of a ferroelectric, respectively. However, the treatment of such restructuring processes and also of the problems associated with the nucleation of new domains and the motion of domain boundaries (the latter topic has been touched upon only partly here) would take us far beyond the scope of this book.

11. Ferroelectrics with an Incommensurate Phase

11.1 Phase Transitions into an Incommensurate Phase

In a series of ferroelectrics ($NaNO_2$, $(NH_4)_2BeF_4$, etc.) the transition into the polar phase is preceded by a transition into a rather peculiar intermediate phase called the incommensurate phase, i.e., when the the temperature falls off, there is observed the following sequence of phases: symmetric/incommensurate/polar (commensurate) (Fig. 11.1).

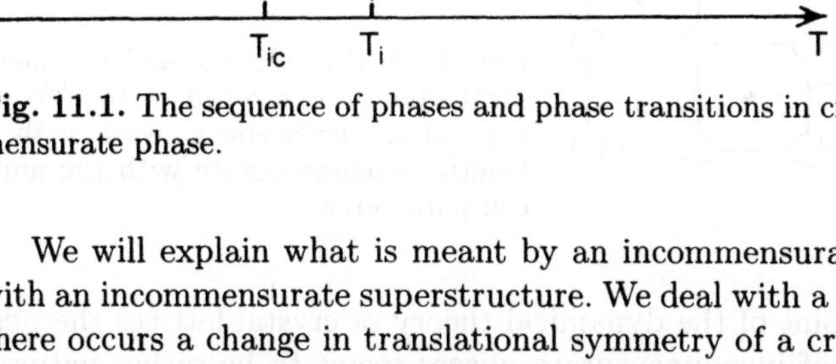

Fig. 11.1. The sequence of phases and phase transitions in crystals with an incommensurate phase.

We will explain what is meant by an incommensurate phase or a phase with an incommensurate superstructure. We deal with a superstructure when there occurs a change in translational symmetry of a crystal during a phase transition, i.e., an increase of the number of atoms in the unit cell. In Chaps. 1 and 5 we considered examples of such transitions associated with the doubling of the unit-cell volume. Figure 11.2 reproduces the model of Fig. 1.3; where the dashed lines indicate that in this particular case we may speak of a frozen wave of displacements with a period twice as large as the unit-cell parameter of the symmetric phase. This is tantamount to the formation of a superstructure with a period doubled with respect to the original one. It is evident that the parameter of such a transition may be assumed to be a normal coordinate corresponding to $k = \pi/a$ (see Chap. 10), and so the dashed lines in Fig. 11.2 exactly represent a wave with such a wave vector.

Suppose now that the frozen wave itself is in no way tied up with the principal lattice period. The ratio of the wavelength λ to the unit-cell parameter

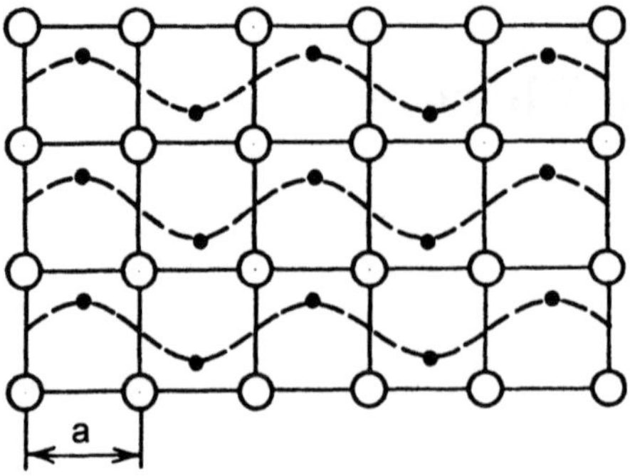

Fig. 11.2. Formation of a superstructure with a period twice as large as the main period as a result of the "freezing" of a displacement wave of length $\lambda = 2a$.

may be any number, even an irrational one. In such a case there is formed an incommensurate phase – a phase with a superstructure, whose period is incommensurate with the principal period (Fig. 11.3). The first characteristic feature of a crystal in the incommensurate phase is that it has no translational symmetry, but there is a strictly ordered arrangement of atoms, such that the electron density may be represented as the sum of two independent periodic functions.

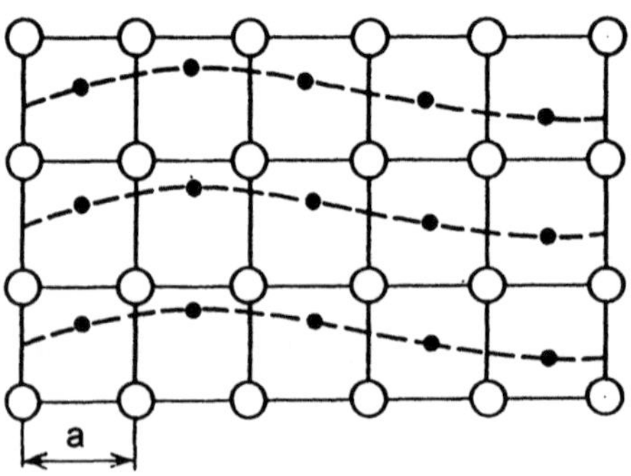

Fig. 11.3. Formation of an incommensurate phase as a result of the "freezing" of a displacement wave with a length incommensurate with the unit-cell parameter.

From the standpoint of the dynamical theory of crystal lattices the possibility of formation of incommensurate phases seems to be rather natural. This may be explained by considering displacive phase transitions. While examining transitions involving no changes in translational symmetry (see Chap. 10), we established that the soft mode corresponds to the wave vector $k = 0$, since the minimum of the soft optic branch corresponds to the center of the Brillouin zone. On the other hand, the doubling of the number of atoms in the unit cell occurs when the minimum of the soft optic branch corresponds to the edge of the Brillouin zone. What is the factor that enables one to find the minimum of the dependence $\omega^2(\boldsymbol{k})$ at this or that point? In a general form in the dynamical theory of crystal lattices it is only possible to prove that in the case of a nondegenerate optic branch in the center and at the boundary of the Brillouin zone there are extrema in the $\omega^2(\boldsymbol{k})$ plot.

Whether these extrema are maxima or minima and also whether there are other extrema in the $\omega^2(\boldsymbol{k})$ plot is determined by the specific form of the interaction of atoms in the crystal. The existence of mimima at arbitrary points of the Brillouin zone is possible though it does not follow from symmetry considerations. There is a strong possibility that one of these minima will prove to be absolute, i.e., the lowest one, and the point $k = 0$ and the boundaries of the Brillouin zone will correspond to the maxima of the function $\omega^2(\boldsymbol{k})$ (Fig. 11.4). Then, the frequency which vanishes at the phase transition temperature will correspond to a certain nonspecified point in k-space and it is exactly to this point that there will correspond the frozen displacement wave that arises upon transition, i.e., an incommensurate superstructure, which is incommensurate by virtue of this nonspecificity.

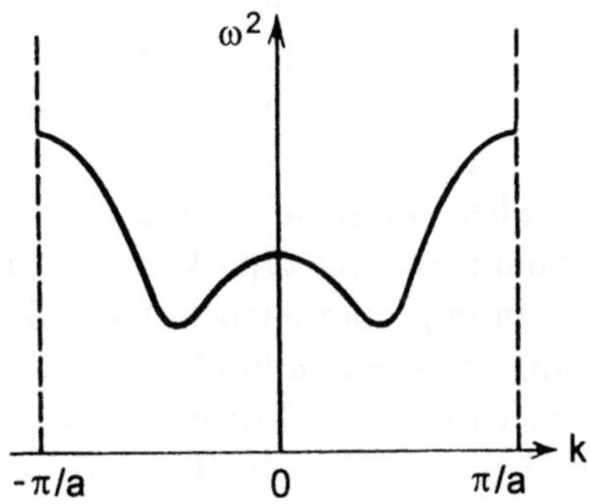

Fig. 11.4. Dependence of the frequencies of the soft optic branch on the wave vector near the temperature of a phase transition into an incommensurate phase.

Since in the case under discussion the position of the minimum of the function $\omega^2(\boldsymbol{k})$ is not due to symmetry, it is natural to expect that the point representing the wave vector of the resultant superstructure will be shifted with a change in temperature within the Brillouin zone of the main structure, i.e., the $\boldsymbol{k}$ vector will be changed. It is clear that the change of $\boldsymbol{k}$ with respect to the reciprocal-lattice parameter is essential. It may be expected that eventually the vector $\boldsymbol{k}$ will arrive at a "commensurate" point (of the type $k = (m/n)(\pi/a)$, where m and n are small integers) where it will stay. There occurs a phase transition: an incommensurate phase $\rightarrow$ a commensurate phase. If the latter phase is polar, we speak of a ferroelectric with an incommensurate phase. The latter exists in a certain temperature range above the point of the ferroelectric phase transition; this has already been mentioned above. The phase transition "an incommensurate phase $\rightarrow$ a ferroelectric phase" has a number of specific features as compared with ordinary (proper and improper) ferroelectric transitions. This topic will be discussed below.

In the case under consideration the formation of an incommensurate phase at a temperature higher than the ferroelectric transition temperature occurred by chance, i.e., did not follow from symmetry considerations. Such a variant is realized, for example, in $NaNO_2$ crystals. Of great interest is also the case where the second-order phase transition directly from the symmetric into the polar phase is forbidden, and the formation of an incommensurate phase can be predicted on the basis of symmetry considerations.

We have just discussed the case of the nondegenerate branch of vibrations where the $\omega^2(\boldsymbol{k})$ curve has necessarily an extremum at least along one of the directions in k-space. However, if the soft vibration branch is degenerate, i.e., for a single frequency corresponding to the point $k = 0$ or to the boundary of the Brillouin zone there are several normal coordinates,[1] the situation may change in a cardinal way. As will be shown in Sect. 11.5, with certain transformation properties of the components of the order parameter for the transition from the symmetric into the polar phase (in the presence of the Lifshitz invariant in the thermodynamic potential) we can draw quite definite conclusions concerning the form of the $\omega^2(\boldsymbol{k})$ curves near the degeneracy point. To be more specific, this form must be analogous to that shown in Fig. 11.5, i.e., the symmetric point (degeneracy point) is approached by two branches with slopes equal in magnitude but opposite in sign. If based on the symmetry analysis we found out that the order parameter for the transition from the symmetric into the polar phase must be a normal coordinate corresponding to the edge of the Brillouin zone (of type $k = \pi/a$, in Fig. 11.5), it means that we are dealing with an improper ferroelectric. The softening of the normal coordinate corresponding to $k = \pi/a$ implies that the point A in Fig. 11.5 will descend upon approach to the transition point to the abscissa; it is clear that when the point A is very close to (but has not yet reached!) the abscissa, the $\omega^2(\boldsymbol{k})$ curve will touch the abscissa at another point, namely $k \neq \pi/a$. The location of this point is determined by the constants of the crystal and any position is possible, i.e., the phase formed will be incommensurate.

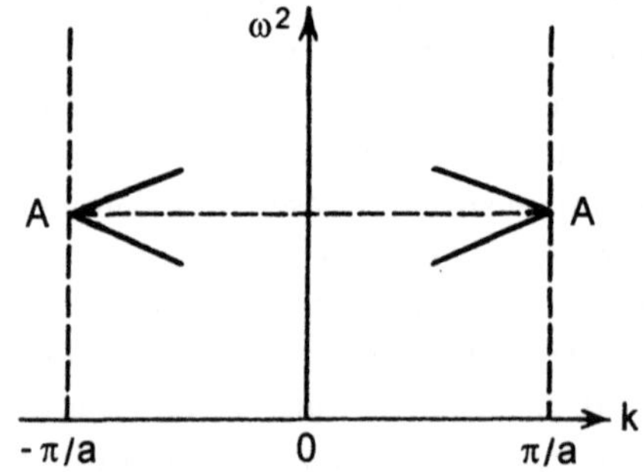

Fig. 11.5. A possible form of dispersion curve near the points of degeneracy A (two-component order parameter) for the case where the crystal symmetry allows for the Lifshitz invariant in the thermodynamic potential.

[1] Such a situation is realized when the order parameter is a multicomponent quantity.

11.2 Phenomenological Theory of Phase Transition into an Incommensurate Phase

We will now consider the phenomenological theory of phase transitions into the incommensurate phase. The question that naturally arises here is: What quantity may be taken in this case as the phase transition parameter? In the context of the dynamical theory of crystal lattices the answer seemse obvious: the parameter in question is the normal coordinate corresponding to the point $k = k_0$, at which the $\omega^2(\boldsymbol{k})$ curve will touch the k axis at the temperature of the transition into the incommensurate phase ($T = T_i$). (More precisely, according to Landau theory, we should say that it is a variable that has the same transformation properties as the normal coordinate mentioned above; however, for the sake of brevity, we will use terminology which is not quite adequate.)

The efficiency of Landau theory consists in the following: knowing only the transformation properties of the order parameter, one can draw a number of conclusions with regard to the properties of the system near the phase transition. In our treatment we will restrict ourselves to the case where the situation depicted in Figs. 11.4 and 11.5 occurs only for one of the coordinates of the crystal (x). It means that the symmetry elements of the crystal do not transform one axis into another; we will assume, for example, that the crystal is rhombic. As can be seen from Figs. 11.4 and 11.5, two points on the k axis are "soft", i.e., we are dealing with two components of the order parameter: with normal coordinates for the points $\boldsymbol{k} = \boldsymbol{k}_0$ and $\boldsymbol{k} = -\boldsymbol{k}_0$ ($Q(\boldsymbol{k}_0)$ and $Q(-\boldsymbol{k}_0)$).

As follows from Chap. 9, an example of the normal coordinate corresponding to the $\boldsymbol{k}$ vector may be the quantity $\eta_{\boldsymbol{k}}$. We use this example in order to establish the transformation properties of the normal coordinate. From (6.3) it follows that

$$\eta_{\boldsymbol{k}} = \frac{1}{v}\int \eta(\boldsymbol{r})\mathrm{e}^{\mathrm{i}\boldsymbol{k}\cdot\boldsymbol{r}}\mathrm{d}\boldsymbol{r}. \tag{11.1}$$

Let us carry out one of the symmetry transformations – the translation by a unit length a along the axis in question, i.e., the following replacement is made in (11.1): $\boldsymbol{r} \to \boldsymbol{r} + \boldsymbol{a}$. Since the function $\eta(\boldsymbol{r})$ is specified only at lattice points which coincide upon translation, i.e., $\eta(\boldsymbol{r}+\boldsymbol{a}) = \eta(\boldsymbol{r})$, it follows that upon translation $\eta_{\boldsymbol{k}}$ is multiplied by $\mathrm{e}^{\mathrm{i}\boldsymbol{k}\cdot\boldsymbol{a}}$. One can also see that if a symmetry operation transforms $\boldsymbol{k}$ into $-\boldsymbol{k}$, then $\eta_{\boldsymbol{k}}$ is transformed into a complex-conjugate guantity $\eta^*_{\boldsymbol{k}}$. This reasoning is quite sufficient to find the form of the thermodynamic potential as a function of $Q(\boldsymbol{k}_0)$ and $Q(-\boldsymbol{k}_0)$ (for brevity, these quantities will be denoted here by Q and Q^*).

Taking account of the requirement of translational invariance, we see that only one second-order invariant is possible; we write it here as $(1/2)AQQ^*$. As regards higher-order terms, one has, strictly speaking, to distinguish cases where $k_0a/(2\pi)$ is a rational or irrational number, i.e., when k_0 enters a "commensurate" or "incommensurate" point of the Brillouin zone. As will be shown below, the commensurability effects manifest themselves only at a certain distance from T_i, which is why we will assume that k_0 corresponds to the "incommensurate" point. In this case, only the powers of QQ^* are invariant with respect to translations, i.e., if we introduce the variables ρ and φ, then $Q = \rho e^{i\varphi}$, and we arrive at the conclusion that

$$\Phi(Q, Q^*) = \Phi(\rho^2) = \Phi_0 + \frac{1}{2}A\rho^2 + \frac{1}{4}\beta\rho^4, \tag{11.2}$$

where, as usual, $A = A'(T - T_i)$ and $\beta = \text{const}$.

Expression (11.2) refers to the spatially homogeneous case. Let us now take into account the possibility of the spatial inhomogeneity of the order parameter. Using the same reasoning as in Chap. 6, we conclude that the expression for the density of the thermodynamic potential must contain a term of the form

$$\frac{1}{2}\delta_{ik}\frac{\partial Q}{\partial x_i}\frac{\partial Q^*}{\partial x_k}. \tag{11.3}$$

Here the set of allowed components of the tensor δ_{ik} depends on the crystal symmetry. For what follows it is sufficient to restrict our consideration to the simplest situation where expression (11.3) reduces to

$$\frac{1}{2}\delta(\nabla Q)(\nabla Q^*) = \frac{1}{2}\delta[(\nabla\rho)^2 + \rho^2(\nabla\varphi)^2]. \tag{11.4}$$

Let us prove that in our case the symmetry allows one more invariant containing spatial derivatives of Q. Consider the expression

$$Q^*\frac{\partial Q}{\partial x} - Q\frac{\partial Q^*}{\partial x}. \tag{11.5}$$

Since the symmetry operation that transforms x into $-x$ also transforms Q into Q^*, expression (11.5) is invariant. It is known as the *Lifshitz invariant.* It is clear that the Lifshitz invariant is allowed for any pair of normal coordinates $Q(\boldsymbol{k})$ and $Q(-\boldsymbol{k})$, where $\boldsymbol{k}$ is located on the axis in question. But $\boldsymbol{k}_0$ is not any wave vector; it is a special one corresponding to the point of tangency – the minimum of the $\omega^2(\boldsymbol{k})$ curve. It means, as will be seen below, that for $\boldsymbol{k} = \boldsymbol{k}_0$ the coefficient of the Lifshitz invariant is equal to zero at $T = T_i$. It is natural that with $T \neq T_i$ the coefficient σ of the Lifshitz invariant is different from zero. Retaining, according to Landau theory, only those terms which are linear in $T - T_i$, we get

$$\sigma = \sigma'(T - T_i). \tag{11.6}$$

Passing over to the variables ρ and φ, we see that the Lifshitz invariant is written in the form

$$\sigma\rho^2\frac{\partial\varphi}{\partial x}. \tag{11.7}$$

So, the expression for the thermodynamic potential assumes the following form:

$$\Phi = \Phi_0 + \frac{1}{2}A\rho^2 + \frac{1}{4}\beta\rho^4 + \frac{1}{2}\delta(\nabla\rho)^2 + \frac{1}{2}\delta\rho^2(\nabla\varphi)^2 + \sigma\rho^2\frac{\partial\varphi}{\partial x}. \tag{11.8}$$

The presence in this expression of a term with the first derivative with respect to φ implies that with $T < T_i$ there is a state with a spatially inhomogeneous order parameter Q. Since expression (11.8) contains the first derivative of the phase φ alone, it is reasonable to assume that at equilibrium only φ (and not ρ) is changed in space. Let us show that the equations for the extremum of the thermodynamic potential, whose density is given by expression (11.8) (i.e., by the Euler equation), does have such a solution. For $\varphi(x)$ we obtain the equation

$$\frac{\mathrm{d}^2\varphi}{\mathrm{d}x^2} = 0, \tag{11.9}$$

the solution of which has the form

$$\varphi = qx + \varphi_0 \tag{11.10}$$

where q and φ_0 are certain constants. Substituting (11.10) into (11.8), we see that the φ-dependent part of the thermodynamic potential takes the following form:

$$\frac{1}{2}\delta\rho^2q^2 - \sigma\rho^2q. \tag{11.11}$$

To its minimum there corresponds

$$q = \frac{\sigma}{\delta} \propto (T - T_i). \tag{11.12}$$

Upon minimization with respect to ρ the contributions of φ-dependent terms may be neglected.

Indeed, as follows from (11.11) and (11.12), they are proportional to $(T - T_i)^2$ and, hence, if they are taken into account in the determination of the equilibrium value of ρ, the accuracy of Landau theory will be exceeded. In other words, for ρ to be determined we use a thermodynamic potential of the same form as in the case of the one-component order parameter.

The result obtained needs some comments. The spatial distribution of the order parameter found by us is as follows:

$$Q(x) = \bar{Q}(k_0)\mathrm{e}^{\mathrm{i}qx},$$

$$Q^*(x) = \bar{Q}^*(k_0)\mathrm{e}^{-\mathrm{i}qx},$$

where the bar indicates that the quantity is independent of x. Let us now carry out the translation and recall that with this transformation $Q(k_0)$ is multiplied by $\mathrm{e}^{\mathrm{i}k_0a}$. We arrive at the conclusion that $Q(x)$ is multiplied upon translation by $\exp[\mathrm{i}(k_0+q)a]$ and $Q^*(x)$ is multiplied by $\exp[-\mathrm{i}(k+q)a]$, i.e., $Q(x)$ and $Q^*(x)$ possess the translational properties of the normal coordinates corresponding, respectively, to $k_0 + q$ and $-k - q$. From this it becomes clear that if at $T = T_i$ the coefficient σ had not vanished, the role of the point $k = k_0$ would have been played by another point, i.e., the vanishing of σ at $T = T_i$ is the reflection of the fact that the point corresponds exactly to the minimum of $\omega^2(k)$ at $T = T_i$. The difference of the quantity q from zero at $T \neq T_i$ and its temperature dependence mean that the wave vector of the "frozen" displacement wave (see Fig. 11.3) varies with temperature; according to (11.12) it does so by a linear law (near T_i). This circumstance may also be treated correctly by stating that with $T < T_i$ the spatially inhomogeneous order-parameter distribution is realized by a sinusoidal law or by noting that as the temperature varies the role of the order parameter is now played by new normal coordinates (implying that the order parameter is homogeneous in space).

The amplitude of the frozen displacement wave varies with temperature in the same way as the equilibrium value of the order parameter during an ordinary phase transition.

Since the thermodynamic variables are "tied up" exactly with this amplitude, the anomalies of all thermodynamic quantities upon transition into an incommensurate phase exhibit (within the range of validity of the Landau theory) the same character as the anomalies of the same quantities in a nonferroelectric and a nonferroelastic transition with a one-component order parameter.

We will show now that none of the components of any material tensor arises upon transition of the normal into an incommensurate phase when, just as in our case, the order parameter has two components, i.e., when only one wave of any material tensor becomes "frozen". Indeed, the new component must be proportional to any combination of the order-parameter components and this combination must be translationally invariant since the material tensor does not change under translation. The only way to construct such combinations in our case is to use QQ^* and its powers. In doing so, we will obtain not only translational but also complete invariants on which there may depend only the components of tensors that are allowed by the symmetry of the symmetric phase. In particular, in the Landau thermodynamic potential an invariant of the type $P\rho^2$ is possible if P is a full-symmetric variable and only an invariant $P^2\rho^2$ if it is not. Thus, if the symmetric phase is nonpolar, then upon transition into an incommensurate phase the dielectric constant experiences a kink on the temperature dependence and if the phase is polar, it displays a jump. This is also valid for other material tensors.

From what has been said above it may seem that upon transition into an incommensurate phase the properties of the crystal change only slightly. As will be shown below, this is not actually the case since the incommensurate phase has rather specific properties. We have already encountered the main factor responsible for this specific form. We are speaking here of the specific feature of the solution obtained for the equilibrium value (or equilibrium distribution) of the order parameter, where the quantity φ_0 (see (11.10)) remains indeterminate, i.e., may assume any arbitrary value. The physical meaning of this circumstance can be easily understood if one takes into account that the change of φ_0 corresponds to the shift of the frozen displacement wave as a whole. Will the energy of the crystal be changed in this case? It can easily be seen that it will not.

Indeed, because of the incommensurability of the length of this wave with the principal lattice period (in Fig. 11.3 this period is determined by the disposition of atoms depicted by open circles) the magnitude of atomic displacements is changed in going from one unit cell to another in such a manner that this value is not repeated, i.e., in an infinite crystal we deal with an infinite set of displacement values. The same character will also be exhibited by the displacement picture for a shifted wave, i.e., the energy of interaction of atoms, which is determined by their relative arrangement, remains unaltered.

This is what accounts for the fact that the thermodynamic potential of the state of equilibriurn is independent of φ_0. More exactly, we should speak of equilibrium states because their number is indefinite (even countless). In this case the system is said to exhibit a *continuous degeneracy*.

It should be noted that any system with a phase transition of the ferroelectric type is degenerate: there are several (two in the simplest case) equilibrium values of the order parameter. But this degeneracy is discrete.

11.3 Specific Features of Crystal Lattice Vibrations in an Incommensurate Phase

It is often emphasized that the incommensurate phase exhibits the following specific feature: the presence of an additional acoustic vibrational branch. Before explaining the implication of this feature, recall that the pecularity of the acoustic branch is the vanishing of the function $\omega^2(k)$ at the value of the wave vector $k = 0$. For optical branches, however, the function $\omega^2(k)$ does not pass through zero (we do not consider the Curie point here). The vanishing of the frequency of the optical branch follows from the fact that with a simultaneous displacement of all atoms of the crystal by the same distance (just as in the case of the acoustic vibration corresponding to $k = 0$) the potential energy of the crystal is not changed since this energy is the interaction energy of atoms, the distances between which do not change upon uniform displacement. We have found in Sect. 11.2 that an analogous situation takes place during a displacement of the frozen wave such that it is not deformed. In other words, in this case too no displacing force originates and the frequency of the corresponding vibration is equal to zero. Since the shift of the dashed line in Fig. 11.3 implies a change of phase of the frozen wave, such a vibration is called a homogeneous *phason*. It is evident that if the displacement of the frozen wave is inhomogeneous, in which case it experiences a strain, the potential energy of the crystal is changed. With a slightly inhomogeneous strain this change will be insignificant, i.e., the frequencies of slightly inhomogeneous phasons or of phasons with small wave vectors are also low. A more detailed study shows that the dependence of the phason frequency on the wave vector is analogous to that for acoustic vibrations, i.e., $\omega \propto k$.

There exists, however, one significant difference between a phason and acoustic vibrations. They differ with respect to damping. From the considerations that were used in the analysis of the change of the potential energy upon displacements corresponding to acoustic vibrations at $k = 0$ it follows that the acoustic vibration energy does not undergo dissipation at $k = 0$ either. It can be shown that the damping constants for acoustic vibrations in the region of small k are proportional to k^2, i.e., the quality factor of acoustic normal vibrations (the ratio of frequency to the damping constant) goes to infinity as $k \to 0$. The situation is different for a homogeneous phason. Here, the homogeneous displacement of the frozen wave too involves the displacement of some atoms relative to others (for example, black atoms relative to white ones in Fig. 11.3), which is always associated with energy dissipation. So, the damping constant of a phason remains finite at $k = 0$ as well, i.e., the quality factor of the phason mode is equal to zero at $k = 0$.

What has been said shows that we may speak of lattice vibrations corresponding to a phason only conditionally. Vibrations for which the damping constants are larger than the natural frequency (so-called overdamped vibrations) are no longer vibrations; they are classified as motions of the relaxational type. Thus, phasons (long-wavelength at least) are thermal fluctuations in a crystal, which are of relaxational nature.

Another type of atomic displacement which does not change the homogeneity of the wave shown in Fig. 11.3 is the change of the amplitude of this wave. Such a vibration is referred to as the *amplitudon*. It is evident that a change of the amplitude of the frozen wave is always associated with a change of the potential energy, so that the frequency of a homogeneous amplitudon is quite finite. Moreover, the amplitudon is analogous to the soft mode for ordinary phase transitions since it is precisely the frozen wave amplitude that characterizes the value of the order parameter.

Let us now check the qualitative reasoning by means of direct calculations. To this end, we have first to write down the equations of motion for the order parameter (Q, Q^*). As a matter of fact, such an equation is already known to us (see (9.37)). For the purposes of this chapter it is convenient to write this equation in the form

$$\frac{\partial}{\partial t}\frac{\partial K}{\partial \dot{\eta}_i} + \frac{\delta \Phi}{\delta \eta_i} = -\frac{\partial R}{\partial \dot{\eta}_i}, \tag{11.13}$$

where K is the kinetic energy; $\delta\Phi/\delta\eta_i = \partial\Phi/\partial\eta_i - (d/dx)\partial\Phi/\partial(\partial\eta_i/\partial x)$; and R is the dissipative function (for (11.13) $R = (1/2)L\dot{\eta}_i^2$). The convenience of the form of (11.13) is that it can be used not only for cases where η_i is a linear combination of atomic displacements but also for any variables.

We will now express the kinetic energy and the dissipative function in terms of Q and Q^* (here it is convenient to assume that $Q = Q(k)$ rather than $Q = Q(k_0)$). Taking account of the requirement of translational invariance, we have

$$K = \frac{\tilde{m}}{2}\frac{\partial Q}{\partial t}\frac{\partial Q^*}{\partial t}, \tag{11.14}$$

$$R = \frac{L}{2}\frac{\partial Q}{\partial t}\frac{\partial Q^*}{\partial t}. \tag{11.15}$$

Passing over to the variables ρ and φ, we get.

$$K = \frac{\tilde{m}}{2}\dot{\rho}^2 + \frac{\tilde{m}}{2}\rho^2\dot{\varphi}^2, \tag{11.16}$$

$$R = \frac{L}{2}\dot{\rho}^2 + \frac{L}{2}\rho^2\dot{\varphi}^2. \tag{11.17}$$

From formulas (11.8) and (11.16)–(11.17) it can be seen that the variables ρ and φ are convenient because the equations for small oscillations of these quantities are independent, i.e., the quantities $\rho' = \rho - \rho_0$ and $\rho_0(\varphi - \varphi_0)$ are the normal coordinates of the crystal in the incommensurate phase. The equation for oscillations of ρ is as follows:

$$\tilde{m}\frac{\partial^2 \rho'}{\partial t^2} + L\frac{\partial \rho'}{\partial t} - 2A\rho' - \delta\nabla^2\rho' = 0. \tag{11.18}$$

If use is made of the Fourier components of $\rho(r)$, we will obtain an equation quite analogous to (9.37). We see that the frequency of homogeneous vibrations of amplitude ρ is

$$\omega_0^2 = -\frac{2A}{\tilde{m}} \propto (T_i - T)$$

which coincides with formula (9.15). The amplitudon is thus quite analogous to the soft mode for ordinary phase transitions.

The equation of motion for φ takes the form

$$\tilde{m}\frac{\partial^2 \varphi}{\partial t^2} + L\frac{\partial \varphi}{\partial t} - \delta\nabla^2\varphi = 0. \tag{11.19}$$

Assuming that $\varphi = \varphi_{\omega,\boldsymbol{q}}\exp(\mathrm{i}\omega t - \mathrm{i}\boldsymbol{q}\cdot\boldsymbol{r})$, we obtain

$$-\tilde{m}\omega^2 - \mathrm{i}L\omega - \delta q^2 = 0. \tag{11.20}$$

From this expression it can be seen that if we put $L = 0$, then

$$\omega = q\sqrt{\frac{\delta}{\tilde{m}}}, \tag{11.21}$$

that is, if there were no damping, then the same dispersion law would be valid for vibrations of the phase as in the case of acoustic vibrations. However, as can be seen from (11.20) the damping constant for phase vibrations is equal to $L/\tilde{m}$, i.e., is independent of q. This means that at rather small q when the phason frequency appears to be equal to or lower than the damping coefficient, this vibration will be overdamped.

11.4 The Incommensurate Phase in a Real Crystal

Here we will discuss other consequences of the continuous degeneracy of the incommensurate phase. One consequence of discrete degeneracy is already known to us. This is the possibility of formation of domain structure which arises exactly because an ideal crystal has several states of equilibrium. But what is an analog of the domain structure when the number of equilibrium states is infinite, i.e., in the case of continuous degeneracy?

Consider first a certain analog of the domain wall. Suppose we have a specimen in the form of a plate of thickness L and let us assume that a phase φ_0 ($\varphi(x = -L/2) = \varphi_0$) is given on one surface of the plate and a phase φ_1 ($\varphi(x = L/2) = \varphi_1$) on the other. How will the phase change throughout the plate thickness? The answer to this question is provided by (11.10). According to this equation, the phase varies linearly in x. The corresponding proportionality coefficient is determined not from the equation itself but from the minimum condition for the thermodynamic potential calculated with account taken of the solution obtained. Naturally, the minimum energy will correspond to the minimum rate of variation of the phase, i.e., in our case $d\varphi/dx = (\varphi_1 - \varphi_0)/L$. So, we found that the "domain wall" in the case of the incommensurate phase has a width equal to the specimen width. We may also say that the domain wall width is infinite since it is infinite specimens that are usually considered, without special reservations. We now calculate the surface energy of the wall in the same way as we did for an ordinary domain wall in Chap. 10, i.e., let us calculate the difference in thermodynamic potential between the inhomogeneous and the homogeneous state per unit surface (of a plate in our case). This difference will be equal to $(\delta/2)[\rho_0^2(\varphi_1 - \varphi_0)^2/L^2]L \propto 1/L$.

We see that as $L \to \infty$, then what corresponds to the surface energy of the wall vanishes.

Let us recall in addition another familiar characteristic of the domain structure in discretely degenerate systems: the number of walls of different types. This is equal to the number of different pairs of adjacent domains, i.e., it is determined by the number of energetically equivalent equilibrium states of an ideal crystal. In a continuously degenerate system this number is infinite, i.e., what is an analog of the number of walls of different types is equal to infinity.

Thus, the answer to the question of the domain structure of the incommensurate phase should have been as follows: it has an infinite multitude of domain walls of infinite width and infinitely low energy.[2] It is natural that under these conditions the very concept of the domain structure becomes meaningless. Whereas in a macroscopically inhomogeneous structure of a crystal in a phase with discrete degeneracy we can distinguish the regions corresponding to domains of different types, nothing of the kind may be said of a crystal in the incommensurate phase and we have to elucidate the function $\varphi(r)$, which is multiple-valued, since the phase has been determined to within $2\pi n$ (the amplitude ρ may also be spatially inhomogeneous, but since the correlation radius of the amplitude is a "normal" quantity, i.e., of the same order as the domain wall width in discretely degenerate systems, then while speaking of macroscopic inhomogeneities, we mean the phase φ alone).

Note also the analogy between this situation and the one which occurs in the theory of the mechanical properties of crystals with account taken of dislocations. The departure of the crystal structure from the ideal one may be described in that theory by specifying the displacement of each point relative to its position in the ideal structure – the function $u(\boldsymbol{r})$. This function is ambiguous since the displacement by the lattice constant transforms the structure into itself. This ambiguity is reflected in the existence of dislocations. Something like this must also be observed in the incommensurate phase. This topic cannot be discussed in more detail here, but we can state that the problems concerning the incommensurate phase are very complicated.

Let us compare once more the properties of discretely and continuously degenerate systems. Recall that whereas the transition into a discretely degenerate phase is nonferroelectric and nonferroelastic, the domain structure is formed because of the imperfection of the crystal or of its prehistory, i.e., because the crystal has not yet attained a state of equilibrium. Naturally, this also refers to what is considered to be an analog of the domain structure for the incommensurate phase – the distribution $\varphi(\boldsymbol{r})$ the only difference being that since the energy of the "domain wall" in the incommensurate phase is close to zero, these "walls" must be formed much more readily than the domain walls in ordinary phases. To put it differently, if the crystal is in the incommensurate phase, its boundaries, defects and internal stresses lead to much more significant (than is usually the case) distortions of its structure. This statement can be formulated in a more clear-cut manner. In Section 10.6 we have shown for a doubly degenerate system that the incorporation of a small concentration of randomly arranged and randomly oriented defects of the "random field" type does not lead to a splitting of the crystal into domains, i.e., to the virtual elimination of the nonsymmetric phase

[2] Recall that here we are dealing with an incommensurate phase, in which the frozen wave of displacements is possible only along one direction. If there are several such directions, it may happen that waves of different directions are "frozen" in different regions of the crystal. Such regions may, of course, be called domains of the incommensurate phase, but the meaning of the term will differ from that implied in our treatment.

(the destruction of long-range order). We will now apply analogous reasoning to a degenerate system. It should first be noted that the defect depicted in Fig. 10.10 is a defect of the random-field type with respect to the phase φ as well. Indeed, the change of the energy of the crystal, into which an impurity atom has been incorporated, depends on the position of the "black" atom. The more strongly displaced is the atom from the center, the easier is the incorporation of the impurity atom into the lattice, i.e., the more energetically favorable is the situation when at the point where the impurity atom is located the frozen wave has a maximum or a minimum. This means that the impurity atom fixes the phase of the frozen wave. If we denote the "desired" phase by φ, then the energy of the impurity site may be written in the form $-h_1\rho\cos(\varphi - \varphi_0)$. For the impurity atom located at another point, the "desired" phase will be, generally speaking, different, which is why the energy of the system of impurity atoms should be written as

$$\sum_i h_1\rho\cos[\varphi - \varphi_0(\boldsymbol{r}_i)].$$

If the impurity atoms are arranged randomly, the quantity $\varphi(\boldsymbol{r})$ will also assume random values.

Suppose now that we have reduced the value of h_1 to zero (that we have "turned off" the defects), as a result of which the incommensurate phase has become homogeneous. Now we will gradually increase h_1. If an inhomogeneity of the incommensurate phase does arise, it will be weak at first. Let us presume that over the length L the phase will change by an amount of the order of unity (from π up to 2π). We divide the crystal into regions of size L. Since the crystal tends to a minimal energy, in a given region the number of impurity atoms, for which $\cos(\varphi - \varphi_0)$ is positive, is larger than that of atoms for which this quantity is negative. The difference between the numbers of two types of defects is estimated in the same way as in Sect. 10.6, i.e., in order of magnitude it is equal to $(NL^3)^{1/2}$, where N is the concentration of defects. The energy gain (per unit volume) is estimated, because of the readjustment of the "frozen wave" to defects, as $h_1\rho(NL^3)^{1/2}/L^3 = h_1\rho N^{1/2}L^{-3/2}$. The loss of specific energy associated with the inhomogeneity of φ is naturally evaluated as $1/2\delta\rho^2(d\varphi/dx)^2 \approx \delta\rho^2(1/L^2)$. Thus, the overall change of the energy per unit volume upon incorporation of defects is given by

$$U(L) = -h_1\rho N^{1/2}L^{-3/2} + \delta\rho^2\frac{1}{L^2}. \tag{11.22}$$

The graph of this function is given in Fig. 11.6. The energy minimum corresponds to

$$L = L^* \sim \frac{\rho^2}{h_1^2 N}. \tag{11.23}$$

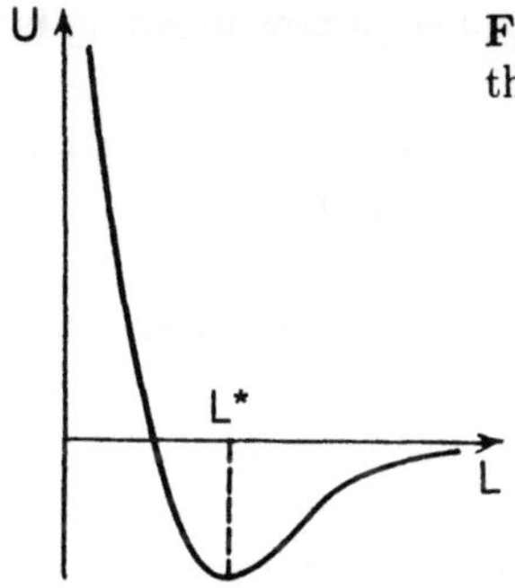

Fig. 11.6. Dependence of the total unit-volume energy on the size of the inhomogeneity of the order parameter.

Thus, in contrast to a doubly degenerate system, at a low concentration of defects the size of an inhomogeneity ("domain") is large (it lies in the physical region; see Sect. 10.6), i.e., the crystal does become spatially inhomogeneous, being split into regions having a size of the order of L with random values of the phase φ. In other words, there is no long-range order in the incommensurate phase with defects of the type under consideration. This treatment clearly confirms our expectation that the "domain structure" of the incommensurate phase originates much more readily than in a phase with discrete degeneracy.

11.5 The Commensurate–Incommensurate Phase Transition: A Special Type of Phase Transition

At the beginning of this chapter it was noted that in ferroelectrics the incommensurate phase is usually observed as an intermediate phase between the symmetric (nonpolar) and the polar phase. A low-temperature phase may be nonferroelectric as well, say ferroelastic, or neither, but, as a rule, its structure is such that it could have been formed as a result of a second-order transition: from the symmetric to the nonsymmetric (commensurate) phase if the appearance of an incommensurate phase had not passed ahead of this event.

Since the incommensurate phase is intermediate, it can be entered in two ways, by lowering or raising the temperature. In Sects. 11.2–11.4 we were dealing with the transition from the symmetric into the incommensurate phase and found that in order to understand the main properties of a crystal in the vicinity of this transition it is sufficient to use the Landau theory with account taken, of course, of the specific features of the case considered.

An entirely different situation arises when one is dealing with the commensurate–incommensurate phase transition. In this section we will consider this kind of transition, which is of a totally different nature. Some of the concepts and definitions that we have successfully used in the preceding chapters prove inapplicable in this case. For example, the question of what physical quantity is the order parameter becomes pointless unless, of course, a new meaning is imparted to the term "order parameter".

The main feature of the commensurate–incommensurate phase transition consists in the statement that it is associated with the spontaneous nucleation of domain walls.[3]

Just as in Chap. 1, we assume that when the temperature of the polar phase is increased, there is reached a temperature ($T = T_{ic}$) at which the domain-wall energy tends to zero (though the spontaneous polarization is still finite). In Section 11.6 we will examine a concrete model and demonstrate this possibility in a clear-cut manner, but for the present we limit ourselves to the realistic supposition that at $T > T_{ic}$ the domain-wall energy becomes negative and that it is unfavorable for the crystal to remain monodomain; the formation of a domain wall lowers its energy. It might seem that the larger the number of domain walls formed, the stronger the reduction of the crystal energy. However, this may not be the case: such a treatment neglects the interaction between the walls. The origin of this interaction will be considered separately; at this point we will discuss a certain simple and basically speculative situation. Suppose that the domain walls repel one another and that this repulsion decreases exponentially with increasing distance between them. This presumption is natural if it is assumed that the walls act on one another and the degree of this interaction is equal to the overlap of the "tails" of inhomogeneous distributions of the order parameter corresponding to these walls. As shown in Chap. 4, since the inhomogeneity of the order parameter caused by the wall decreases far from the wall by an exponential law, it is clear that the energy of interaction between the walls will also fall off by the same law.

It is easy to show that in the case in question the transition with a spontaneous nucleation of domain walls at $T = T_{ic}$ will be continuous. Indeed, at a temperature very close to (but higher than) T_{ic} the energy gain due to the nucleation of domains is infinitely small and an infinitely low repulsive energy is required per wall in order to stabilize the structure, i.e., the distance between the walls is infinitely large. But if at large distances the walls attract one another (at small distances the interaction will still be repulsive), the transition will occur in a jumpwise manner. In any case, when the temperature increases from $T = T_{ic}$, the wall density increases and, in the long run, the distance between them will become comparable to their width. We can no longer draw a distinction between the domain boundary and its width, and the order-parameter distribution assumes the character of the frozen wave (see Fig. 11.3), i.e., we are getting into the region whose properties have been discussed earlier.

---

[3] In the literature such domain walls, which originate in the incommensurate phase near the point of transition into the commensurate phase and which have a negative energy, are called *solitons*.

Consider now the properties of a crystal in the vicinity of $T = T_{\rm ic}$; suppose that the domain walls repel one another. For the energy of such repulsion we may write in accordance with what has been said above:

$$W \approx a_2 \exp(-l/\varkappa), \tag{11.24}$$

where l is the distance between adjacent walls; $\varkappa$ is a quantity of the order of the domain-wall width; and a_2 is a coefficient which must not depend appreciably on temperature in the neighborhood of $T = T_{\rm ic}$. In this vicinity the thermodynamic potential of the unit volume may be written as

$$\Phi = a_1(T_{\rm ic} - T)n + a_2 n \exp\left(-\frac{1}{n\varkappa}\right). \tag{11.25}$$

Here $n = 1/l$ is the number of walls per unit length; $a_1(T_{\rm ic} - T)$ the surface energy of the wall. Minimizing expression (11.25) with respect to n, we find

$$\frac{a_1}{a_2}(T - T_{\rm ic}) = \left(1 + \frac{l}{n\varkappa}\right) \exp\left(-\frac{l}{n\varkappa}\right). \tag{11.26}$$

Since we are interested in the case where $l \gg \varkappa$ $(n \to 0)$, the 1 in parentheses on the right-hand side of (11.26) may be neglected. Taking the logarithm of both parts of (11.26), we obtain

$$\ln \frac{a_1}{a_2}(T - T_{\rm ic}) = -\frac{l}{\varkappa} + \ln \frac{l}{\varkappa}. \tag{11.27}$$

In the zeroth approximation we may neglect $\ln(l/\varkappa)$ as compared to $(l/\varkappa)$ and, as a result, we find for the equilibrium value of l:

$$l_0 = \varkappa \ln \frac{a_2/a_1}{T - T_{\rm ic}}. \tag{11.28}$$

We see that, as expected, the equilibrium distance between domain walls tends to infinity as $T \to T_{\rm ic}$. The approximate expression (11.28) yields a correct notion of the temperature dependence of l, but it is not sufficiently accurate in order to find anomalies of the thermodynamic quantities. Indeed, if l from expression (11.28) is substituted into (11.25), we will obtain for the thermodynamic potential a value equal to zero, i.e., in our approximation the incommensurate phase even with $T > T_{\rm ic}$ does not appear to be energetically more favorable than the commensurate phase. This inference, of course, indicates only that our approximation is insufficient. Let us now solve (11.27) with greater accuracy.

We put $l = l_0 + l'$. Neglecting l' as compared to l_0 under the logarithmic sign, we get

$$l' = \varkappa \ln \frac{l_0}{\varkappa} = \varkappa \ln \ln \frac{a_2/a_1}{T - T_{\rm ic}}. \tag{11.29}$$

Thus, the more accurate expression for the equilibrium value l_0 has the following form:

$$l_0 = \varkappa \ln \frac{a_2/a_1}{T - T_{\rm ic}} + \varkappa \ln \ln \frac{a_2/a_1}{T - T_{\rm ic}} + \dots, \tag{11.30}$$

where the ellipsis represents terms that exhibit a weaker temperature dependence (these terms can be found in the successive approximations). From (11.26), (11.25) and (11.30) it follows that the equilibrium value of the thermodynamic potential is given by

$$\Phi = \frac{a_2}{\varkappa} \exp\left(-\frac{l}{\varkappa}\right) = \frac{a_1}{\varkappa}(T - T_{\rm ic}) \left[\ln \frac{a_2/a_1}{T - T_{\rm ic}}\right]^{-1}. \tag{11.31}$$

Using formula (2.6), we find for the heat capacity of the incommensurate phase:

$$C \approx \frac{T a_1}{\varkappa} \left[\ln \frac{a_2/a_1}{T - T_{\rm ic}}\right]^{-2} \frac{1}{T - T_{\rm ic}}. \tag{11.32}$$

We see that the heat capacity of the incommensurate phase sharply increases as $T \to T_{\rm ic}$ by a law similar to the Curie–Weiss law (Fig. 11.7).

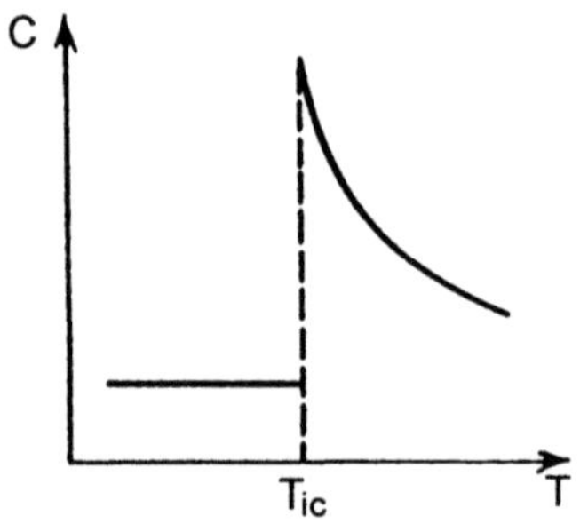

Fig. 11.7. Temperature dependence of the heat capacity of the incommensurate phase near $T_{\rm ic}$.

Consider now the case where the commensurate phase is polar and the sign of the energy of the wall separating the domains with opposite directions of the polarization vector $\boldsymbol{P}_{\rm s}$ changes at $T = T_{\rm ic}$ (note that different domains with the same polarization are possible in the polar phase of an improper ferroelectric). The incommensurate phase will represent a regular domain structure, in which there are neighboring domains with oppositely directed polarizations. It is not difficult to show that the anomaly of the dielectric constant in an incommensurate–commensurate phase transition must be similar to the heat-capacity anomaly considered above. Indeed, the displacement of domain walls upon application of an electric field occurs the more freely, the farther are these walls from one another, i.e., the weaker is the interaction of the walls. Therefore, upon approach (from above) of T to $T_{\rm ic}$ the dielectric constant of the crystal must increase.

Let us test this reasoning by a direct calculation. Suppose that upon application of an electric field E the width of domains with a positive polarization (i.e., directed along the field) has increased by Δl and, accordingly, the width of domains with a negative polarization has diminished by the same amount. The change of the thermodynamic potential occurs at the expense of that part of it which corresponds to the interaction of domain walls and due to the work performed by the electric field. This change referred to the unit length is given by

$$\Delta\Phi = \frac{a_2}{2l}\left[\exp\left(-\frac{l_0+\Delta l}{\varkappa}\right) + \exp\left(-\frac{l_0-\Delta l}{\varkappa}\right) - 2\exp\left(-\frac{l_0}{\varkappa}\right)\right] - \frac{2P_s\Delta l E}{2l_0}. \tag{11.33}$$

Minimizing expression (11.33) with respect to Δl, we find

$$\Delta l^* = \frac{2P_s\varkappa^2}{a_2}\exp\left(\frac{l_0}{\varkappa}\right)E. \tag{11.34}$$

The polarization associated with the displacement of domain walls is

$$P = \frac{2P_s\Delta l^*}{2l} = \frac{P_s}{l}\Delta l^* = \frac{P_s^2\varkappa^2}{a_2 l}\exp\left(\frac{l_0}{\varkappa}\right)E. \tag{11.35}$$

Using expression (11.30) for l_0, we obtain

$$\chi = \frac{\mathrm{d}P}{\mathrm{d}E} = \frac{P_s^2\varkappa}{a_1(T-T_{\mathrm{ic}})} + \dots, \tag{11.36}$$

where the ellipsis represents terms having a weaker singularity as $T \to T_{\mathrm{ic}}$. Thus, the dielectric constant of the incommensurate phase tends to infinity as $T \to T_{\mathrm{ic}}$ if the transition into the commensurate phase is continuous. Note that the dielectric constant of the commensurate phase experiences no anomaly – it is the dielectric constant of a monodomain ferroelectric far from the Curie point. Figure 11.8 shows the temperature dependence of the dielectric constant and polarization in the region of the commensurate–incommensurate phase transition. A remarkable point is that the polarization varies in a jumpwise manner, though the transition is continuous. This

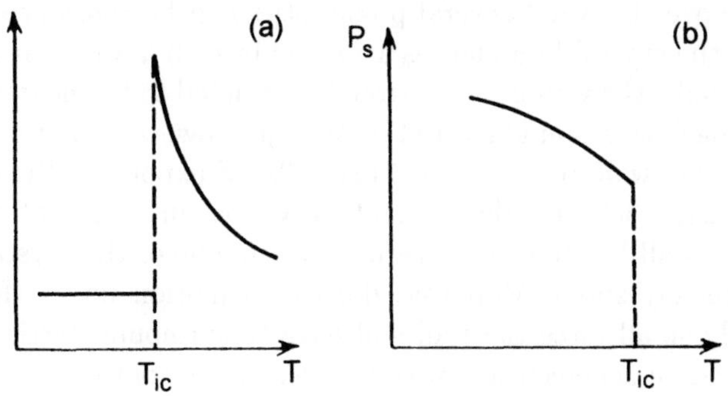

Fig. 11.8. (**a**) Temperature dependence of the dielectric constant and (**b**) spontaneous polarization in the region of the phase transition from the incommensurate into the commensurate phase.

is quite natural since we are dealing, in fact, with the transition from the single-domain state into a state with a regular domain structure, in which the number of domains with opposite polarization directions is the same.

Having concluded the reasoning concerning the expected anomalies of various quantities in the commensurate–incommensurate phase transition within the framework of a model with an exponential repulsion of domain walls, it is natural to pose the question: In what measure does this model correspond to the real situation and to available experimental data? The problem is that, as has been shown by a more rigorous theory, the decisive factor in the interaction of domain walls at large distances is attraction rather than repulsion. The outcome is that the commensurate–incommensurate phase transition becomes discontinuous. This is associated with the vanishing of the domain-wall energy at the temperature T_{ic}; it is natural that here the equilibrium concentration of domain walls will be determined by their interaction alone and in the case of repulsion it will tend to zero. On the contrary, in the presence of attraction it is advantageous to diminish the interwall distance until the attraction at small distances between them is replaced by repulsion. As a result, the concentration of walls at the point of the commensurate–incommensurate phase transition will change in a jumpwise manner. However, with a further rise of temperature the distance between domain walls diminishes approximately by the same law as in the case of the speculative situation considered above: at small distances between the walls the crucial factor in their interaction is still repulsion.

As to the comparison of theory with experiment, it may be said at once that significant discrepancies might be expected. As a matter of fact, what has so far been considered is, in essence, the temperature evolution of an equilibrium domain structure. However, under experimental conditions we practically never have an equilibrium domain structure: the time of relaxation of the domain structure to equilibrium is extremely long.

Consider, for example, a crystal with a single domain wall, whose energy is positive. It is advantageous for the crystal to "get rid" of this wall, i.e., there is no doubt that the wall will eventually leave the crystal. But how can it do so? What forces will move the wall? Several possibilities can be considered. One possibility is that the crystal boundary is arranged in such a way that it attracts the wall. Naturally, the wall is more strongly attracted to the nearest boundary: it will approach this boundary and disappear. However, the force of interaction between the wall and the boundary falls off exponentially as a function of the distance between them; in other words, in a crystal of macroscopic size with a wall located somewhere in the middle of the crystal this force will be extremely small. Moreover, during its motion under the action of this negligibly small force the wall will have to surmount barriers associated with the presence of defects and with the discreteness of the crystal lattice. Of course, with $T \neq 0$ the wall will eventually overcome these barriers due to thermal motion by way of thermal activation, but this will take a very long time.

Let us now focus our attention on the incommensurate phase. We begin with the temperature dependence of the domain-wall density n. Suppose that prior to the experiment the crystal had been at a fixed temperature for so long that the domain-wall density had attained an equilibrium value. We start cooling the crystal. The equilibrium wall density decreases, i.e., from the standpoint of energy favorableness part of the domain walls must leave the crystal and the remaining ones must move apart. However, since all these displacements and, possibly, disappearances of the walls at the crystal boundary are associated with the overcoming of energy barriers, their velocity will be very low until the difference between the equilibrium and the actual density of domain walls becomes sufficiently high. Stated differently, upon cooling of the crystal we will not record at the outset changes in domain-wall density; it is only upon attainment of a certain degree of overcooling that the density begins to change noticeably. Of course, overcooling depends on the rate of variation of the temperature and also on the quality of the crystal: the larger the number of defects, the larger is the number of obstacles to any reconstructions of the domain structure. Thus, upon cooling the $n(T)$ plot in our experiment will be similar to the curve *abc* in Fig. 11.9, where the point *b* marks the onset of "appreciable" changes in the domain structure.

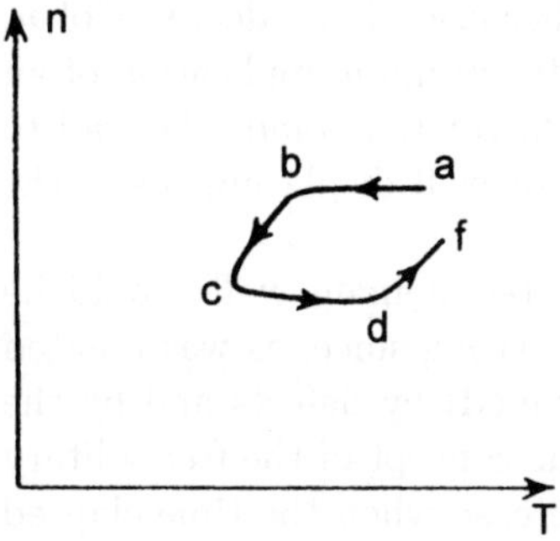

Fig. 11.9. Temperature dependence of the domain-wall density in a real crystal with an incommensurate phase.

Imagine that, having reached the point *c*, we stopped cooling the crystal and began to heat it, i.e., we began increasing the equilibrium value of the domain-wall density. At first this increase will only lead to a decrease of overcooling, which occurred when the point *c* was reached, i.e., new walls are not generated in advance. Neither does it happen immediately after the crystal is heated above the temperature at which the domain-wall density existing in the crystal at the point *c* is at equilibrium, i.e., when overcooling is replaced by overheating. Under these conditions it is advantageous that new walls are nucleated; it is clear, however, that there exists a barrier to nucleation and it is not the entire wall (or a group of walls) that is formed immediately; at first a small portion (or a group of portions) is nucleated, which then grows and forms a whole wall. It is also clear that this stage of formation of a wall is also associated with the overcoming of barriers. Thus, no new domain walls appear immediately after cooling is replaced by heating; a certain degree of overheating is required for their density to begin changing noticeably. To put it differently, the $n(T)$ plot on heating will be represented by a, curve of the *cdf* type in Fig. 11.9. So, experimentally there is observed a hysteresis in the $n(T)$ plot in the incommensurate phase. Naturally, a hysteresis will also be observed for other quantities and not only on temperature plots. As an example, we will consider the temperature dependence of the dielectric constant for the case where the commensurate phase is ferroelectric. We begin with the $\varepsilon(T)$ plot for cooling. As follows from the preceding reasoning, as the temperature falls off the dielectric constant ε increases (since the domain-wall density is diminished), but upon attainment of the temperature T_{ic} the dielectric constant remains finite because the domain-wall density will not be equal to zero, as would be the case at equilibrium. It also remains finite when the crystal is cooled down to $T < T_{\mathrm{ic}}$. Upon further cooling the wall energy (which is now positive) increases more and more and the crystal tends to get rid of them. It should be expected that upon cooling to temperatures substantially lower than T_{ic} the wall density will become very low. What will be the form of the $\varepsilon(T)$ plot upon cooling in the region of temperatures $T < T_{\mathrm{ic}}$? For this to be found, one has to recall the discussion of the dependence $\varepsilon(T)$ for the incommensurate phase. It has been established that the increase of ε as $T \to T_{\mathrm{ic}}$ is associated with the fact that, because of the decrease of n, the domain walls are displaced more and more freely upon application of an electric field, i.e., it was believed that the only factor that binds the wall to its site and creates the "rigidity" with respect to wall displacements is the interaction betweem the walls.

Within the scope of such reasoning an isolated domain wall should be regarded as absolutely free. At first sight this is wrong since, as was pointed out in Chap. 10, the wall becomes clamped (pinned) by defects and by the periodic potential relief of the crystal lattice. The concept of the free solitary wall is valid for strictly static fields, i.e., for the case when the time elapsed from the moment the measuring field is turned on to the moment when the experiment is started is infinitely long. In this case, the wall has time enough to surmount any barriers and to move infinitely far away from the original position. However, in reality no static fields exist, i.e., the measuring field always corresponds to a nonzero frequency. This may well happen and, as a rule, it does happen so that during the period the field is varying (or is switched on) the wall has no time to overcome the barriers, i.e., it should be regarded as being clamped, naturally not "dead fast" but as if it were attached by a spring to the initial position. The action of this spring will be called here the effect of individual pinning.

Let us return to the $\varepsilon(T)$ plot upon cooling. If on entering the temperature region of the commensurate phase the domain-wall density has not yet become so small that the interaction between the walls still limits their displacement more strongly than does the individual pinning, then upon cooling, i.e., when n falls off, the dielectric constant ε will increase. However, when the mean distance between the walls increases so that the decisive factor that interferes with displacements of the walls is the individual pinning rather than their interaction, the quantity ε will diminish with decreasing temperature, i.e., with decreasing n.

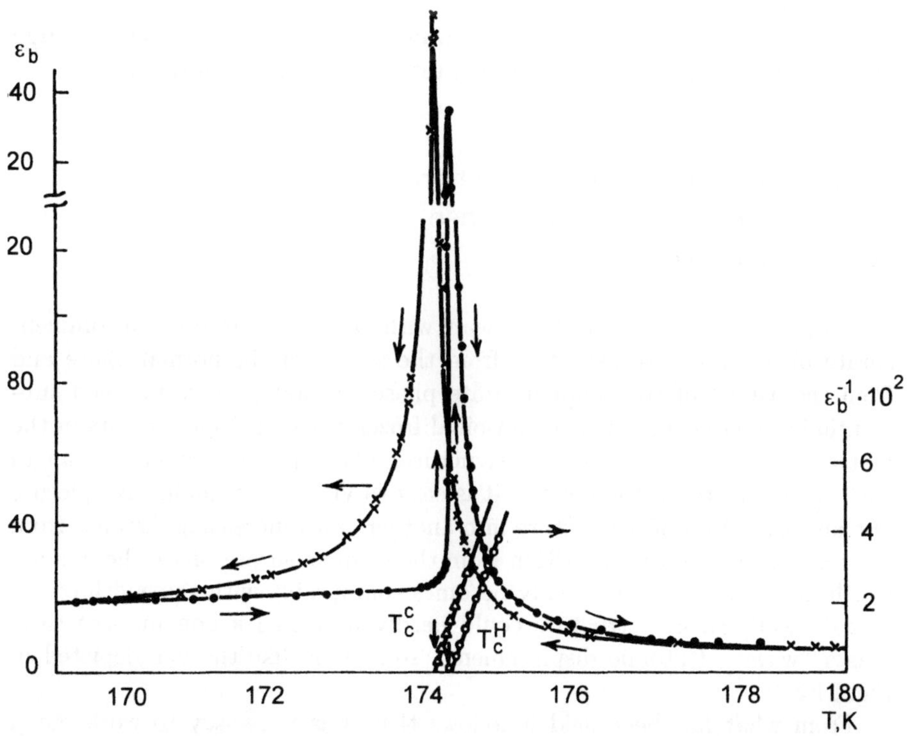

Fig. 11.10. Temperature dependence of the dielectric constant of the $(NH_4)_2BeF_4$ crystal in the region of transition from the incommensurate into the commensurate phase [16].

Suppose now that after the crystal has been deeply cooled into the commensurate phase its temperature increases. As long as the crystal is in this phase the domain-wall energy is positive and there is no reason for nucleation. The dielectric constant therefore practically remains constant and relatively low – this is the dielectric constant of a single-domain ferroelectric far away from the transition point. In the region of temperatures $T > T_{\mathrm{ic}}$, where the domain-wall energy is negative, the nucleation of new domain walls does not practically occur as long as the difference $T - T_{\mathrm{ic}}$ is small and becomes appreciable only after a certain degree of overheating is attained. As the first walls are formed the dielectric constant sharply increases and then diminishes with their increasing density. The experimentally observed dependences $\varepsilon(T)$ are in good agreement with the scenario outlined above (see Fig. 11.10).

Concluding this section, it should be pointed out once more that we have been mainly concerned with the continuous commensurate–incommensurate phase transition, i.e., we presumed that the domain walls, whose energy change sign at $T = T_{\text{ic}}$, repel one another. In experiment, however, this transition is discontinuous, as a rule; upon cooling the domain-wall density decreases in a jumpwise manner and upon heating there immediately appears their finite density n.

In fact the interaction between the walls is of alternating sign and at large distances between the walls attraction rather than repulsion takes place.

11.6 Evolution of the Structure of the Incommensurate Phase (General Picture)

In the preceding section we have dealt with transitions into the incommensurate phase from two aspects – from the aspect of the normal phase and from the aspect of the commensurate phase. In doing so we revealed substantially different pictures: a sinusoidal frozen wave of displacements in the first case and a regular domain structure, whose period regularly changes with temperature, in the second. It is easy to visualize a qualitative picture of transformations of one picture into another: with increasing distance from the temperature of the transition from the commensurate phase the domain width diminishes and eventually becomes comparable with the width of the domain wall. Such a structure could hardly be called a domain structure; it is a "wave" of atomic displacements, something like the one depicted in Fig. 11.3.

From what has been said it follows that it is necessary to work out a theory which could be used to trace out the evolution of the structure of the incommensurate phase and to describe both phase transitions (normal → incommensurate and incommensurate → commensurate) within the framework of a unified approach. To this end, one should consider the Landau thermodynamic potential as a function of the order parameter for the normal–commensurate phase transition (this transition is not observed in reality, but nothing prevents us from considering its parameter). For temperatures corresponding to the region of existence of the commensurate phase, with the minimum of the thermodynamic potential there must occur a spatially homogeneous distribution of the order parameter, and for temperatures corresponding to the incommensurate phase there is a spatially inomogeneous distribution, a domain structure, a "wave" of the order parameter (it should be noted once again that we are speaking of the order parameter of the transition that is not realized in practice). Since the temperature region of existence of the incommensurate phase is usually rather small, it is hoped that the entire evolution of the structure of the incommensurate phase may be regarded as a consequence of the temperature dependence of a single thermodynamic potential coefficient.

An advantage of this theory is its simplicity and a shortcoming consists in that it ignores a very important circumstance. This circumstance has not been taken into account either in the preceding treatment: we mean the discreteness of the crystal structure. This was pointed out in Sect. 11.2. We presumed that the wave vector formed in the phase transition of a "frozen" wave corresponds to the incommensurate point, i.e., that its ratio to the corresponding reciprocal lattice vector ($k_0 a/2\pi$) is an irrational number. True, we immediately found that k_0 varies with a change in temperature, and this means that it covers all values, both "incommensurate" (irrational values of the quantity $k_0 a/2\pi$) and "commensurate" (rational ratios ($k_0 a/2\pi$)). The latter circumstance (i.e., commensurability effects) was ignored in Sect. 11.2, but we promised to show that near T_i these effects are unimportant. We will now try to demonstrate that this is really the case.

To do so, we have to pose the question: How will the expression for the thermodynamic potential in which $Q(\boldsymbol{k})$ is the order parameter with the commensurate wave vector $k = (2\pi/a)(m/n)$ (m and n are integers) differ from formula (11.2)? While (11.2) is an expansion series in powers of $Q(\boldsymbol{k})Q^*(\boldsymbol{k})$ since it is the only invariant for the incommnensurate $\boldsymbol{k}$, in our case there are additional invariants. Under a translation transformation $x \to x + a$ the quantity $Q^n(\boldsymbol{k})$ is multiplied by $[\exp(\mathrm{i}(2\pi/a)(m/n)]^n$, i.e., it remains invariant. Each of the expressions $Q^n(\boldsymbol{k}) + Q^{*n}(\boldsymbol{k})$ and $Q^n(\boldsymbol{k}) - Q^{*n}(\boldsymbol{k})$ and the difference of their squares may serve as an invariant with respect to all transformations of symmetry groups (it is only such invariants that enter into the thermodynamic potential). As a result, in the thermodynamic potential there appear terms which depend not only on ρ but also on φ, for example $Q^n(\boldsymbol{k}) + Q^{*n}(\boldsymbol{k}) = 2\rho^n \cos(n\varphi)$ or $[Q^n(\boldsymbol{k}) + Q^{*n}(\boldsymbol{k})]^2 - [Q^n(\boldsymbol{k}) - Q^{*n}(\boldsymbol{k})]^2 = 4\rho^{2n}\cos(2n\varphi)$.

Now, in order to determine the equilibrium value of the order parameter, the thermodynamic potential has to be minimized with respect not only to ρ but also to φ . Let the quantity $\rho^n \cos(n\varphi)$ be invariant. Then, depending on the sign of the coefficient in front of this invariant, the value of $\cos(n\varphi)$ equal to $+1$ or -1 will correspond to the minimum of the thermodynamic potential. In any case, a decrease of the thermodynamic potential due to the "commensurability effect" is proportional to ρ^n, i.e., $(T - T_i)^{n/2}$ or $(T - T_i)^n$ if $\rho^{2n}\cos(2n\varphi)$ is the only invariant. Recall that the contribution of the terms taken into account in formula (11.2) to the thermodynamic potential is proportional to $(T - T_i)$. So, if n is large and $(T - T_i)$ is small, there is only a slight difference between the thermodynamic potentials of the two phases – the commensurate and the incommensurate phase – with practically coinciding vectors. The foregoing is also valid with respect to other properties, say the vibrations of the phase φ. Repeating the reasoning of Sect. 11.3, we find that the phason frequency in the commensurate phase does not vanish for wave vectors close to zero and remain finite. This frequency, however, is proportional to ρ^{n-2} (or to ρ^{2n-2}), i.e., is very small, and at large n the corresponding mode is practically undetectable in experiment.

Nevertheless, let us focus our attention once again on one of the remarks made above. We saw that if we compare the thermodynamic potentials of two phases, for which the wave vectors of the frozen wave are infinitely close to each other, but one of the wave vectors is "incommensurate" and the other is "commensurate" (recall that in any small neighborhood of an irrational number there is an infinite set of rational numbers), then the thermodynamic potential of the commensurate phase is found to be lower than that of the incommensurate phase. It thus follows that it is not advantageous for the wave vector of the "frozen" wave to assume incommensurate values: there will always be found a value infinitely close to it, though commensurate. Thus, while varying with temperature, the wave vector will "avoid" incommensurate values, assuming only commensurate ones. In other words, what we called the incommensurate phase is in fact a sequence of commensurate phases, this sequence being infinite since to each rational value of $ka/(2\pi)$ there corresponds a possible phase.

Having found its way into the commensurate phase, the crystal is retained in it in a certain temperature range. Indeed, suppose that we "switched off" the commensurability effects. The wave vector of the frozen wave $\boldsymbol{k}$ will then choose the value dictated by formula (11.2). Let this value be commensurate by chance, i.e., $ka/2\pi = m/n$. Now we "switch on" the commensurability partly, i.e., we assume that the coefficient at the term $\rho^n \cos(n\varphi)$ is different from zero. The crystal energy will then decrease by a certain (finite) amount. We then change the value of σ. Had there been no commensurability effects, the vector $\boldsymbol{k}$ would have been immediately changed. Due to an energy gain provided by commensurability effects the vector $\boldsymbol{k}$ remains the same as before, being commensurate even if the value of σ is different; from the standpoint of the thermodynamic potential (11.12) the former vector $\boldsymbol{k}$ is no longer favorable. Of course, this is valid only as long as the energy loss resulting from the fact that the vector $\boldsymbol{k}$ is not the most favorable, is more than compensated by an energy gain due to commensurability effects. We have just compared the energies (thermodynamic potentials) corresponding to the commensurate and incommensurate wave vectors. Now we will compare the "original" commensurate and other commensurate wave vectors. Could it so happen that with a change of σ it is advantageous for the wave vector to assume without delay a different commensurate value? The answer should be negative (more exactly, it is negative, as a rule). Indeed, if we deal with rational numbers very close to the given rational number, they will have larger denominators, as a rule, than the initial number. Therefore, because of commensurability effects, the energy gain will be smaller for them.

We come to the conclusion that the wave vector of the frozen wave varies in an extremely complicated way with a change in temperature. It takes only commensurate values, retaining them in a finite temperature range. In characterizing this dependence, use is made of a "devil's staircase". This mathematical term represents a dependence $y(x)$ when with a change of x the quantity y remains constant within a certain range of values for each rational value of x. From the reasoning given above it follows that the wave vector $\boldsymbol{k}$, while varying with a change of temperature, does not necessarily cover all steps of the "devil's staircase", which does not prevent the dependence $\boldsymbol{k}(T)$ from being quite "devilish" in its complexity. True, as has already been noted, these complexities are mostly outside the limits of accuracy of experiments. Just the same, one has to keep them in mind: the possibility of encountering these complexities is not ruled out at all.

The attentive reader has probably noticed that, while discussing commensurability effects, we have implicitly assumed that the frozen wave is sinusoidal. As a matter of fact, this is not the case, of course: commensurability effects alter not only the energy of the frozen wave but its form as well. Moreover, we have already dealt with a nonsinusoidal frozen wave: the domain structure considered in Sect. 11.5 is exactly such a wave if the crystal is taken to be a continuous medium. Now we shall take into account commensurability effects (discreteness of the crystal lattice) for the case in question. To do this, we first consider an isolated domain wall. If the crystal were a continuous medium, the energy of this wall would not depend on its position. However, in actuality this energy will be different and depend on whether the middle of the wall (in our simplest case, a point where the order parameter vanishes in the continuous-medium approximation (see Fig. 10.1)) is midway between atomic layers or coincides with an atomic layer. One of these positions must correspond to the energy minimum. This means that the wall can change its position not continuously but in a jumpwise manner – by an amount not smaller than the length of primitive translation along the direction perpendicular to the wall. This also means that the superlattice of domain walls displays the tendency (other conditions being equal) to form a commensurate structure with a period being a multiple of the translation mentioned above. The phrase "other conditions being equal" means that such a location is favorable from the standpoint of the wall interaction energy. Suppose this is really the case at a certain temperature. Let us now change the temperature slightly. The interaction of the walls will be changed and, accordingly, the optimal interwall distance will also change: it will no longer be a multiple of a primitive translation. But how should the walls be located in order to achieve the energy minimum? They have to be located at a distance most favorable from the standpoint of the interaction energy: they have to be driven out of commensurate positions, which implies a large energy loss. Another way is to change the distance (by the length of a primitive translation), for example, between each millionth and a million-plus-one wall.

The mean distance will then change and a new configuration will probably be energetically more favorable than the previous one. In order to answer the question of how and what distance exactly should be changed, we have to take into account the interaction not only of nearest-neighbors (which was done in Sect. 11.5) but also of walls located far away from each other. These interactions are evidently very weak and, like the attendant effects, are hardly accessible to experimental observation. It should also be added that we have had in mind the minimum of the incomplete thermodynamic potential which describes the distribution of displacements in a "frozen" wave, neglecting thermal fluctuations and randomness effects in the arrangement of domain walls. A situation could, in principle, be possible where it is not advantageous for the system to maintain a long-range order in the arrangement of domain walls, i.e., the equilibrium state of the system corresponds to their random arrangement. Such conclusions are arrived at by some authors, while speaking of a "chaotic phase" near $T = T_{\rm ic}$. The key point from the viewpoint of the experimentalist is that we discussed here the equilibrium configurations of the system and recognized that the transition between them (for example, with a change in temperature) is exceedingly difficult. For instance, in order to change the distance between occasional pairs of walls, with the remaining distances being unaltered, we have to move all walls! Stated in a different manner, in experiment we will actually deal with nonequilibrium states, whose relaxation time to equilibrium is extremely long. Overall, the situation is rather complicated, a fact which leads us to take a different view on the theory employing the continuous-medium approximation. Its approximate nature ceases to be a drawback and becomes an advantage. But one should not forget that the theory describes equilibrium states, which may appear to be difficult to attain in experiment. Note that if nonequilibrium effects (say, hysteresis) are clearly observed in experiments, it does not at all mean that for the experiment to be explained we have to take into account the discreteness of the crystal. As has been pointed out above, the delays involved in the establishment of equilibrium are inevitable if there are defects in the crystal. The incomplete agreement between experiment and a theory that treats an ideal crystal as a homogeneous continuous medium may thus arise mainly not because the crystal is, in fact, discrete but because it is, as a rule, far from being ideal.

11.7 Evolution of the Structure of the Incommensurate Phase (the Continuum Approximation)

We begin with the first of the cases of transition into the incommensurate phase considered in Sect. 11.1, where the location of the minimum on the soft optic branch is not governed by symmetry (see Figs. 11.3 and 11.4). We assume that the commensurate phase here is a phase in which the displacement of atoms marked by filled circles is the same in all unit cells. This displacement is denoted by η. Now we are to write the thermodynamic potential as a function of η with account taken of spatially inhomogeneous states. In fact, we have already derived an expression of this kind – formula (6.9). As follows from Chap. 6, $\eta_{\boldsymbol{k}}$ are normal coordinates corresponding to the vicinity of the minimum of the soft branch, in which case $|\boldsymbol{k}|$ is the distance to the minimum and the quantity $A + \delta k^2$ is proportional to the square of the optical frequency of vibrations. To describe tbe case of interest, which corresponds to Fig. 11.4, it is evidently necessary to alter the character of the dependence on k of the coefficient at $\eta_{\boldsymbol{k}}\eta_{-\boldsymbol{k}}$; for example, we may need to assume that this coefficient is equal to $A + \delta k^2 + gk^4$. So, expression (6.9) becomes

$$\Phi(p, T, \eta_{\boldsymbol{k}}) = \varphi_0 v + \frac{v}{2} \sum_{\boldsymbol{k}} (A + \delta k^2 + gk^4) \eta_{\boldsymbol{k}} \eta_{-\boldsymbol{k}}$$

$$+ \frac{v\beta}{4} \sum_{\boldsymbol{k}_1 + \ldots + \boldsymbol{k}_4 = 0} \eta_{\boldsymbol{k}_1} \eta_{\boldsymbol{k}_2} \eta_{\boldsymbol{k}_3} \eta_{\boldsymbol{k}_4}, \tag{11.37}$$

where now, as can be seen from Fig. 11.4, the coefficient $\delta < 0$. If the minimum of the function $\omega(\boldsymbol{k})$ lies sufficiently close to the point $\boldsymbol{k} = 0$, the equation $\omega^2 = A + \delta k^2 + gk^4$ will describe the course of the function $\omega^2(\boldsymbol{k})$ rather accurately, its minimum corresponding to the value

$$k_0^2 = -\delta/2g \tag{11.38}$$

and being equal to

$$A - \delta^2/4g = A - \alpha_0. \tag{11.39}$$

It is exactly this quantity that vanishes at the point of transition from the symmetric to the incommensurate phase, i.e., at $T = T_i$. We assume, as before, that in (11.37) $A = \alpha(T - T_c)$ and the remaining coefficients are temperature-independent; note that from the fact of the existence of a phase transition it only follows that expression (11.39) tends to zero at the transition point and, of course, it is quite unclear whether this occurs due to the temperature dependence of A or of δ and g. It is natural that T_c here no longer has the meaning of the phase transition temperature. Thus, our assumption implies something more significant than the assumptions that are usually made in the Landau theory. Accordingly, certain results will also go beyond the framework of the Landau theory; however, as will be shown below, they are in agreement with experimental data, say with respect to $NaNO_2$ crystals. As a matter of fact, our assumption implies that the optical branch in Fig. 11.4 descends as a whole as $T \to T_i$ (in the region of small k at least); it touches to the axis of wave vectors at

$$T_i - T_c = \delta^2/(4g\alpha). \tag{11.40}$$

In order to find out which of the normal coordinates $\eta_{\boldsymbol{k}}$ is the order parameter, let us compute the thermodynamic potential corresponding to a certain frozen wave with a wave vector $\boldsymbol{q}$:

$$\eta(\boldsymbol{r}) = \eta_{\boldsymbol{q}} \mathrm{e}^{\mathrm{i}\boldsymbol{q}\cdot\boldsymbol{r}} + \eta_{-\boldsymbol{q}} \mathrm{e}^{-\mathrm{i}\boldsymbol{q}\cdot\boldsymbol{r}}. \tag{11.41}$$

To this end, it is first necessary to determine the number of terms in each of the sums appearing in (11.37) that contain only $\eta_{\boldsymbol{q}}$ and $\eta_{-\boldsymbol{q}}$. It is obvious that the number of such terms in the first sum is only two and they correspond to the values of $\boldsymbol{k} = \boldsymbol{q}$ and $-\boldsymbol{q}$. As regards the second sum, it represents the Fourier representation of the quantity $(\beta/4)\int \eta^4 \mathrm{d}\boldsymbol{r}$; it is therefore expedient to raise both parts of the equality (11.41) to the fourth power and to calculate the coefficient at $(\eta_{\boldsymbol{q}}\eta_{-\boldsymbol{q}})$. The outcome is

$$\Phi(\eta_{\boldsymbol{q}}\eta_{-\boldsymbol{q}}) = \Phi_0 + (A + \delta q^2 + g q^4)\eta_{\boldsymbol{q}}\eta_{-\boldsymbol{q}} + \frac{3\beta}{2}(\eta_{\boldsymbol{q}}\eta_{-\boldsymbol{q}})^2. \tag{11.42}$$

Let us now find the value of the amplitude of the frozen wave corresponding to the minimum of the thermodynamic potential. From (11.40) we obtain

$$\eta_{\boldsymbol{q}}\eta_{-\boldsymbol{q}} - A(q)/(3\beta) \quad \text{for} \quad A(q) < 0, \tag{11.43}$$

where $A(q) = A + \delta q^2 + g q^4$. Substitution of (11.43) into (11.42) yields for the thermodynamic potential of the incommensurate phase:

$$\Phi = \Phi_0 - A^2(q)/(6\beta), \tag{11.44}$$

from which it can be seen that the freezing of such a displacement wave which corresponds to the largest absolute negative value of $A(q)$ is energetically favorable, i.e., it corresponds to the minimum of this function. As has already been shown, at the point of the minimum $q = k_0$ (see (11.38)), i.e., in the approximation used the equilibrium value of $\eta_{\boldsymbol{q}}$ (let us denote it as $\eta_{\boldsymbol{q}}^0$) is given by the following expression:

$$\eta_{\boldsymbol{q}}^0\eta_{-\boldsymbol{q}}^0 = A(T - T_i)/(3\beta), \tag{11.45}$$

and the thermodynamic potential of the incommensurate phase may be written in the form

$$\Phi = \Phi_0 - \alpha^2(T - T_i)^2/(6\beta). \tag{11.46}$$

Let us compare this expression with the expression for the thermodynamic potential of the commensurate (polar) phase, which is easy to obtain by substituting the equilibrium value of the order parameter into the incomplete thermodynamic potential. As follows from Chap. 2, we have

$$\eta_0^2 = -\alpha(T - T_{\mathrm{c}})/\beta, \tag{11.47}$$

$$\Phi = \Phi_0 - \alpha(T - T_{\mathrm{c}})^2/(4\beta). \tag{11.48}$$

The graphs of the temperature dependences (11.46) and (11.48) are presented in Fig. 11.11. One can see that at a certain temperature these plots intersect and that the thermodynamic potential of the commensurate phase becomes lower than that of the incommensurate phase. Thus, according to our results, at $T = T_{\mathrm{ic}}$ there takes place a first-order transition from the incommensurate into the commensurate phase (see Fig. 11.1) and the transition temperature is determined by the condition

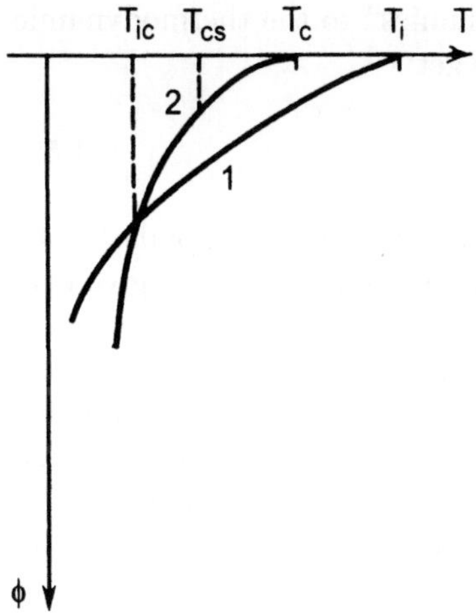

Fig. 11.11. Temperature dependence of thermodynamic potentials: (*1*) incommensurate phase; (*2*) commensurate phase.

$$\frac{1}{6}(T_{\rm ic}-T_i)^2=\frac{1}{4}(T_{\rm ic}-T_{\rm c})^2. \tag{11.49}$$

Of course, this formula is approximate, just as in the above treatment.

Our appxoximation consisted of the transformation from (11.37) to (11.42), i.e., of the assumption of the possibility of a solution of equations that determines the minimum of the thermodynamic potential, for which all the Fourier components of the order parameter, except one, are equal to zero. We will show that this assumption is really an approximation, i.e., that there arise other Fourier components. Indeed, in the thermodynamic potential (11.37) the second sum contains, for example, terms of the type $\eta_{3q}\eta_{-q}\eta_{-q}\eta_{-q}=\eta_{3q}\eta_{-q}^3$. If the essential terms containing η_{3q} are retained in expression (11.37)

$$A(3q)\eta_{3q}\eta_{-3q}+\beta(\eta_{3q}\eta_{-q}^3+\eta_{-3q}\eta_q^3), \tag{11.50}$$

and the thermodynamic potential is minimized with respect to η_{3q}, η_{-3q}, we find that

$$\eta_{3q}^0=-\beta\eta_q^{0^3}/A(3q). \tag{11.51}$$

From (11.45) it follows that $\eta_{3q}^0\propto(T-T_i)^{3/2}$ and $\eta_{3q}^0/\eta_q^0\propto(T-T_i)$. One can see that other Fourier components do appear, but at least near T_i the amplitude of the "third harmonic" is much lower than that of the "fundamental tone" of the incommensurate structure. This is valid even to a greater extent in the case of higher harmonics, which also appear in the incommensurate phase. For example, due to the presence of a term of the type $\eta_{5q}\eta_{-3q}\eta_{-q}^2$ in (11.37), the quantity η_{5q}^0 is different from zero; as can easily be seen, $\eta_{5q}\propto(T-T_i)^{5/2}$.

We arrive at the conclusion that formula (11.46) is sufficiently accurate only at small values of $T-T_i$; as $T-T_i$ increases it becomes necessary to take into account the contributions of "higher harmonics" to the thermodynamic potential. Thus, taking account of (11.51), we get

$$\Phi = \Phi_0 - \frac{1}{6\beta}A^2(\boldsymbol{q}) + \frac{1}{216\beta}\frac{A^3(\boldsymbol{q})}{A(3\boldsymbol{q})}. \tag{11.52}$$

The structure realizable in the incommensurate phase corresponds to the value of q at which expression (11.52) is at a minimum. We do not give here the results of the cumbersone (though simple) procedure of minimization.

From formula (11.52) it can be seen, in the first place, that the thermodynamic potential of the incommensurate phase is lower than that which follows from (11.46). This is quite natural. Our first approximation consisted in minimizing the potential only with respect to one Fourier component of the order parameter, the remaining ones being forbidden to change. Now we allowed the quantity $\eta_{3\boldsymbol{q}}$ to change. Naturally, this provides additional possibilities for the system in attaining a state with the lowest thermodynamic potential. We emphasize that fairly far from T_i the additional contribution becomes appreciable, as a result of which the curves for the thermodynamic potential of the two phases will lie near the point of intersection closer to each other. Taking account of higher harmonics will only increase this closeness. The situation for an accurate solution which would take into account any desired higher harmonics is not known to us, strictly speaking. However, there is ground to expect (considering, in particular, the results of numerical calculations) that for the continuum model discussed by us the transition into the incommensurate phase is a first-order transition.

We now presume that the appearance of an incommensurate phase is associated with the instability at $q = k_0$ in the polar optical branch (see Fig. 11.4), i.e., $\eta \propto P$. We determine the temperature dependence of the dielectric constant in the region of $T \approx T_i$. In accordance with what has been said in Sect. 11.2, the temperature dependence of the dielectric constant must experience a kink at $T = T_i$.

Let us see whether this is really the case. To this end, for the incomplete thermodynamic potential we substitute $\eta(\boldsymbol{r}) = \eta_{q=0} + \eta_{\boldsymbol{q}}\mathrm{e}^{\mathrm{i}\boldsymbol{q}\cdot\boldsymbol{r}} + \eta_{-\boldsymbol{q}}\mathrm{e}^{-\mathrm{i}\boldsymbol{q}\cdot\boldsymbol{r}}$ and add the term $-a\eta_{q=0}E$. We obtain the following result:

$$\begin{aligned}\Phi(\eta_{q=0}, \eta_{\boldsymbol{q}}, \eta_{-\boldsymbol{q}}, E) &= (A(\boldsymbol{q}) + 3\beta\eta_{q=0}^2)\eta_{\boldsymbol{q}}\eta_{-\boldsymbol{q}} + \frac{3}{2}\beta(\eta_{\boldsymbol{q}}\eta_{-\boldsymbol{q}})^2 \\ &\quad + \frac{1}{2}A\eta_{q=0}^2 - a\eta_{q=0}E.\end{aligned} \tag{11.53}$$

In the symmetric phase $\eta_{\boldsymbol{q}} = \eta_{-\boldsymbol{q}} = 0$ and

$$\Phi = \frac{1}{2}A\eta_{q=0}^2 - a\eta_{q=0}E. \tag{11.54}$$

For the nonsymmetric phase we have to substitute expressions (11.43) into (11.53) with the replacement of $A(\boldsymbol{q})$ by $A(\boldsymbol{q}) + 3\beta\eta_{q=0}^2$. As a result, we get

$$\Phi = \frac{1}{2}[A - 2(A - \alpha_0)]\eta^2_{q=0} - a\eta_{q=0}E. \qquad (11.55)$$

From (11.54), (11.55) and the results of Chap. 3 it follows that near T_i the dielectric constant varies by the Curie–Weiss law with the same Curie constant in both phases; at $T = T_i$ it is finite and continuous. Thus, as was expected, the dependence $\varepsilon(T)$ undergoes a kink at $T = T_i$ (Fig. 11.12). But our model gave us something more than simply the presence of a kink: we came to our conclusion concerning the Curie–Weiss law and the equality of the Curie constants for $T > T_i$ and $T < T_i$. In principle, our model also involves the behaviour of $\varepsilon(T)$ in the vicinity of the temperature of transition into the commensurate phase. However, for this region a simple treatment, which takes account of only the sinusoidal distribution of the quantity η, does not provide complete agreement with experiment, which reveals a sharp peak of ε at $T = T_{ic}$ (the curve in Fig. 1.22a). The cause of the discrepancy may, in principle, consist of the inaccuracy of the theory (all higher harmonics are not adequately taken into account) and also of the fact that our model is not quite realistic. It is not possible, however, to discuss this topic in detail here.

Let us now turn our attention to the second case dealt with in Sect. 11.1, where the soft vibration branch is degenerate. For clarity, we consider a concrete example (though it does not correspond to any of the known ferroelectric crystals). We discuss a proper ferroelectric transition involving a spontaneous polarization in the xy plane perpendicular to the four-fold axis z. The parameter of such a transition is a two-component one; its components η_1 and η_2 are transformed as the x- and y-components of the vector (which may, of course, be regarded as the polarization vector (P_x, P_y)). The transformation properties of the order parameter and of some of its combinations are given in Table 11.1. Using this table, we write down the Landau thermodynamic potential for this case:

$$\begin{aligned}\Phi(P_x, P_y) &= \Phi_0 + \frac{A}{2}(P_x^2 + P_y^2) + \frac{1}{4}B_1(P_x^2 + P_y^2)^2 \\ &\quad + 2B_2P_x^2P_y^2 + B_3(P_x^2 - P_y^2)P_xP_y \\ &\quad + \sigma\left(\frac{\partial P_x}{\partial z}P_y - P_x\frac{\partial P_y}{\partial z}\right) + \frac{\delta}{2}\left[\left(\frac{\partial P_x}{\partial z}\right)^2 + \left(\frac{\partial P_y}{\partial z}\right)^2\right]. \qquad (11.56)\end{aligned}$$

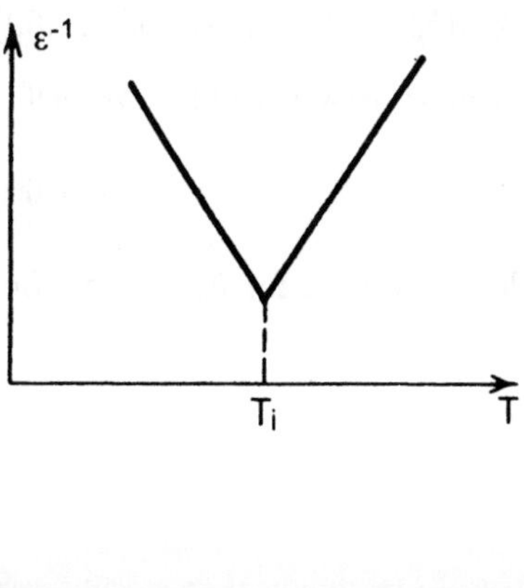

Fig. 11.12. Temperature dependence of the inverse dielectric constant ε^{-1} near the point of transition T_i from the symmetric into the incommensurate phase.

Table 11.1. Transformation properties of the order parameter.

1	4	2	4^{-1}
P_x	P_y	$-P_x$	$-P_y$
P_y	$-P_x$	$-P_y$	P_x
$P_x^2 - P_y^2$	$-P_x^2 + P_y^2$	$P_x^2 - P_y^2$	$-P_x^2 + P_y^2$
$P_x P_y$	$-P_x P_y$	$P_x P_y$	$-P_x P_y$
$\frac{\mathrm{d}P_x}{\mathrm{d}z} P_y - \frac{\mathrm{d}P_y}{\mathrm{d}z} P_x$	$\frac{\mathrm{d}P_x}{\mathrm{d}z} P_y - \frac{\mathrm{d}P_y}{\mathrm{d}z} P_x$	$\frac{\mathrm{d}P_x}{\mathrm{d}z} P_y - \frac{\mathrm{d}P_y}{\mathrm{d}z} P_x$	$\frac{\mathrm{d}P_x}{\mathrm{d}z} P_y - \frac{\mathrm{d}P_y}{\mathrm{d}z} P_x$

As usual we set $A = \alpha(T - T_c)$ and assume the remaining coefficients to be constant for the present. Let us now be clear that in our case T_c no longer means the transition temperature since because of the presence of the Lifshitz invariant (the penultimate term on the right-hand side of (11.54)) in the thermodynmic potential the second-order phase transition $C_4 \to C_1$ proves to be impossible.

Let us consider the spatially inhomogeneous fluctuations of P_x and P_y, assuming that these quantities depend on z alone. Using, just as in Chap. 6, the Fourier expansion for the incomplete thermodynamic potential, we have

$$\Phi = \varphi_0 v + \frac{v}{2} \sum_{\boldsymbol{k}} [(A+\delta k^2)(P_{x,\boldsymbol{k}} P_{x,-\boldsymbol{k}} + P_{y,\boldsymbol{k}} P_{y,-\boldsymbol{k}} + 4\mathrm{i}\sigma k P_{x,\boldsymbol{k}} P_{y,-\boldsymbol{k}}]. \quad (11.57)$$

In order to calculate the fluctuations corresponding to the given quantity $\boldsymbol{k}$, we pick out in (11.57) the terms corresponding to $\boldsymbol{k}$ values which are equal in magnitude but opposite in sign:

$$\begin{aligned} v[(A+\delta k^2)(P_{x,\boldsymbol{k}} P_{x,-\boldsymbol{k}} + P_{y,\boldsymbol{k}} P_{y,-\boldsymbol{k}}) \\ +4\mathrm{i}\sigma k P_{x,\boldsymbol{k}} P_{y,-\boldsymbol{k}} - 4\mathrm{i}\sigma k P_{x,-\boldsymbol{k}} P_{y,\boldsymbol{k}}] \end{aligned} \quad (11.58)$$

($\boldsymbol{k}$ now assumes only positive values). We introduce, just as in Chap. 6, the real and imaginary parts of the quantities $P_{x,\mathbf{k}}$ and $P_{y,\mathbf{k}}$:

$$P_{x,\boldsymbol{k}} = a_{\boldsymbol{k}} + \mathrm{i}b_{\boldsymbol{k}}, \qquad P_{y,\boldsymbol{k}} = c_{\boldsymbol{k}} + \mathrm{i}d_{\boldsymbol{k}},$$

$$P_{x,-\boldsymbol{k}} = a_{\boldsymbol{k}} - \mathrm{i}b_{\boldsymbol{k}}, \qquad P_{y,-\boldsymbol{k}} = c_{\boldsymbol{k}} - \mathrm{i}d_{\boldsymbol{k}}.$$

With the new variables expression (11.58) becomes

$$v[(A+\delta k^2)(a_{\boldsymbol{k}}^2 + d_{\boldsymbol{k}}^2 + 4\sigma k a_{\boldsymbol{k}} d_{\boldsymbol{k}}) + (A+\delta k^2)(b_{\boldsymbol{k}}^2 + c_{\boldsymbol{k}}^2) - 4\sigma k b_{\boldsymbol{k}} c_{\boldsymbol{k}}]. \quad (11.59)$$

This expression can be simplified by passing over to new variables. We set

$$a_{\boldsymbol{k}} = \frac{a'_{\boldsymbol{k}} + d'_{\boldsymbol{k}}}{\sqrt{2}}, \qquad d_{\boldsymbol{k}} = \frac{a'_{\boldsymbol{k}} - d'_{\boldsymbol{k}}}{\sqrt{2}}. \quad (11.60)$$

An analogous substitution must also be made, of course, for $b_{\boldsymbol{k}}$ and $c_{\boldsymbol{k}}$. So, instead of (11.58), we obtain

$$v\left\{\left[(A+\delta k^2)-2\sigma k\right]\left(d_{\boldsymbol{k}}'^2+b_{\boldsymbol{k}}'^2\right)\right.$$
$$\left.+\left[(A+\delta k^2)+2\sigma k\right]\left(a_{\boldsymbol{k}}'^2+c_{\boldsymbol{k}}'^2\right)\right\}. \tag{11.61}$$

Now, in order to calculate the fluctuations, we use formulas (6.1) and (6.2). We find

$$\langle a_{\boldsymbol{k}}'^2\rangle = \langle c_{\boldsymbol{k}}'^2\rangle = \frac{k_{\mathrm{B}}T}{v(A+\delta k^2+2\sigma k)}, \tag{11.62}$$

$$\langle d_{\boldsymbol{k}}'^2\rangle = \langle b_{\boldsymbol{k}}'^2\rangle = \frac{k_{\mathrm{B}}T}{v(A+\delta k^2-2\sigma k)}. \tag{11.63}$$

If $\sigma > 0$, the fluctuations of $d_{\boldsymbol{k}}'$ and $b_{\boldsymbol{k}}'$ are greater than those of $a_{\boldsymbol{k}}'$ and $c_{\boldsymbol{k}}'$. They have the largest value at the minimum of the denominator in (11.63); this value is $k = k_0 = \sigma/\delta$. We see that at $A = \sigma^2/(2\delta)$ the fluctuations of $d_{\boldsymbol{k}_0}'$ and $b_{\boldsymbol{k}_0}'$ go to infinity. At lower values of A (i.e., at lower values of $T - T_{\mathrm{c}}$) formula (11.63) becomes meaningless: the a priori positive quantity – the quadratic fluctuation – is negative according to this formula. This means, in fact, that the expression for the thermodynamic potential (11.57) which was used by us to calculate fluctuations and in which it was assumed that the equilibrium values of $P_x(z)$ and $P_y(z)$ are equal to zero (it is for this reason that higher-order terms were omitted) has become incorrect. In other words, the equilibrium values of $P_x(z)$ and $P_y(z)$ are no longer equal to zero, i.e., a certain phase transition has taken place at the temperature

$$T_i = T_{\mathrm{c}} - \frac{\sigma}{2\delta\alpha}. \tag{11.64}$$

In Chap. 6 we saw that the fluctuation of the order parameter becomes infinite at the phase transition point. We have now become convinced of the validity of the opposite (in a certain sense) statement: when a certain fluctuation becomes infinite, the phase reaches the limit of its stability. Any further slight change in external conditions results in the transition into a different phase. The change in structure that occurs in this case is determined by an analysis of the thermodynamic potential, but it can be predicted by finding out which of the fluctuations does go to infinity. In Chap. 6 it was a spatially homogeneous fluctuation of η and the change in structure during a phase transition consisted of the appearance of a nonzero order parameter, which was the same at all points in space. In our case the fluctuations of the combinations of Fourier components ("wave amplitudes") P_x and P_y build up, i.e., a spatially inhomogeneous distribution of P_x and P_y should be expected to occur in the new phase. Stated in a different way, we come to a certain frozen wave, i.e., to an incommensurate phase. This phase is sometimes called spatially inhomogeneous. However, one should ascertain that this term is strictly conditional. It originated from the description of the normal–incommensurate phase transition with the use of the parameter of the phase transition that "could not have occurred" because of the presence of the Lifshitz invariant in the thermodynamic potential.

In order to determine the structure of the new phase, one has to find the distribution of $P_x(z)$ and $P_y(z)$ corresponding to the minimum of the thermodynamic potential, i.e., one has to find the solutions of Euler equations and to choose the most appropriate one. For what follows it is convenient to introduce the polar coordinates in the P_xP_y plane:

$$P_x = P\cos\psi, \qquad P_y = P\sin\psi. \tag{11.65}$$

The formula for the thermodynamic potential will then become

$$\begin{aligned}\Phi = \varphi_0 v + \frac{A}{2}P^2 + \frac{\tilde{B}_1^2}{4}P^4 + \tilde{B}_2 P^4 \sin^2[2(\psi-\psi_0)] \\ - \sigma P^2 \frac{\mathrm{d}\psi}{\mathrm{d}z} + \frac{1}{2}\delta\left[\left(\frac{\mathrm{d}P}{\mathrm{d}z}\right)^2 + P^2\left(\frac{\mathrm{d}\psi}{\mathrm{d}z}\right)^2\right],\end{aligned} \tag{11.66}$$

where $\tilde{B}_1 = B_1 + B_2 - (B_2^2 + B_3^2)^{1/2}$, $\tilde{B}_2 = 1/2(B_2^2 + B_3^2)^{1/2}$, and $\psi_0 = \frac{1}{4}\arctan(-B_3/B_2)$. An exact solution of the Euler equation cannot yet be obtained for the thermodynamic potential and we have to limit ourselves to a qualitative analysis.

It should first be noted that at $\tilde{B} = 0$ we obtain the thermodynamic potential that has already been considered (see Sect. 11.2). Using the results obtained, we find ($k_0 = |\sigma|/\delta$):

$$P^2 = \frac{A_0 - A}{B} \propto (T_i - T), \qquad \psi = k_0 z. \tag{11.67}$$

This solution describes a "screw" (or, as sometimes said, helicoidal) "frozen" wave of the polarization vector with a period of $L = 2\pi/k_0$, i.e., with a change of z the polarization vector rotates in the plane uniformly and with no change of its length at an "angular velocity" equal to k_0. Naturally, with $\tilde{B}_2 \neq 0$ the rotation ceases to be uniform. Let us evaluate the conditions under which the nonuniformity becomes noticeable. To this end, we compare the volume-averaged values of the terms $\tilde{B}_2 P^4 \sin^2[2(\psi - \psi_0)]$ and $\sigma P^2 \partial\psi/\partial z$ upon substitution of the expression for ψ from formula (11.67). Since the average value of $\sin^2[2(\psi-\psi_0)]$ is equal to 1/2, the first term is estimated as $\frac{1}{2}\tilde{B}_2 P^4$ and the second is naturally equal to $\sigma P^2 k_0 = \sigma^2/(2\delta)P^2$. One can see that we, in fact, compare $\tilde{B}_2 P^2$ with $\sigma^2/(2\delta)$. Clearly, near T_i, when P is small, the first term may be neglected, i.e., the "helicoidal" solution proves to be a good approximation.

But what will happen when $\tilde{B}_2 P^2$ increases up to $\sigma^2/(2\delta)$? The rotation will, of course, become nonuniform, but it is hoped that the length of the vector P will remain nearly constant if the anisotropy is low, i.e., $\tilde{B}_2 \ll \tilde{B}_1$. Recall that such a case has already been dealt with in the discussion of the domain-wall structure (see Sect. 10.3).

If we compare formulas (10.22) and (11.66), we see that they are nearly identical with an accuracy of up to the replacement of ψ by $\psi - \psi_0$ and the absence of the Lifshitz invariant in formula (10.22). But if P (or ρ) does not change in space, neither the Lifshitz invariant nor ψ appears in the Euler equation, i.e., in the case under consideration we have the Euler equation with ψ being replaced by $\psi - \psi_0$. The analogy with the pendulum helps us to run over all solutions of this equation – all the possible movements of the pendulum. We have to choose a motion which will correspond to the minimum of the thermodynamic potential. It is exactly this motion that will provide the equilibrium distribution of $\psi(z)$. In this choice the Lifshitz invariant will now play an important role. One can see from formula (11.66) that if, for example, $\sigma > 0$, it will be advantageous that the quantity $\mathrm{d}\psi/\mathrm{d}z$ be positive, i.e., such a state is favorable when the pendulum rotates in a single direction. This is easily attainable near $T = T_i$ since, as can be seen from (10.25), the "force of gravity" vanishes at $T = T_i$. Note that this rotation cannot be too fast; otherwise, the contribution of the last term in formula (11.66) would be increased. In moving away from $T = T_i$ towards lower temperatures the "force of gravity" increases and interferes with the tendency of the pendulum to rotate (by virtue of the fact that the potential energy of the pendulum is the spatially homogeneous part of the thermodynamic potential with the minus sign (see Chap. 10); to the equilibrium position of the pendulum there corresponds the "upside-down" position). After all, because of the increased force of gravity it will be favorable for the pendulum to come to a stop. This is exactly the transition into a commensurate phase. If we compare the equilibrium state of the incommensurate phase (the pendulum is constantly in the "upside-down" position) with a state in which the pendulum performs a single rotation (in the same direction as in the incommensurate phase) and then stops, the latter state will naturally appear energetically less favorable. It corresponds to a commensurate phase with a domain wall. The difference in energy (thermodynamic potential) between these two states is the wall energy. We have already calculated the domain-wall energy for the case of the thermodynamic potential (10.22). For the case corresponding to (11.66) we have, in addition, to take into account the contribution of the term involving the Lifshitz invariant. For the unit area of the wall we then have

$$\sigma P^2 \int_{-\infty}^{\infty} \frac{\mathrm{d}\psi}{\mathrm{d}z}\mathrm{d}z = \sigma P^2(\psi(+\infty) - \psi(-\infty)) = \sigma P^2 \frac{\pi}{4}. \tag{11.68}$$

Here we have taken into account that the complete rotation of the pendulum is a change of ψ from 0 to 2π and treated the wall with a direction of rotation such that the contribution of the Lifshitz invariant is negative. The overall energy of the domain wall is now evaluated as

$$\frac{2\pi}{4}\left(\frac{\delta}{\tilde{B}_2 P^2}\right)^{1/2} \tilde{B}_2 P^4 - \sigma P^2 \frac{\pi}{2}. \tag{11.69}$$

This energy will vanish at

$$P^2 = \frac{\sigma}{\delta \tilde{B}_2}. \tag{11.70}$$

It should be recalled that the expression for the first term in formula (11.69) was obtained not rigorously (but as an estimation). A mathematically rigorous calculation (under the assumption of the quantity P being constant) leads to a formula different from (11.70) by a factor of 1/2. Thus, the transition from the commensurate to the incommensurate phase occurs at the temperature

$$T_{\text{ic}} = T_{\text{c}} - \frac{\sigma^2 \tilde{B}_1}{2\delta \tilde{B}_2 \alpha}. \tag{11.71}$$

In the derivation of this formula we have taken into account that $\tilde{B}_2 \ll \tilde{B}_1$. Let us now rewrite expression (10.26) for the domain-wall width in the commensurate phase in the form

$$l = \frac{a}{\sqrt{|T - T_{\text{c}}|}}, \tag{11.72}$$

where

$$a = \frac{\pi}{2} \left(\frac{\tilde{B}_1}{\tilde{B}_2} \right)^{1/2} \frac{\delta^{1/2}}{\alpha^{1/2}}.$$

Using (11.71), we find that

$$k_0^2 a^2 \cdot \frac{2}{\pi} = T_{\text{c}} - T_{\text{ic}}. \tag{11.73}$$

Since, as can be seen from a comparison of formulas (11.64) and (11.71), $T_{\text{c}} - T_{\text{ic}} \gg T_i - T_{\text{c}}$ (provided, of course, that the low-anisotropy condition that has been used in all computations is satisfied), it follows that $T_i - T_{\text{ic}} \approx T_{\text{c}} - T_{\text{ic}}$. We have thus established the relationship between three quantities which can be experimentally measured: the coefficient that characterizes the temperature dependence of the domain-wall width in the incommensurate phase; the length of the wave that arises upon transition from the normal into the incommensurate phase; and the temperature range of existence of the incommensurate phase. Of course, this relationship has been obtained due to certain simplifications: only the coefficient A in formula (11.66) was assumed to be temperature dependent. The validity of the neglect of the temperature dependence of the remaining coefficients is not a priori obvious and has, in principle, to be tested by special investigations. We will not, however, plunge into this problem, noting only that for displacive-type transitions, for which the characteristic scale of temperature dependences is T_{at} (see Chap. 7), the neglect of the temperature dependence of all the coefficients, except A, is quite legitimate if σ is small under these particular conditions, i.e., the length of the frozen wave is much larger than the lattice constant. It is exactly this situation that is realized for the majority of transitions into the incommensurate phase.

What has been said does not exhaust the results of our model that do not follow from the general treatments in Sects. 11.2 and 11.5. We arrive at the conclusion, for example, that domain walls are nucleated in the commensurate phase by groups of four. For example, let the equilibrium state of the commensurate phase correspond to $\psi = 0$. Suppose now that somewhere there is formed a nucleus of the incommensurate phase. The possibility that this is an insular domain wall is ruled out since the wall with $\mathrm{d}\psi/\mathrm{d}z > 0$ separates domains with $\psi = 0$ and $\psi = \pi/2$. The change of ψ from zero to 2π for $\mathrm{d}\psi/\mathrm{d}z > 0$ is possible if there is a sequence of four domain walls, as shown in Fig. 11.13. The points A and B are projections of dislocation lines onto the plane of the figure: the trip of each of these points around the closed contour leads to a change of the phase ψ by 2π. Naturally, within the framework of our model we will obtain for the structure of the incommensurate phase not only a helicoidal wave (near T_i) and a sequence of domains (near T_{ic}) but also the spatial distributions of ψ which are intermediate between these extreme cases. We can, of course, refuse to confine ourselves to the mathematical simplicity of theory and, with the use of computers, abandon the conditions of low anisotropy and constancy of P and obtain more realistic theoretical inferences concerning the properties of the incommensurate phase.

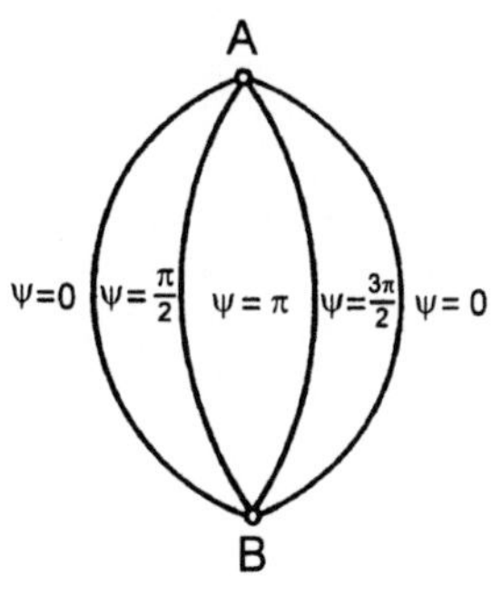

Fig. 11.13. A nucleus of the incommensurate phase for the case discussed in the text.

12. Ferroelectric Liquid Crystals

12.1 Basic Types of Orientational Ordering in Liquid Crystals

In the preceding chapters we have dealt with the phenomena associated with the origination of a spontaneous polarization in crystalline media. The basic mechanisms leading to ferroelectricity involved spontaneous displacements of charged particles in unit cells or redistributions of the probabilities of their localization at certain positions. A characteristic feature of such transitions is the change of the crystal symmetry into a group with a special polar direction, with the original and polar phases belonging to one of the 32 discrete crystallographic groups.

There arises the following question: Can ferroelectric properties exist in systems possessing the symmetry of one of limiting groups? In other words, can there exist liquid ferroelectrics? A search should evidently be made among liquids composed of polar molecules, i.e., molecules having a permanent dipole moment. An isotropic liquid consisting of randomly oriented polar molecules may be treated as possessing a spherical symmetry – an infinite number of infinite-fold axes and an infinite number of symmetry planes, i.e., a $\infty\infty m$ group. It is quite possible that in the absence of an external electric field the molecules of the liquid, which are involved in a certain interaction, are "aligned" at a certain temperature in such a manner that there arises a dipolar ordering of the "head-to-tail" pattern. As a result, a spontaneous polarization occurs in the liquid and the symmetry of the system is lowered: $\infty\infty m \rightarrow \infty m$.

However, such a possibility has not yet been verified by experiment. It turns out that as the temperature falls off liquids crystallize before the conditions required for the realization of such a phase transition are fulfilled; the orientational interaction of polar molecules is insufficient for overcoming the disordering action of thermal motion. One may say that at present no phase transition associated with the spontaneous occurrence of a macroscopic electric polarization is known to exist in polar liquids.

Rather specific phenomena associated with unusual types of dipolar ordering occur, however, in "nontraditional" media – *liquid crystals* – substances which are intermediate with respect to physical properties between liquids and anisotropic crystals. While possessing the ability to flow and to form drops and exhibiting no resistance, like liquids, to shearing stresses, liquid crystals display the well-pronounced anisotropy of optical, electrical and magnetic properties, which is associated with certain types of molecular ordering.

Liquid crystals belong, as a rule, to organic compounds, whose main feature is the strongly extended shape of molecules, which is due, for example, to the sequence of benzene rings as their structural units. Upon melting such substances do not immediately pass from the ordinary crystalline state into the state of an isotropic liquid; one or more anisotropic intermediate phases are successively formed (mesophases), in which there occurs a specific ordering in the arrangement of molecules. Each liquid-crystalline phase exhibits a certain macrosymmetry, which is due to both the type of ordering and the symmetry of the molecules themselves. An essential point is the presence or absence in molecules of planes of symmetry: molecules with no planes of symmetry may be "right-handed" and "left-handed" (Fig. 12.1) and are called *chiral.* Accordingly, the point macrosymmetry of liquid crystals composed of such molecules does not have planes of symmetry either.

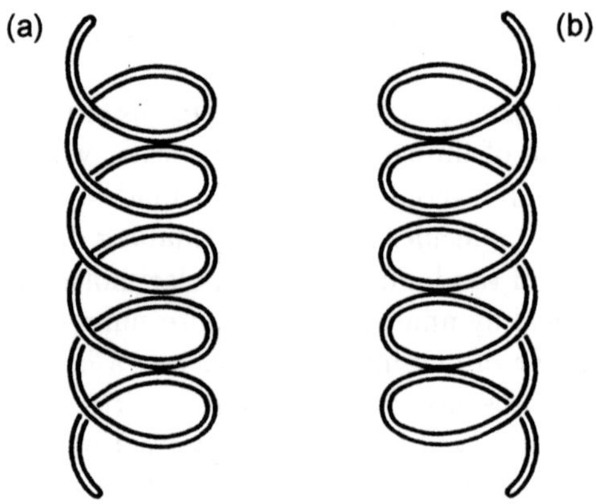

Fig. 12.1. (**a**) The right-handed and (**b**) left-handed chiral molecules of a liquid crystal (schematic).

Certain types of orientation ordering of molecules in liquid crystals are presented in Table 12.1. The first row shows a disordered isotropic liquid, which is devoid of symmetry planes if the molecules are chiral: the reflection in the plane would convert "right-handed" to "left-handed" molecules and vice versa. In Table 12.1 to the absence of such planes there conditionally corresponds a geometrical object in the form of a sphere, whose radius rotates in a clockwise or counterclockwise direction (if one looks along the radius in the direction of the sphere center). Such a sphere has a ∞/∞ symmetry.

The second row corresponds to liquid crystals, which are characterized by a predominant orientation of the long axes of molecules along a single direction determined by the unit vector of the director $\boldsymbol{n}$. Such liquid crystals

Table 12.1. Orientation ordering of molecules in liquid crystals.

	Kind of ordering a	Symmetry: Nonchiral molecules b	Symmetry: Chiral molecules c
I		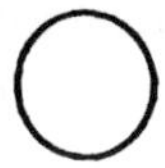 $\infty\infty m$	 ∞/∞
II		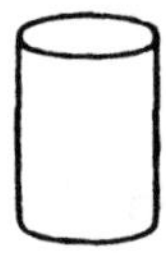 ∞/mm	 $\infty/2$
III	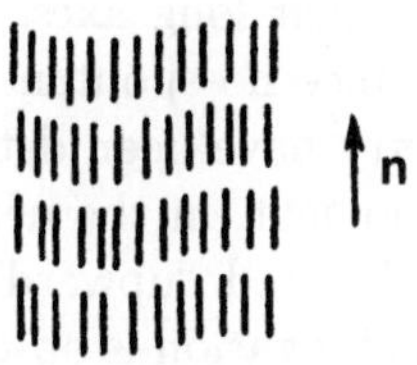	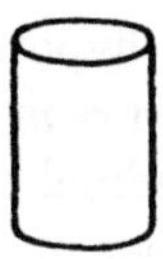 ∞/mm	 $\infty/2$
IV	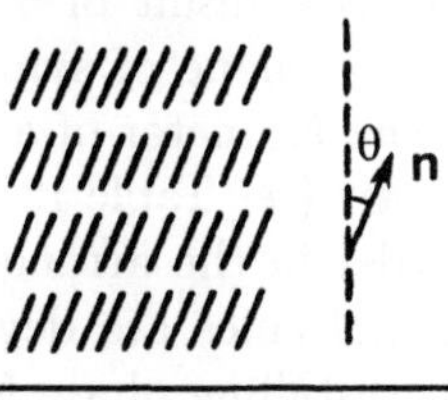	 $2/m$	 2

are called *nematic*; to a drop of a nematic consisting of achiral molecules may be ascribed the symmetry of a cylinder at rest (∞/mm) and there is one symmetry axis of an infinite order, which coincides in direction with the vector n, which is not polar, however, since the plane perpendicular to it renders the ends of this axis physically equivalent. On the contrary, in a nematic liquid crystal composed of chiral molecules there are no symmetry planes and there remain only an infinitely large number of two-fold axes perpendicular to the ∞ axis ($\infty/2$ symmetry). As can be seen from Table 12.1, in nematic liquid crystals the centers of gravity of molecules are arranged in space quite randomly and anisotropy arises due to the prevailing orientation of molecular axes, i.e., is of exceptionally orientational character. The third and fourth rows shows liquid crystals which are called *smectic*. We see that in these substances there is an additional ordering in the form of a layered arrangement of the centers of mass of molecules with a nematic orientational order being preserved; the smectic phases may differ in the orientation of the director vector. In the smectic A-phase (smectic A) the molecules are oriented, on an average, perpendicular to the planar layers (the third row), whereas in the smectic C-phase (smectic C) the director vector is inclined relative to the normal to the planes of layers at a certain angle θ (the fourth row). Note the following important circumstance: the smectic C liquid crystal consisting of chiral molecules has a special polar direction which lies in the plane of smectic layers.

12.2 Conditions for Existence of Dipolar Ordering in Liquid Crystals

In analyzing variants of dipolar ordering in liquid crystals we should discuss the already mentioned possibility of occurrence in nematic or smectic A-phases of liquid crystals of a homogeneous electric polarization associated with the dipole moments of molecules directed along their long axes. However, here too this possibility is purely hypothetical since no spontaneously polarized liquid-crystalline systems with ∞m symmetry have been detected so far: even if molecules have a longitudinal dipole moment, in the nematic and smectic A-phase the average number of dipoles directed "upward" and "downward" is always the same and no spontaneous polarization is observed.

From the symmetry standpoint this means that as a result of a phase transition into the nematic state there is preserved the plane of symmetry perpendicular to the direction of the ∞ axis, and the parameter of a phase transition involving a symmetry change $\infty\infty m \to \infty mm$ (see Table 12.1) will be a tensor quantity characterizing the degree of order in the arrangement of molecular long axes. It should be noted that if molecules are devoid of symmetry planes, i.e., if they are chiral, the liquid-crystalline phase formed has, in addition, a spiral ordering and a $\infty/2$ symmetry. Here the direction of the director vector $\boldsymbol{n}$ undergoes a regular change in space so that along the helical axis z we have $n_x = \cos(q_0 z + \varphi_0)$, $n_y = \sin(q_0 z + \varphi_0)$ and $n_z = 0$. Helicoidal structures of this type, which are periodic along the z axis with a period of $L = 2\pi/q_0$, are called *cholesteric*.

As pointed out above, from the standpoint of the possibility of a dipolar ordering in liquid-crystalline systems the following phase transitions may be of interest:

Chiral smectic A (SmA^*) $\to$ chiral smectic C (SmC^*)

which involve a symmetry change $\infty/2 \to 2$. The latter group is polar. It is easy to see that the ∞ axis is lost in such a transition as a result of the departure, at a certain temperature, of the director vector $\boldsymbol{n}$ from the direction normal to the plane of smectic layers.

Here the symmetry of a liquid crystal in the SmC^*-phase is polar, irrespective of the presence or absence of a permanent dipole moment in molecules. For a spontaneous polarization to occur along the polar axis 2 it is necessary that the molecules have a transverse dipole moment lying in the plane of a smectic layer.

At present about 20 organic compounds are known to exist, which undergo phase transitions of the type indicated. The most thoroughly studied compound is one with the abbreviated name DOBAMBC. The molecule of this organic compound is shown in Fig. 12.2 and Fig. 12.3 gives the sequence of phase transitions that occur upon cooling.

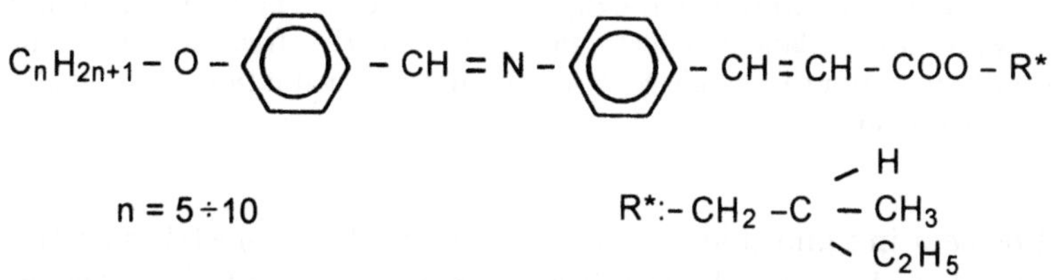

Fig. 12.2. The formula representation of the molecule of the ferroelectric liquid crystal DOMAMBC.

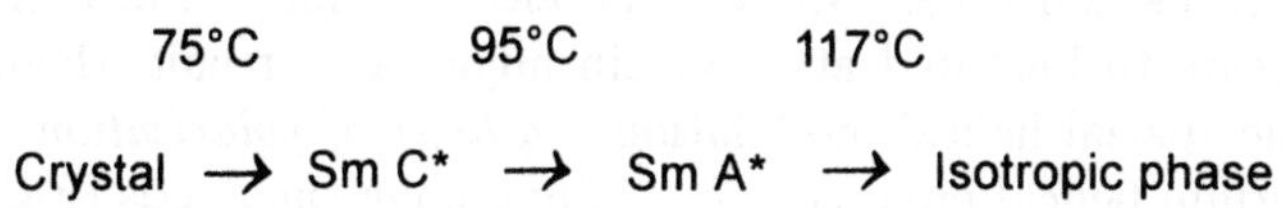

Fig. 12.3. Sequence of phase transformations in the chiral liguid crystal DOBAMBC.

The characteristic features are the strongly extended shape of the molecule, and the presence of two benzene rings, a long alkyl "tail" C_nH_{2n+1} and a chiral group R^*, which eliminates the plane of symmetry. The transverse dipole moment of the molecule is tied up with –C=O, –C=N and CH_3–C groups, being distributed along the length of the molecule. Though this moment is of the order of 1 D, the contribution of each molecule to the total spontaneous polarization of a liquid crystal ($P_s \approx 10^{-9}$–10^{-8} C·cm^{-2}) does not exceed 0.01 D. This means that, in contrast to the smectic A^*-phase, where there exists a free rotation of molecules about their long axes, in the C^*-phase there occurs only a partial redistribution of the probability of different directions of the transverse dipole moment (Fig. 12.4), such that one of the short axes of the molecule along which its dipole moment is aligned spends a somewhat longer time in the azimuth corresponding to the axis of the smectic layer.

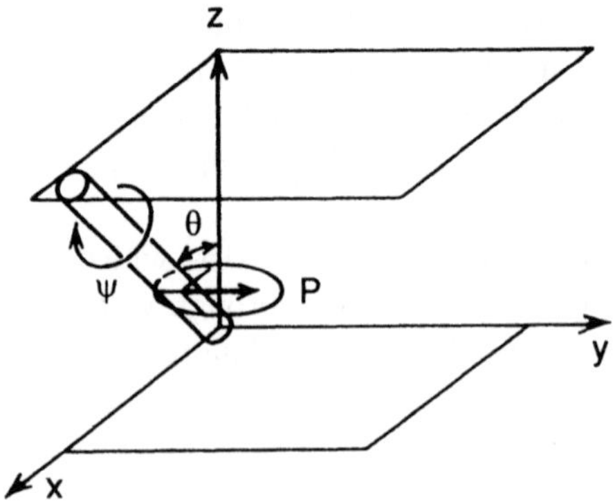

Fig. 12.4. Predominant orientation of the dipole moment of the molecule of a chiral SmC^* liquid crystal in the plane of smectic layers xy: "nearly" free rotation about the long axis with an azimuthal angle ψ; θ is the angle of inclination of molecules in the plane of smectic layers.

It should be kept in mind that the majority of chiral ferroelectric smectics have their achiral analogs, which exhibit the same sequence of phase transitions and which, however, do not undergo a spontaneous polarization in the "inclined" C-phase; in general, there is practically no difference in the thermodynamic properties of simple (achiral) and ferroelectric (chiral) smectics. A characteristic feature of dipolar ordering in chiral SmC^* liquid crystals, is that in the phase transition SmA^*→SmC^* the electric moment of each smectic layer xy appears to be tilted at a certain angle; as a result, there appears a macroscopic spatial helical modulation – *a helix of polarization* – with a certain equilibrium helical pitch $p_0 = 5 - 10\,\mu$m along the z axis (Fig. 12.5). The result is that the volume-averaged polarization of a liquid crystal becomes equal to zero, and in this sense the SmC^*-phase is an analog of the incommensurate phase in proper ferroelectrics (see Chap. 11). The only difference is that in the case under consideration the pitch of a spatially modulated structure has the above-indicated macroscopic values and the question of whether the interlayer spacing is a multiple of the helical pitch or not is devoid of any real meaning.

Thus, we state that a spontaneous polarization occurs here only within each of the smectic layers; on the whole, the modulated incommensurate phase displays the symmetry of the high-temperature SmA^*-phase and exhibits no macroscopic spontaneous polarization. It is therefore clear that the SmC^* phase of a liquid crystal may be called ferroelectric only conditionally. There is some justification: a polarization helix can be "unwound" by applying a relatively low electric field along the planes of smectic layers. As a result, the liquid-crystalline specimen passes into a state with a homogeneous electric polarization, which is characteristic of ferroelectrics. In low-frequency variable fields the dependence of electric polarization on electric field in the SmC^* phase has the form of a hysteresis loop since the "winding up" of the

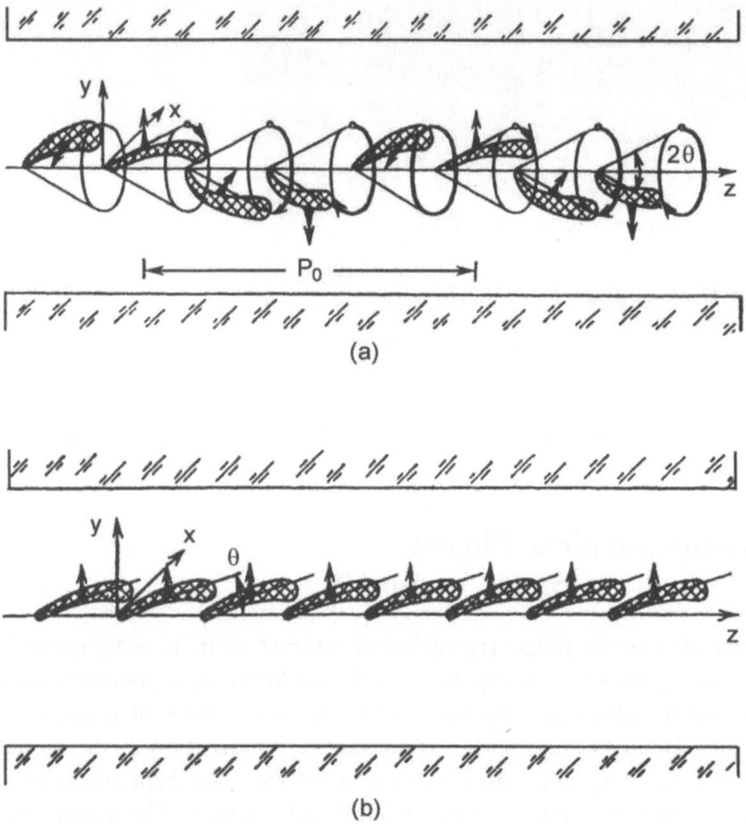

Fig. 12.5. Arrangement of molecules and their transverse dipole moments in the SmC^*-phase of a liquid crystal (smectic layers are parallel to the xy plane; inside the layers the long axes of molecules are rotated by an angle θ relative to the normal and smectic layers); (**a**) for the case $E = 0$; (**b**) homogeneous polarization in an external electric field directed along the plane of smectic layers.

helix and the disappearance of polarization with the electric field being reduced to zero are relatively slow processes (Fig. 12.6). Phase transitions are accompanied by anomalies in dielectric permittivity, which will be considered after the discussion of the currently available versions of the Landau theory of phase transitions worked out for describing phase transitions of the type Sm$A^* \rightarrow$ SmC^* in liquid crystals.

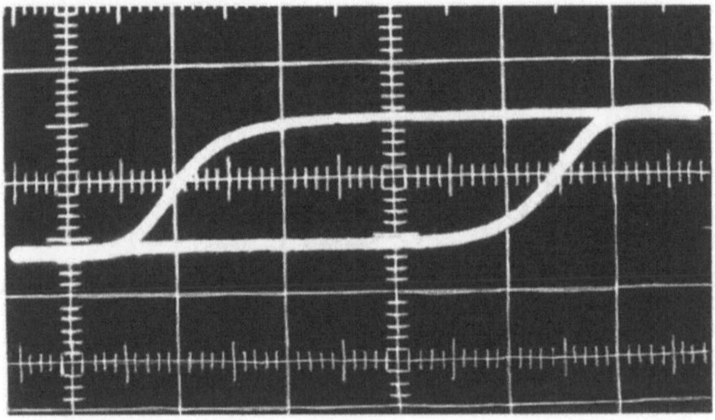

Fig. 12.6. Dependence of polarization on electric field in the SmC^*-phase of a liquid crystal [17].

12.3 Phenomenological Theory of Phase Transition SmA^* → SmC^*

As pointed out above, the phase transition of interest to us is accompanied by a hypothetical symmetry change $\infty/2 \to 2$; the lowering of symmetry in the low-temperature phase is associated with the appearance of a slope of molecules in smectic layers and with a spontaneous polarization in the plane of smectic layers. To analyze phenomena connected with this type of second-order phase transition, use may be made of the Landau theory; the parameter of such a phase transition must be a physical quantity, whose transformation properties provide the required symmetry change. A specific feature of the situation in question is that the symmetry of a high-temperature phase is continuous (limiting) and the hypothetical symmetry of a low-temperature phase is discrete. The first contains, as symmetry elements, rotations by an arbitrary angle φ about an infinite-fold axis and rotations by an angle π about twofold axes which are perpendicular to the ∞ axis.

Formally, the phase-transition parameter may be, in this case, the two-component vector quantity $\boldsymbol{P} = \boldsymbol{i}P_x + \boldsymbol{j}P_y$, which has the meaning of the polarization vector lying in the plane of a smectic layer. From Fig. 12.7 one can see that the orientation of the director vector is specified by the angles θ and φ, so that $n_1 = \sin\theta\cos\varphi$, $n_2 = \sin\theta\sin\varphi$, $n_3 = \cos\theta$ and $\theta = [\boldsymbol{k} \times \boldsymbol{n}]$ ($\boldsymbol{i}$, $\boldsymbol{j}$ and $\boldsymbol{k}$ are unit vectors directed along the axes x, y and z, respectively). The vector $\boldsymbol{P}$, like θ, is normal to the plane in which the vectors $\boldsymbol{k}$ and $\boldsymbol{n}$ lie; its projections on to the axes x and y are equal to:

$$P_x = -P\sin\varphi, \qquad P_y = P\cos\varphi. \tag{12.1}$$

Under symmetry operations these quantities are transformed as the components of the director vector n_2 and n_1. Analogous transformations are experienced by quadratic combinations of the components of the director vector

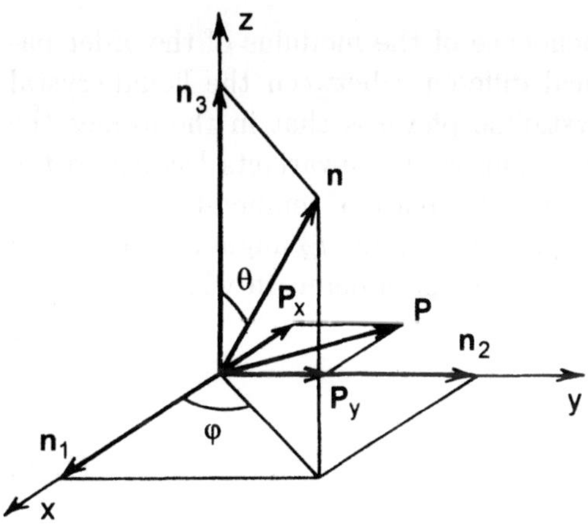

Fig. 12.7. Disposition of the director vectors and spontaneous polarization in a SmC^* liquid crystal (xy is the plane of a smectic layer).

n_3n_2 and n_3n_1, which determine its slope relative to the normal to the planes of smectic layers, i.e., the orientation of the long axes of molecules.

Thus, according to the logic of the Landau theory, the parameter of the phase transition SmA_* →SmC^* may be either the quantities P_x and P_y or

$$\xi_1 = n_3n_1 = \frac{1}{2}\sin(2\theta)\cos\varphi, \tag{12.2}$$

$$\xi_2 = n_3n_2 = \frac{1}{2}\sin(2\theta)\sin\varphi. \tag{12.3}$$

In the foregoing, while characterizing the symmetry of the SmC^*-phase, we used the word "hypothetical". It was implied that the Landau theory forbids a second-order phase transition involving a symmetry change $\infty/2 \to 2$ since the components of the order parameter for such a transition may be used to construct the Lifshitz invariant

$$P_x\frac{\partial P_y}{\partial z} - P_y\frac{\partial P_x}{\partial z},$$

or

$$\xi_1\frac{\partial \xi_2}{\partial z} - \xi_2\frac{\partial \xi_1}{\partial z}.$$

Using the inferences of the preceding chapter, we come to the conclusion that the presence of such invariants must lead in this particular case to the appearance of an incommensurate modulation below the phase-transition temperature – the helical winding of the vector $\boldsymbol{P}$ about the z axis, which is observed experimentally (see Fig. 12.5).

It is obvious that the expression for the thermodynamic potential may be written in this case in the form given by (11.8), from which the fourth term on the right-hand side has to be dropped with the substitution $\rho \to P$. As has been shown, the routine analysis of the expression obtained points to properties that are characteristic of the SmC^* phase, such as the helical winding of polarization about the z axis with a constant "angular velocity"

$q_0 = \sigma/\delta$ and the temperature dependence of the modulus of the order parameter $P^2 \propto (T_i - T)$. The typical difference between the liquid-crystal incommensurate phase and the crystalline phase is that in the former the thermodynamic potential contains no anisotropic invariants leading to the nucleation of narrow domain walls with a decrease of temperature.

It should be pointed out that the quantities ξ_1 and ξ_2 might also be chosen equally well as the components of the order parameter, in which case their invariant combinations are defined as

$$\xi_1^2 + \xi_2^2 = \frac{1}{4}\sin^2(2\theta) \approx \theta^2, \tag{12.4}$$

$$(\xi_1^2 + \xi_2^2)^2 \approx \theta^4, \tag{12.5}$$

$$\xi_1 \frac{\partial \xi_2}{\partial z} - \xi_2 \frac{\partial \xi_1}{\partial z} = \theta^2 \frac{\partial \varphi}{\partial z}, \tag{12.6}$$

$$\left(\frac{\partial \xi_1}{\partial z}\right)^2 + \left(\frac{\partial \xi_2}{\partial z}\right)^2 = \theta^2 \left(\frac{\partial \varphi}{\partial z}\right)^2, \tag{12.7}$$

and the density of the thermodynamic potential is given by

$$\Phi = \Phi_0 + \frac{1}{2}A\theta^2 + \frac{1}{4}\beta\theta^4 + \frac{1}{2}\delta\theta^2 \left(\frac{\partial \varphi}{\partial z}\right)^2 + \sigma\theta^2 \frac{\partial \varphi}{\partial z}. \tag{12.8}$$

The last expression coincides, with an accuracy of up to the substitution $\rho \to \theta$, with (11.8) and, naturally, all the conclusions arrived at above remain valid.

We have demonstrated that the choice of any physical quantity with identical transformation properties as the phase-transition parameter does not (and cannot) change the general picture of the phase transition Sm$A^* \to$ SmC^*. This picture can be extended if we introduce both two-component order parameters into the thermodynamic potential. Thus, it is suggested that the smectic C^* phase be treated as a pseudoproper phase (see Chap. 4). In such a case, the role of the "true" order parameter can be played by the deflection of the long axes of molecules from the direction of the normal to the planes of smectic layers and the attendant polarization may be treated as a "secondary" effect. Such a treatment may be justified by the obvious circumstance that the absence of polarization (the condition $P = 0$) is not an obstacle for a phase transition and the appearance of a spontaneous inclination of molecules, whereas the "constant-inclination" condition leads to the elimination of the transition. The relationship between the two order parameters is determined in this case by two possible manifestations of a peculiar "piezoeffect" – *an electroclinic effect*, when the artificial creation of an inclination of molecules leads to the polarization of a liquid crystal ($P \propto \theta$), and *a flexoelectric effect* when an electric polarization occurs as a result of an inhomogeneous distribution of the director vector in space: its bending and winding. In this case, the components of the polarization vector appear to be proportional to the space derivatives of the orientational order parameter:

$$P_x \propto \frac{\partial \xi_1}{\partial z}, \qquad P_y \propto \frac{\partial \xi_2}{\partial z}. \tag{12.9}$$

With account taken of the identical transformation properties of the "primary" and "secondary" order parameters, their cross-terms in the thermodynamic potential may be written as

$$P_x\xi_2 - P_y\xi_1 = \theta P, \tag{12.10}$$

$$P_x\frac{\partial \xi_1}{\partial z} + P_y\frac{\partial \xi_2}{\partial z} = \theta P\frac{\partial \varphi}{\partial z}. \tag{12.11}$$

The incomplete thermodynamic potential of a liquid crystal will thus have the following form:

$$\begin{aligned}\Phi = \Phi_0 &+ \frac{1}{2}\alpha(T - T_0)(\xi_1^2 + \xi_2^2) + \frac{1}{4}\beta(\xi_1^2 + \xi_2^2)^2 \\ &+ \sigma\left(\xi_1\frac{\partial \xi_2}{\partial z} - \xi_2\frac{\partial \xi_1}{\partial z}\right) + \frac{1}{2\chi_0}(P_x^2 + P_y^2) \\ &- \mu_1(P_x\xi_2 - P_y\xi_1) - \mu_2\left(P_x\frac{\partial \xi_1}{\partial z} + P_y\frac{\partial \xi_2}{\partial z}\right) \\ &+ \frac{1}{2}\delta\left[\left(\frac{\partial \xi_1}{\partial z}\right)^2 + \left(\frac{\partial \xi_2}{\partial z}\right)^2\right].\end{aligned} \tag{12.12}$$

Here α, β, δ, and χ_0 are constants and μ_1 and μ_2 are the piezomoduli which describe the electroclinic and flexoelectric effects, respectively. In order to analyze the behaviour of the system in an external electric field, it is necessary to add the term $-\boldsymbol{E}\cdot\boldsymbol{P}$ to (12.12). Expression (12.12) in variables θ and φ becomes (with account taken of the smallness of the quantity θ):

$$\begin{aligned}\Phi = \Phi_0 &+ \frac{1}{2}\alpha(T - T_0)\theta^2 + \frac{1}{4}\beta\theta^4 + \sigma\theta^2\frac{\partial \varphi}{\partial z} \\ &+ \frac{1}{2\chi_0}P_2 - \mu_1\theta P - \mu_2\theta P\frac{\partial \varphi}{\partial z} + \frac{1}{2}\delta\theta^2\left(\frac{\partial \varphi}{\partial z}\right)^2.\end{aligned} \tag{12.13}$$

As a matter of fact, expression (12.13) differs from (12.8) only in that the two parameters considered are two-component quantities and that the space derivatives of the components of the orientational order parameter are taken into account in explicit form.

The magnitude and character of the temperature and coordinate (z) dependences of θ, P and φ are determined by the minimum of the functional

$$\int_L \Phi\left(\theta, P, \varphi, \frac{\partial \varphi}{\partial z}\right)\mathrm{d}z. \tag{12.14}$$

Writing, as before, the Euler equation for the function $\varphi(z)$, for example, we have $\mathrm{d}^2\varphi/\mathrm{d}z^2 = 0$, which means that $\varphi = qz + \varphi_0$. The corresponding equations for the determination of the equilibrium values of q, P and θ reduce to the following equations:

$$\frac{\partial \Phi}{\partial q} = \sigma\theta_0^2 - \mu_2\theta_0 P_0 + \delta\theta_0^2 q_0 = 0; \tag{12.15}$$

$$\frac{\partial \Phi}{\partial P} = \chi_0^{-1} P_0 - \mu_1\theta_0 - \mu_2\theta_0 q_0 = 0; \tag{12.16}$$

$$\frac{\partial \Phi}{\partial \theta} = \alpha(T - T_0)\theta_0 + \beta\theta_0^3 - \mu_1 P_0$$

$$+ 2\sigma\theta_0 q_0 - \mu_2 P_0 q_0 + \delta\theta_0 q_0^2 = 0. \tag{12.17}$$

From (12.15) we find

$$q_0 = \frac{\mu_2\theta_0 P_0 - \sigma\theta_0^2}{\delta\theta_0^2}, \tag{12.18}$$

and (12.16) yields

$$P_0 = \chi_0(\mu_1 - \mu_2 q_0)\theta_0, \tag{12.19}$$

from which we obtain

$$q_0 = \frac{\chi_0\mu_1\mu_2 - \sigma}{\delta - \chi_0\mu_2^2} \tag{12.20}$$

which is the equilibrium value of the "wave number" of the helix. We see that in the approximation adopted this quantity is temperature independent. Substituting (12.19) and (12.20) into (12.17), we obtain an equation for the determination of the equilibrium value of the angle of inclination of molecules to the plane of layers:

$$\alpha(T - T_c)\theta_0 + \beta\theta_0^3 = 0 \tag{12.21}$$

where

$$T_c = T_0 + \frac{1}{\alpha}[(\delta - \chi_0\mu_2^2)q_0^2 + \chi_0\mu_1^2] \tag{12.22}$$

is the temperature of the phase transition SmA^* →SmC^*.

Equation (12.21) immediately yields the sought-for solution:

$$\theta_0 = 0 \qquad \text{for the Sm}A^* \text{ phase} \qquad (T > T_c), \tag{12.23}$$

$$\theta_0^2 = \pm\left[-\frac{\alpha(T - T_c)}{\beta}\right] \qquad \text{for the Sm}C^* \text{ phase} \qquad (T < T_c). \tag{12.24}$$

It is easy to see that the thermodynamic potential (12.16) provides a qualitative description of the basic regularities of the thermodynamic behavior of a chiral smectic liquid crystal in the region of the phase transition into a spontaneously polarized state: the appearance of the angle of inclination of molecules to the normal to the planes of smectic layers and a proportional polarization which lies in these planes; a helicoidal "winding" of the director vector and a spontaneous polarization about the z axis, which is normal to the planes of smectic layers; and the root dependence (characteristic for the Landau theory) of the angle of inclination of molecules θ and, accordingly, of the spontaneous polarization on $T_c - T$.

The helical pitch p_0 (Fig. 12.5) equal to $2\pi/q_0$ is found to be independent of temperature.

It should be noted that the helix-like ordering of the dipole moments of molecules corresponds to the minimum of the thermodynamic potential of the system and in this sense the helix pitch is an analog of the period of a spatially modulated structure in the incommensurate phase of crystalline ferroelectrics. In contrast to the latter, in liquid crystals the helix pitch is large, being equal to 5–10 μm, i.e., thousands of smectic layers; thus, the presence of a helicoidal structure can easily be registered by direct optical observations. Moreover, the angle of inclination of molecules determines the orientation of the optical indicatrix in a liquid crystal with a homogeneous polarization, i.e., in an external electric field directed along the smectic layers and sufficient for the helix to be "unwound". Thus, using the measurements of the value of P_s from the hysteresis loops (see Fig. 12.6), the orientation of the optical indicatrix (by means of conventional crystalloptic methods) and also of the period of space modulation, the helix pitch (by direct observations or by using diffraction phenomena), we can determine the temperature dependences of these quantities.

We do not try here to offer a detailed analysis of rather specific methodological problems associated with a liquid-crystal experiment; we only note that their specific nature lies in the necessity to place the specimen between the containing planes – ordinarily in a planar capillary made mostly of glass. The distance between the supporting planes is provided by spacers with a calibrated thickness and the orientation of the director vector is specified by special (chemical, mechanical) treatment of the containing surfaces or by the action of external magnetic or electric fields. Thus, it is possible to provide on characteristic spacer thicknesses of 4–100 μm the specified orientation of the vector $\boldsymbol{n}$ parallel to the containing surfaces (planar) or perpendicular to them (homeotropic). The object of investigation is usually a liquid-crystalline cell schematically depicted in Fig. 12.8.

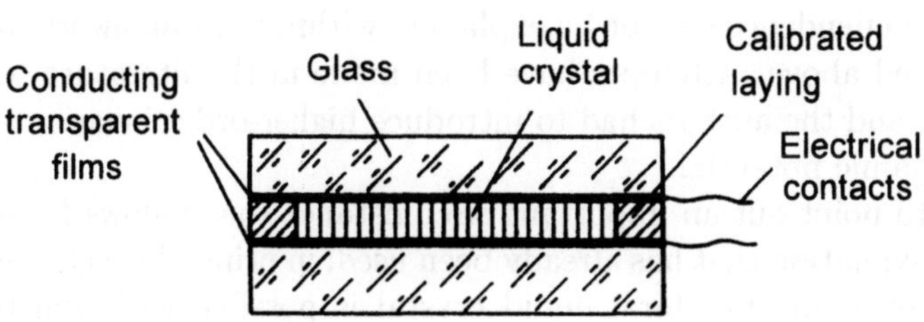

Fig. 12.8. A liquid-crystalline cell for the study of the optical and electrical properties of a liquid-crystal object.

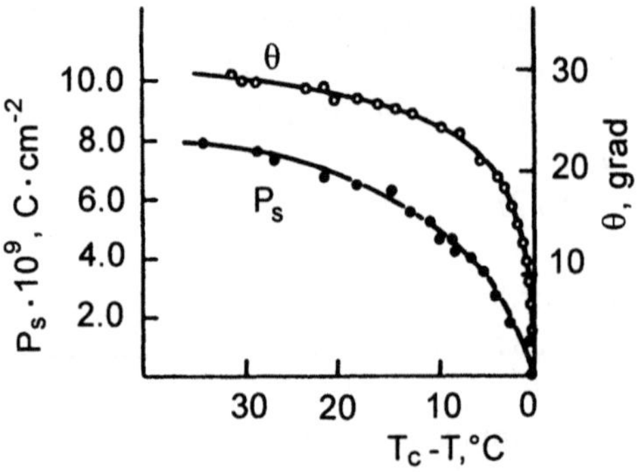

Fig. 12.9. Temperature dependence of P_s and θ in the liquid DOBAMBC crystal [18].

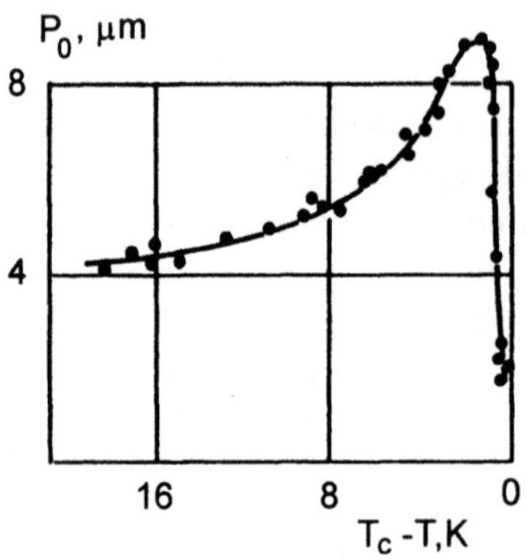

Fig. 12.10. Temperature dependence of the helicoid pitch of the liquid DOBAMBC crystal [19].

Figure 12.9 shows the temperature dependence of the spontaneous polarization P_s and the angle θ for the liquid crystal DOBAMBC near the temperature of the phase transition SmA^* →SmC^*; it is obvious that they are qualtitatively similar, though there are experimental data indicating that the proportionality coefficient that establishes the relationship between these two quantities depends on temperature and displays an anomaly near T_c. Experiment shows also that the helical pitch is far from being a constant quantity, as follows from (12.10). Figure 12.10 shows the variation of the quantity p_0 with temperature. It can be seen that the helical pitch increases upon approach from below to the phase-transition point, reaches a rather sharp peak at $T \sim T_c - 1\,\mathrm{K}$ and then diminishes sharply at the transition point. Evidently, this kind of dependence cannot be explained within the framework of the theory considered above. Attempts have been made in the literature to describe this effect, and the authors had to introduce higher-order invariants into the thermodynamic potential.

It is necessary to point out an important circumstance that follows from (12.20) in the approximation that has already been used: in principle, a choice of the parameters of a smectic chiral liquid crystal is possible with which $\chi_0\mu_1\mu_2 - \sigma \sim 0$ and the quantity $p_0 = 2\pi/q_0$ becomes infinitely large, which corresponds to the homogeneous polarization of the liquid crystal. The realization of this possibility is extremely important for numerous practical applications of ferroelectric liquid crystals.

12.4 The Behavior of a Ferroelectric Smectic Liquid Crystal in an External Electric Field

In this section we will consider the manner in which the above-described specific features of the structure of smectic chiral liquid crystals determine their behavior in an electric field; the basic characteristics here are the dependences of the electric polarization on the field $P(E)$ and the dielectric "response" – the temperature and frequency dependences of the dielectric constant measured in a low electric field.

We have already pointed out above that in the SmC^* phase the dependence of the polarization on the electric field directed along the plane of a smectic layer perpendicular to the helical axis has the form of a hysteresis loop characteristic of ferroelectrics; to the total polarization of the sample there corresponds the complete "unwinding" of the helix i.e., its "pitch" going to infinity.

In what follows we will use the coordinate system adopted in Figs. 12.5 and 12.7; the electric field is directed along the axis y perpendicular to the helical axis. It is clear that the dielectric response of the system is associated with the behavior of the helix in an electric field: the macroscopic polarization $\langle P\rangle$ that appears can be determined by two types of distortion of the helix: the first is caused by a perturbation in the distribution of the azimuthal angle $\varphi(z,E)-\varphi_0(z)$ at constant value of the angle of inclination of molecules θ_0 relative to the z axis and the second by a periodic perturbation $\theta_1(z,E)$ of the angle of inclination of molecules $\theta(z,E)=\theta_0+\theta_1(z,E)$ with the helix pitch being unperturbed.

In order to analyze the situation quantitatively, we can use the thermodynamic potential in the form of (12.13) with the addition of two terms: $-\boldsymbol{P}\cdot\boldsymbol{E}$, which takes into account the interaction of the polarization with the electric field $\boldsymbol{E}(0,E,0)$ and $(1/2)g(\partial\theta/\partial z)^2$, which describes the increase of the thermodynamic potential with the appearance, in the electric field, of a spatially inhomogeneous distribution of the angle of inclination of molecules – an elastic energy associated with the perturbation of this angle.

Thus, we should analyze the thermodynamic potential that has the following form:

$$\begin{aligned}\Phi =&\Phi_0+\frac{1}{2}\alpha(T-T_0)\theta^2+\frac{1}{4}\beta\theta^4+\frac{1}{2\chi_0}(P_x^2+P_y^2)-(P_x\sin\varphi-P_y\cos\varphi)\\ &\times\left(\mu_1-\mu_2\frac{\partial\varphi}{\partial z}\right)\theta+\sigma\theta^2\frac{\partial\varphi}{\partial z}+\frac{1}{2}\delta\theta^2\left(\frac{\partial\varphi}{\partial z}\right)^2+\frac{1}{2}g\left(\frac{\partial\theta}{\partial z}\right)^2-\boldsymbol{E}\cdot\boldsymbol{P}.\end{aligned} \qquad (12.25)$$

It is clear that at $E = 0$, $\varphi = q_0 z$ and $\theta(z) = \theta_0 = \text{const}$. We return to the case that has already been considered. At this point it is necessary to minimize the functional

$$\int_0^L \Phi\left(\theta, \varphi, P_x, P_y, \frac{\partial\varphi}{\partial z}, \frac{\partial\theta}{\partial z}, E_y\right) \mathrm{d}z.$$

The Euler equations for P_x and P_y reduce to $\partial\Phi/\partial P_x = 0$ and $\partial\Phi/\partial P_y = 0$, from which we have

$$P_x = \chi_0\theta\left(\mu_1 - \mu_2\frac{\partial\varphi}{\partial z}\right)\sin\varphi, \tag{12.26}$$

$$P_y = \chi_0 E_y - \chi_0\theta\left(\mu_1 - \mu_2\frac{\partial\varphi}{\partial z}\right)\cos\varphi. \tag{12.27}$$

Substituting these values into (12.25) and minimizing the functional with respect to $\varphi(z)$ and $\theta(z)$, we arrive at the following system of equations:

$$\delta\theta_0^2\frac{\partial^2\varphi}{\partial z^2} + \chi_0 E\left(\mu_1\theta_0\sin\varphi - \mu_2\frac{\partial\theta}{\partial z}\cos\varphi\right) = 0, \tag{12.28}$$

$$-2\tilde{\alpha}\theta_1 - g\frac{\partial^2\theta}{\partial z^2} + \tilde{\delta}\theta_0\left(\frac{\partial\varphi}{\partial z} - q_0\right)^2 + \chi_0 E + \left(\mu_1 - \mu_2\frac{\partial\varphi}{\partial z}\right)\cos\varphi = 0, \tag{12.29}$$

where

$$\theta = \theta_0 + \theta_1, \quad \theta_1 \ll \theta_0, \quad \tilde{\delta} = \delta - \chi_0\mu_2^2$$
$$\tilde{\alpha} = \alpha(T - T_0) - \chi_0\mu_1^2 - \tilde{\delta}q_0^2.$$

In low electric fields ($E \ll E_\mathrm{u}$, where E_u is the field which causes a complete unwinding of the helix) these equations may be written in the simplified form:

$$\tilde{\delta}\theta_0\frac{\partial^2\varphi}{\partial z^2} + \chi_0 E\mu_1\sin\varphi = 0, \tag{12.30}$$

$$-2\tilde{\alpha}\theta_1 + g\frac{\partial^2\theta_1}{\partial z^2} - \chi_0 E\left(\mu_1 - \mu_2\frac{\partial\varphi}{\partial z}\right)\cos\varphi = 0. \tag{12.31}$$

It is easy to see that with $E \ll (\delta_3 q_0^2\theta_0)/(2\chi_0\mu_1)$ the system has the following periodic solutions:

$$\varphi = q_0 z + \frac{\chi_0\mu_1 E}{\tilde{\delta}_3 q_0^2\theta_0}\sin(q_0, z), \tag{12.32}$$

$$\theta_1 = \frac{\chi_0(\mu_1 - \mu_2 q_0)}{2\tilde{\alpha} - g q_0^2}\cos(q_0 z), \tag{12.33}$$

These solutions can be used to find the macroscopic (averaged) polarization $\langle P\rangle$ and the dielectric constant.

Using (12.32) and (12.33) in combination with (12.27), we find for $T < T_c$:

$$\langle P_y \rangle = (\chi_0 + \delta\chi)E,$$

$$\delta\chi|_{T<T_c} = \frac{\chi_0^2}{2}\left[\frac{\mu_1^2}{\tilde{\delta}q_0^2} + \frac{(\mu_1 - \mu_2 q_0)^2}{g q_0^2 - 2\tilde{\alpha}}\right]. \tag{12.34}$$

Above the phase-transition point $\theta_0 = 0$; the homogeneous external electric field E_y is responsible for a homogeneous deflection of the vector $\boldsymbol{n}$ in low fields:

$$P_y = \frac{\chi_0 \mu_1^2 E}{\tilde{\alpha} + \tilde{\delta}q_0^2}, \tag{12.35}$$

from the direction of the normal to the plane of a smectic layer. In this case the dielectric constant is given by

$$\delta\chi|_{T>T_c} = \frac{\chi_0^2 \mu_1^2}{\tilde{\delta}q_0^2 + \tilde{\alpha}}. \tag{12.36}$$

Thus, in the paraelectric phase the quantity $\delta\chi$ is determined by the piezoelectric coefficient μ_1 alone since there is no macroscopic twisting of the orientation of molecules and, hence, the flexoelectric effect is not observed. From (12.34) and (12.36) it can be seen that the diectric constant does not go to inifinity as $T \to T_c$ (where $\tilde{\alpha} = 0$) since the low-temperature phase is spatially modulated and, hence, only the response of the system to the electric field modulated accordingly may be infinitely large. Only with

$$\tilde{\alpha} \gg \tilde{\delta}q_0^2$$

will the relations (12.34) and (12.36) be close to the Curie–Weiss law; upon approach to the Curie point the medium is no longer orientationally homogeneous and the contribution associated with the macroscopic twisting of the structure begins to be dominant. In the polar phase this contribution associated with polar (angle θ) and azimuthal (angle φ) perturbations of the helix remains predominant. If we eliminate in some way the helical winding of the structure (say, by applying, apart from the low variable field, a high constant displacing field, which completely "unwinds" the helix or by choosing a certain concentration of "right-handed" and "left-handed" chiral molecules of a liquid crystal providing the infinitely large helix pitch and uniformity of polarization), the dielectric response will be determined by a change of the polar angle and by the corresponding polarization induced by the the action of an electric field.

Experiment confirms, on the whole, the above considerations. Figure 12.11 presents the temperature dependence of the dielectric constant measured in a low variable electric field at different frequencies in the liquid crystal DOBAMBC. It ean be seen that the dielectric response of the liquid crystal sharply increases upon transition into the helicoidal phase. There is observed a strong frequency dependence of the dielectric constant, indicating

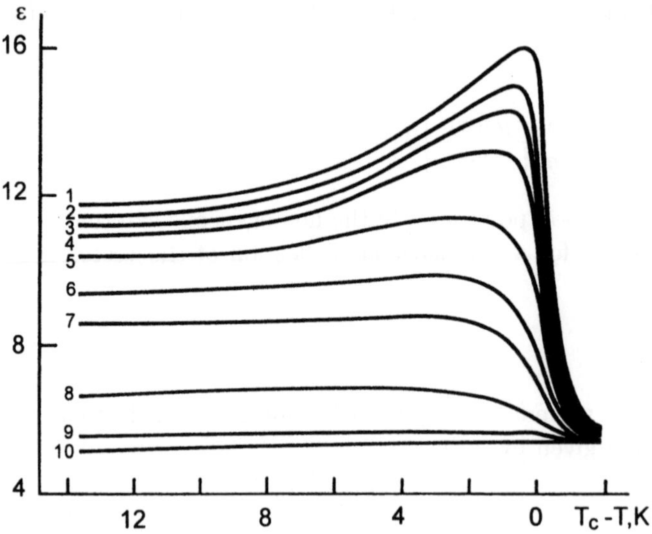

Fig. 12.11. Temperature dependence of dielectric constant at different frequencies of the measuring field in DOBAMBC, Hz: (*1*) 40; (*2*) 60; (*3*) 80; (*4*) 120; (*5*) 200; (*6*) 300; (*7*) 400; (*8*) 800; (*9*) 3000; (*10*) 8000 [19].

the occurrence of relaxation processes, which manifest themselves at very low frequencies ($f_{\rm rel} \sim 10^2\,{\rm Hz}$). As has been shown above, the interaction of a variable electric field with the helicoidal structure involves a periodic variation of the helical parameters; as the frequency of the measuring field increases these processes fail to occur because of the high dynamic viscosity of the liquid crystal. Application of a high constant electric field leads to a sharp decrease of ε in the ferroelectric phase, in which case the dielectric dispersion disappears completely when $E > E_u$ and the helix "unwinds" (Fig. 12.12).

The liquid ferroelectric crystal DOBAMBC is a model object, which has been used for carrying out important experimental investigations. These investigations have revealed all the most characteristic properties of these nonconventional materials. As has been noted earlier, the spontaneous polar-

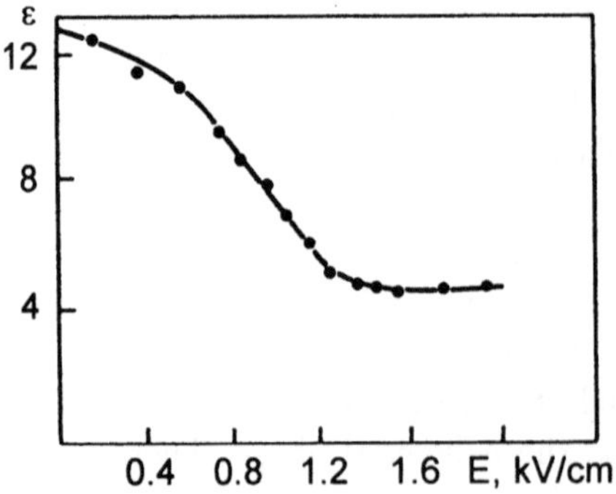

Fig. 12.12. Dielectric constant as a function of the electric field for the liquid DOBAMBC crystal ($T_c - T = 2.10°\,$C, $f = 80\,$Hz, $E_u = 1.28\,{\rm kV\cdot cm^{-1}}$ [20].

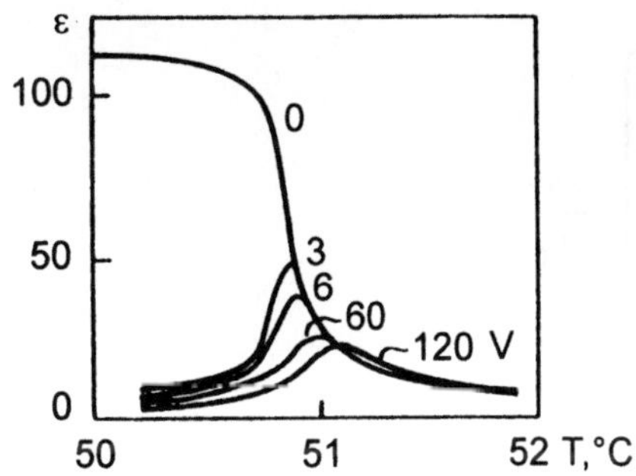

Fig. 12.13. Temperature dependence of the dielectric constant of the ferroelectric liquid 3M2CPOOB crystal at different electric voltages applied to a liquid-crystalline unit cell of thickness 250 μm, $f = 300$ Hz [21].

ization of DOBAMBC has a value of $\sim 10\,\mathrm{nC{\cdot}cm^{-2}}$ and $\varepsilon \sim 15$. There have lately been detected new classes of liquid crystals, for which $P_s \sim (1-5)\times 10^2$ $\mathrm{nC{\cdot}cm^{-2}}$ and $\varepsilon \sim 10^3$–10^4. Figure 12.13 is a plot of ε against temperature for such a material, which is known under the abbreviated name 3M2CPOOB ($T_c \sim 51°$ C) at various values of constant voltage applied to the liquid crystal. We see that at $U = 0$ the $\varepsilon(T)$ plot is qualitatively analogous to the one given in Fig. 12.11 for the DOBAMBC crystal; $E_u = 1.2\,\mathrm{kV{\cdot}cm^{-1}}$ for this liquid crystal and even a voltage of $U = 1$ V ($E \sim 10^4$ $\mathrm{V{\cdot}cm^{-1}}$) is sufficient for the helix to be "unwound", but, in contrast to DOBAMBC, in this case we see that in a homogeneously polarized liquid crystal in the region of a phase transition there is observed a peak of the dielectric constant, which does not disappear up to frequencies of ~ 50 kHz. Moreover, the $\varepsilon(T)$ plot obeys the $|T-T_c|^{-1}$ law, i.e., the Curie–Weiss law above and below the phase-transition point. This behavior follows from the reasoning given above.

Let us now consider the behavior of a chiral smectic C^* in high electric fields, when the approximate solutions of (12.30) and (12.31) in the form of (12.32) and (12.33) prove inapplicable and the distortions of the helix are rather strong. Of special interest are the properties of a liquid crystal when $E > E_u$. In this case, (12.30) and (12.31) are not solved in elementary functions and numerical methods are employed for their analysis. The results of numerical integration show that under the action of an electric field the harmonic dependence of the quantity P_y on z is substantially distorted since the angle φ is no longer a linear function of this coordinate. As the electric field increases there occurs a predominant reorientation of molecular dipoles in the direction of the field. Figure 12.14 shows the plot of $\cos\varphi$ versus $q_0 z$ obtained by means of numerical methods at various values of E/E_u; in such a case, the field of the complete "unwinding" of the helicoid was assumed to be equal to

$$E_p = \frac{\pi^2}{16}\frac{q_0\tilde{\delta}\theta_0}{\chi_0\mu_1} \tag{12.37}$$

and to correspond to a state with a homogeneous polarization ($\varphi = \varphi_0 =$ const).

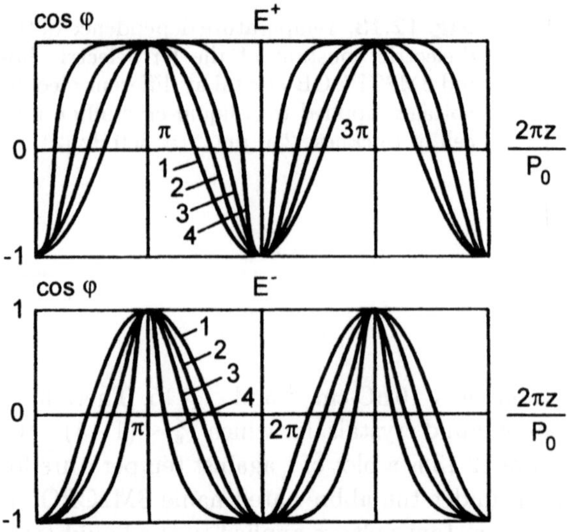

Fig. 12.14. Variation of the distribution of the cosine of the angle of rotation φ of molecules in the xy plane, which characterizes the reorientation of molecules in the helicoidal smectic C at various values of E/E_u: (1) 0; (2) 0.54; (3) 0.92; (4) 0.99 [22].

From Fig. 12.14 we see clearly that as E increases (more exactly, as the ratio E/E_{u} increases) there occurs a predominant orientation of dipoles in the direction of the field (the number of such dipoles is proportional to the area under the $\cos\varphi(z,E)$ curve). At $E = \pm E_{\mathrm{u}}$ the function $\cos\varphi$ assumes values of ± 1, which exactly corresponds to states with a homogeneous polarization, in which case the direction of the external electric field is reversed.

Figure 12.15 presents the thus obtained dependence of the reduced macroscopic polarization $\langle P\rangle/(\chi_0\mu_1\theta_0) = \langle\cos\varepsilon(z,E)\rangle$. Note the sharp increase of this quantity as $E \to E_{\mathrm{u}}$ and the subsequent constant value of polarization at large values of the electric field; moreover, the dependence of $\langle P\rangle$ on the electric field is reversible, i.e., does not display a hysteresis. But, as can be seen from Fig. 12.6, a hysteresis is observed experimentally and its form is very similar to the corresponding plots for crystalline ferroelectrics. Presumably in this particular case the occurrence of hysteresis phenomena is associated with large relaxation times for the transition from a homogeneous into a helicoidal state and vice versa. The complete reversibility of the switching of the polarization of the SmC^* liquid crystal can, in principle, be attained with a very gradual change of the electric field; because of the finite conductivity of liquid-crystal materials, polarization charges are very difficult to measure in this case.

The $P(E)$ plot for ferroelectric liquid crystals must, in principle, be hysteresisless since the original symmetry – the symmetry of the paraelectric smectic A^* phase – is described by a limiting point group $\infty/2$. In this case, the reorientation of dipoles must not be accompanied by energy losses (as

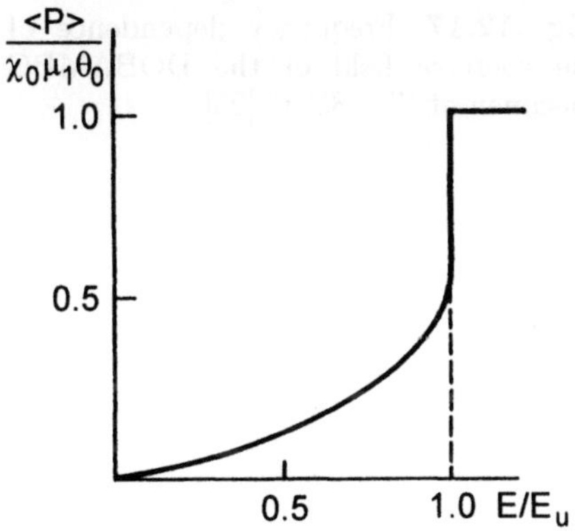

Fig. 12.15. Dependence of the reduced macroscopic polarization in the SmC^*-phase on the quantity E/E_u [23].

is the case with ferroelectric crystals) associated with the overcoming of a potential barrier. It may be said that in a liquid crystal with an infinite-fold axis the potential relief is a surface of revolution (Fig. 12.16) and that the polarization switching can occur by way of the movement of particles along the potential minimum, bypassing the barrier. Experiment shows that the "coercive field" E_c of DOBAMBC crystals decreases with diminishing frequency f of the electric field, which causes the switching of polarization, in which case the $E_c(f)$ plot is almost linear (Fig. 12.17). Extrapolation of this dependence to $f = 0$ results in a zeroth "coercive field". This fact confirms the considerations outlined above.

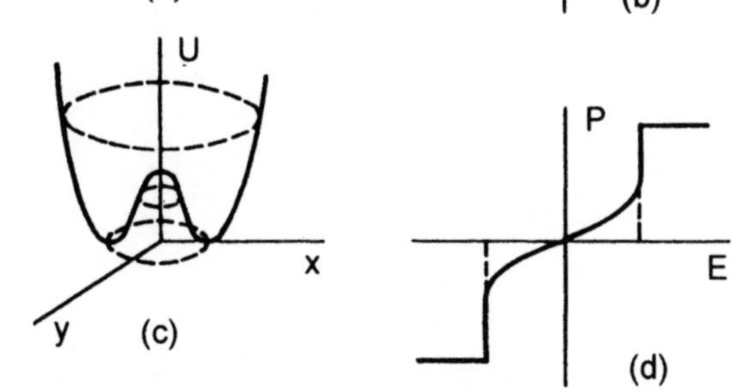

Fig. 12.16a–d. Diagrams illustrating the processes of polarization switching in ferroelectrics: (**a**) potential relief for a crystalline ferroelectric; (**b**) a hysteresis loop for a crystalline ferroelectric; (**c**) potential relief of a smectic C^*; (**d**) the curve of polarization switching for a smectic C^* [24].

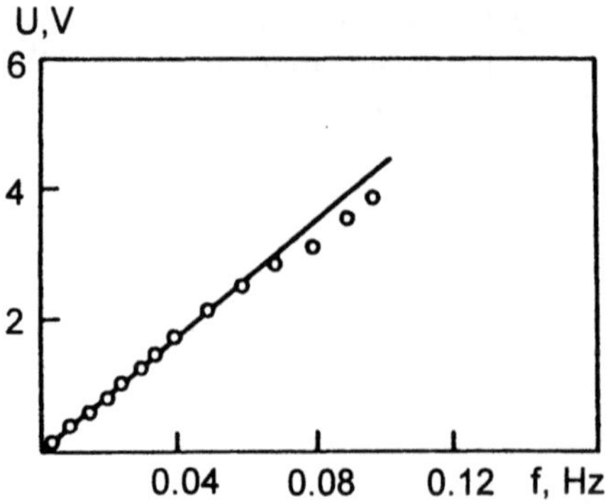

Fig. 12.17. Frequency dependence of the coercive field of the DOBAMBC specimen at $T = 85°$ C [25].

Concluding this chapter, we note that ferroelectric liquid crystals and materials based on such crystals become increasingly important for a more thorough understanding of the processes of electrical ordering in dielectrics and also for diverse practical applications. We have not touched upon such interesting problems as the electro-optic properties of these materials, the plausible interpretations of the anomalous temperature dependence of the helicoid pitch, the practical possibilities of creation of homogeneously polarized liquid-crystalline phases, the effect of the boundaries of a solid phase being in contact with a liquid crystal, and also the thickness of liquid-crystal films. It must, however, be kept in mind that the physics of ferroelectrics and the physics of liquid crystals are quite independent and self-contained branches of the physics of the condensed state of matter. The rather exotic liquid-crystal materials that have been described above serve as a "bridge" between these two branches. The purpose of our treatment was to demonstrate the interesting and unusual physical manifestations of ferroelectric phenomena in liquid crystals and the possibilities that are provided by the Landau theory for their analysis.

13. Crystallochemical Aspects of the Theory of Ferroelectric Phenomena

13.1 Calculation of the Constants of the Hamiltonians of Some Crystals

In Chaps. 8 and 9, while considering the microscopic theory of phase transitions, we used simple model Hamiltonians: the Hamiltonian (7.1) for displacive-type phase transitions and the Hamiltonian (8.8) for order–disorder phase transitions. The form of these Hamiltonians was specified so as to render the theory as simple as possible. The only requirement was the occurrence of a phase transition in a model system. Naturally, the coefficients in the Hamiltonians remained indeterminate; only certain plausible assumptions have been made as to the order of magnitude of their values. Of course, such an approach is insufficient for explaining the predictions of the properties of particular crystals. The problem therefore is to find Hamiltonians, though not as simple as (7.1) and (8.8), such that they would reflect more accurately the properties of concrete crystals.

The finding of such Hamiltonians and calculations of their coefficients constitute the subject matter of what is usually called the crystallochemical aspect or crystallochemical trend of the theory of ferroelectricity and, in general, of the theory of structural phase transitions. In an ideal case, it is presumed that the coefficients of the model Hamiltonian can be determined if only the chemical composition of a crystal is known. If we knew how to solve this problem, we would, in principle, be able to find out, by means of methods of statistical physics, whether a crystal of a given composition is a ferroelectric or not, to calculate the transition temperature in it, etc. In principle, it would be possible to synthesize a crystal if we knew in advance that it would possess the required properties. In reality, however, the solution of this task is only speculative at present.

The long path of its realization may be conveniently discussed if we divide it into several stages. We will cover these stages from the end by presuming that Hamiltonians of the type (7.1) or (8.8) but more "realistic" have been derived. The meaning of the word "realistic" will be discussed at a later time; at this point we are interested in the following question: How can the parameters of a crystal, such as the transition temperature, the Curie–Weiss constant, etc., be determined if we know such Hamiltonians? It is this question that was dealt with in Chaps. 7 and 8, but there we had to use certain approximations.

Of course, in order to avoid cumbersome calculations, we used the simplest approximations; we could have tried to attain greater accuracy, but our attempts would not lead us to any significant results (with respect to calculations of the crystal parameters). In general, it is not possible to avoid approximations, even if we are dealing with the simplest Hamiltonians. (An exception, which seems to be a miracle of a sort is the accurate calculation of the statistical sum, i.e., the free energy for a two-dimensional system with the Hamiltonian (8.8) in the case of the interaction with nearest neighbors only.) Of course, using present-day numerical methods and computers, we can achieve an impressive accuracy in calculations of the thermodynamic characteristics of systems with simple Hamiltonians, but if we are to study concrete crystals, it is worthwhile to be concerned with the following question: To what extent are the "realistic Hamiltonians" more sophisticated than those given by formulas (7.1) and (8.8)? In what follows we will deal only with the Hamiltonian (7.1) because, as we saw in Chap. 7, the Hamiltonian (8.8) is also derived in this case as a special case.

Thus, we are to determine the parameters of interest for a crystal consisting of two species of atoms, which is similar to the crystal dealt with in Chap. 7. What simplifying assumptions were made for writing down the Hamiltonian (7.1) for such a crystal? In this Hamiltonian we restricted ourselves to the variables that correspond to the displacements of atoms A along the z axis for terms containing no products of displacements in different unit cells; we took into account only the second and fourth powers of displacements and imposed strict limitations on the form of terms that contain the products of displacements in different unit cells (a model of springs obeying Hooke's law and connecting ions A in adjacent positions only).

We will now try to render this Hamiltonian more realistic, retaining the same set of variables. Naturally, the Hamiltonian for a real crystal will also contain higher-order terms. Is it important for calculations of the transition temperature, for example? If we add to the first parenthesis of (7.1) a term proportional to $u_{\boldsymbol{R}}^6$, then to the coefficient at $\eta_{k=0}^2$ calculated in Chap. 7 there should be added a term proportional to T^2. Suppose that all the coefficients of the Hamiltonian have a normal, atomic order of magnitude (except, of course, the coefficient a), in which case the ratio of the new term to the second one on the left-hand side of (7.20) will be equal to T/T_{at} in order of magnitide, i.e., it may be neglected. Such a conclusion is, of course, valid only for systems with phase transitions of the displacive type. But in most cases phase transitions are something intermediate between displacive and order–disorder transitions. Nevertheless, there are cases when higher-order terms may be ignored.

We have a different situation with fourth-order terms dependent on displacements in various unit cells. Taking account of terms of the type

$u_{\boldsymbol{R}}^2(u_{\boldsymbol{R}-\boldsymbol{a}_1} - u_{\boldsymbol{R}})^2$, we find that the second term in formula (7.18) assumes the following form:

$$\frac{3}{N}\sum_{\boldsymbol{k}}(b + b_1k^2)\langle u_{\boldsymbol{k}}u_{-\boldsymbol{k}}\rangle. \tag{13.1}$$

If we repeat the calculations that were given in Chap. 7, we will obtain on the right-hand side of formula (7.21) two terms and there is no ground for considering the second term to be smaller than the first. Furthermore, an interaction other than the interaction between nearest unit cells, to which there correspond, for example, the terms $u_{\boldsymbol{R}}^2(u_{\boldsymbol{R}-2\boldsymbol{a}_1} - u_{\boldsymbol{R}})^2$, may appear to be appreciable. The outcome is the appearance of new terms in parentheses in expression (13.1). In other words, the expression for the transition temperature must contain a substantially larger number of Hamiltonian constants than three, as is the case with formula (7.21).

At the next stage we begin to lift the limitations on the number of variables in the Hamiltonian. Recall that we have so far taken into account only the displacements of atom A along the z axis. We now lift this limitation, i.e., we "allow" the atom A to move in other directions and we "release" the atoms B. The number of variables in the Hamiltonian will increase sixfold, in which case the Hamiltonian will, of course, be even more complex; it will, in paricular, contain the products of various displacements. In short, the different variables will be strongly "mixed up". We may, of course, return to the former set of variables and integrate in the expression for the statistical sum Z (see (2.11)) over all variables, except for the displacements of atom A along the z axis. Being multiplied by $(-k_{\mathrm{B}}T)$, the logarithm of the partially integrated statistical sum Z (see Chap. 2) will play the role of a Hamiltonian dependent on the former limited set of variables. True, the coefficients of this Hamiltonian will depend on temperature (such a Hamiltonian is called effective: this is, in fact, an incomplete thermodynamic potential (see Chap. 2)) and this circumstance must be taken into account in the determination of the transition temperature. It may happen that the transition temperature will mainly be determined by the temperature dependence of the coefficients of an effective Hamiltonian that was neglected in Chap. 7 and that was due to the coupling of the displacements of atom A along the z axis to other degrees of freedom of the crystal. In the light of this remark, the conditionality of our inferences as to the scope of applicability of the Landau theory to order–disorder and displacive phase transitions becomes clear: these conclusions were made on the basis of simple models, but real systems are significantly more complicated.

We will now continue to lift limitations on the number of variables in our Hamiltonian. Until now we have assumed atoms (or ions) A and B to be solid balls which move as a whole. Of course, this is a serious simplification of the real situation. Atoms consist of nuclei and electrons and we may speak of an atom or ion in a solid body only conditionally: the formation of a crystal involves the sharing of electrons by all nuclei. True, with respect to well-pronounced ionic crystals we may speak of the electron cloud of a given ion (approximately, of course). In this case, we may take into account additional electronic degrees of freedom, visualizing, for example, an ion as an assembly of a hard "core", to which the electron shell is attached by a "spring". The rigidity of this spring can be determined on the basis of experimental data on the electronic polarizability of the ion, i.e., from the contribution of the ion to the refractive index of the medium, but it will be more consistent to pass from classical mechanics, which has been used by us so far, to quantum mechanics (in principle, the thermal motion of atoms and ions should also be treated within the framework of quantum statistical physics, especially when we deal with low temperatures). We have limited ourselves to classical conceptions since ferroelectric phase transitions commonly occur at temperatures comparable to Debye temperatures, whereas the qualitative problems may well be treated by using the formulas of classical statistical physics.

It is evident that crystals exist due to an electron cloud, without which positively charged nuclei would simply fly apart. But the role of electrons does not reduce to the compensation of nuclear charges only.

Electrostatics alone is not sufficient in our case: a system of classical charged particles interacting with one another according to Coulomb's law is always unstable. It turns out that the quantum-mechanical nature of the laws obeyed by electons is the key point in the understanding of the very possibility of the existence of a crystal and, of course, in the elucidation of its possible phases and in the determination of its material constants. The properties of an electron system are studied by using numerous approaches, most of which involve the construction of more or less realistic models and approximations and the use (sometimes, rather ingenious) of experimental data on the system under study and on related systems. These topics cannot be discussed within the scope of this book.

We will only mention calculations based on first principles, when no use is made of experimental data on a crystal (except, of course, its composition).

If we are interested only in the thermodynamic characteristics of a crystal (in particular, the temperatures of phase transitions in it), we may limit ourselves to the determination of the dependence of the crystal energy on the positions of atomic nuclei. The problem thus consists of the determination of the energy of the electron cloud in the electric field of fixed point charges, with the positions of these charges being different. The total energy of the crystal also includes, of course, the energy of electrostatic interaction of these point charges and the energy of interaction of nuclear charges with the electron cloud (the electron-density distribution at given positions of the nuclei must also be found, of course).

The interesting point is that if the electron-density distribution is known, this information will, in principle, suffice for the determination of the energy of the electron cloud. This statement has been strictly proved and is far from being trivial. Indeed, one might think that with the same distribution of the electron-cloud density its energy will be different because of the possibility of various correlation functions of the type $\langle n(\boldsymbol{r})\, n(\boldsymbol{r}')\rangle$, where $n(\boldsymbol{r})$ is the local electron density. Fortunately, this is not so. Of course, the validity of the theorem mentioned does not facilitate things much. There remains unsolved not only the problem of finding the distribution of the electron-cloud density itself but also the determination of the density dependence of the total cloud energy at different points. Various approximations are possible here.

One of the most natural approximations is to assume that the dependence of energy on density at a given point is the same as it would be with the electron gas being homogeneous. The dependence of the energy of a homogeneous electron gas on its density is known sufficiently well as a result of calculations performed by means of numerical methods. As to the distribution of the cloud density, it can be determined, for example, as follows: one can use the results of calculations of the density distribution in electron clouds of isolated ions (a relatively high accuracy has been achieved in such calculations as well) and assume that the combined cloud is formed as a result of the overlap of individual ions without their deformation

Though such calculations sometimes yield results close to experimental data, their accuracy often appears to be insufficient in order to explain the presence of phase transitions in a crystal and, even more so, to calculate the constants of the Hamiltonian required for the subsequent determination of the transition temperature and other parameters of interest. At present other methods are employed, which provide higher accuracy. We cannot, however, consider these methods because we would have to go far beyond the scope of this book.

We only note that as far as the comparison with experimental findings is concerned, we are mainly interested here in finding out, for example, whether the rhombic structure of barium titanate is more energetically favorable at $T = 0$ than the cubic or tetragonal one on the basis of our calculations. As to the calculation of, say, the temperature of phase transitions in this crystal, it is much more complicated: we have already seen that the treatment involves a large number of constants of the effective Hamiltonian. For most of these constants to be reliably calculated, we need a greater accuracy than can be attained at present since the realization of the problem formulated at the beginning of this section is a matter for the future. There is, however, a certain general outcome of such calculations. The point is that the relatively simple approaches that were used at the beginning of the twentieth century for the analysis of the properties of ionic crystals are still valid. They were based on the works of Lorentz published at the end of the previous century. The latest achievements reduce in many respects to the refinement of the corresponding conceptions and the calculation of the constants. It is these approaches that will be discussed below.

13.2 An Approach Based on the Classical Theory of Ionic Crystals

The classical conceptions are especially efficient in the case of ionic crystals. The theoretical foundations of the calculation of their energy were laid before the advent of quantum mechanics. This is associated with the fact that the nature of the cohesive forces that bind atoms in an ionic crystal has been clarified, on the whole, from the standpoint of classical physics. These are, first of all, Coulomb (electrostatic) forces which are operating between charged ions. These forces, however, are insufficient: we have pointed out earlier that a system in which the particles interact only through the Coulomb field is unstable. Additional forces are thus required. In the case of an ionic crystal these are repulsive forces which become strong when the ions begin to be in "contact" with one another (the quotation marks are used because an ion is a nucleus surrounded by an electron cloud, whose density rapidly diminishes with increasing distance from the nucleus, but, strictly speaking, it does not vanish).

The nature of repulsive forces can be elucidated only within the framework of quantum mechanics. It is natural that, just as in any other many-electron problem, calculations here encounter formidable difficulties. Use is more often made of a certain empirical (power or exponential) law for the dependence of repulsive forces on the distance. Apart from the repulsive forces, the non-Coulombic forces also include the van der Waals attractive forces (fluctuationally dipolar forces), which also act between neutral atoms. An example of the potential energy of the non-Coulombic interaction of ions is the Lennard–Jones potential:

$$U(r) = -\frac{\alpha_1}{r^6} + \frac{\alpha_2}{r^{12}}, \tag{13.2}$$

where r is the distance between the centers of interacting ions and α_1 and α_2 are certain coefficients determined empirically.

In order to calculate the Coulomb forces operating between ions, the ion is replaced in some cases by a point charge (the rigid-spherical-ion approximation). A circumstance taken into account more often is that the ion is a charged region rather than a point charge and that it is not spherical at all and the ionic charge distribution is characterized by the dipole and higher multipole (quadrupole, octupole) moments. These moments are considered to be either fixed or varying under the action of the field acting on the ion, which depends, of course, on the position and distribution of charges in other ions. In such a case, one speaks of the approximation of polarizable ions. Naturally, the ionic parameters, which determine the Coulomb interaction of ions (an effective charge and dipolar, quadrupolar, etc., polarizabilities), are themselves subject to calculations by means of quantum mechanics of many-particle systems, which, as pointed out earlier, is a very difficult task. It is for this reason that commonly the parameters that determine the Coulomb interaction of ions are not calculated; they are left as "free" parameters, which are then chosen so as to provide the best fit between theory and experiment. Of course, this diminishes the heuristic value of theory, but nevertheless this approach is justified if we succeed in describing a great deal of experimental data with the aid of a small number of parameters. Usually, only effective charges and dipole moments of ions, i.e., their polarizability, are taken into account for ferroelectrics.

At first sight the polarizability of an ion can be determined by measuring the refractive index of a gas composed of the corresponding ions. However, the ion in the lattice cannot be regarded as being equivalent to a free ion, though one may hope that their polarizabilities do not differ too strongly and that at least the relationships between the polarizabilities of different species of ions are the same as those for free ions. At present there also exist calculations of the polarizabilities of ions in the lattice. It turns out that they may differ appreciably (by 50%) from the polarizability of free ions. Such calculations are especially important when this or that ion cannot be examined in the free state; an example is the ion O^{2-} which enters into the composition of $BaTiO_3$ (and other perovskites) and which, according to a number of authors, plays an important role in manifestations of the ferroelectric properties of these crystals. Naturally, the fact that the ionic charge is not an integer is a specific quantum effect and is associated with the partial covalence of the bonds between atoms in the crystal lattice.

Despite the limitations of the method, which employs semiempirical laws of the interaction of ions, it often succeeds in establishing the microscopic reason responsible for a phase transition. For displacive-type transitions this is tantamount to proving that the symmetrical structure is unstable at $T = 0$ (in the language of the model Hamiltonian (7.1) it corresponds to the condition $a < 0$).

It is not difficult to point out the general qualitative cause for the negative value of a. Evidently, the cause is the total balance of forces that act on ions. We have already seen that the forces operating between ions is of a different nature. Each of the types of forces introduces a contribution to the coefficients of a Hamiltonian of type (7.1), and it is quite reasonable to presume that these contributions may be of different signs. Then, depending on which of the contributions is dominant, the sign of this or that coefficient will be positive or negative.

Let us show schematically how the sign of the coefficient a is determined for the simple model considered in Chap. 1 (see Fig. 1.3). We are interested in the dependence of the potential energy of a model crystal on the displacement of the sublattice of central ions (the function $U(\eta)$). We assume the ions to be nonpolarizable and ignore the non-Coulombic interaction of non-nearest neighbors. This means that in calculations of the non-Coulombic contribution to $U(\eta)$ we may consider the displacement of the central ion in each unit cell, irrespective of the displacement of the ions in other unit cells. In order to determine the contribution of the Coulomb (long-range) interaction, let us recall that the change of the electric charge distribution upon displacement of the ions may be represented as the result of a dipole being positioned in the center of the unit cell (see Chap. 8). The Coulomb contribution to $U(\eta)$ is then the sum of the interaction energy of such dipoles with the charges of the crystal in the undisplaced (symmetrical) configuration and the interaction energy of the dipoles themselves. In determining the first part of $U_{1\mathrm{Coul}}(\eta)$, we may assume that the ion is displaced only in one unit cell, just as in the calculation of the contribution of short-range forces. It is reasonable to presume that $U_{1\mathrm{Coul}}(\eta) > 0$ since each ion is surrounded by ions of opposite sign and the short-range forces are pure repulsive.

Thus, we have

$$U_1(\eta) = U_{1\mathrm{Coul}}(\eta) + U_{\mathrm{sr}}(\eta) = U_0 + \frac{1}{2}a_1\eta^2 + \dots \tag{13.3}$$

where $a_1 > 0$, i.e., in the absence of a dipole–dipole interaction the symmetric configuration would be stable.

Let us now calculate the electrostatic energy of interaction of dipoles with respect to one dipole $\boldsymbol{\mu}_0$. It is known to be equal to

$$U_2 = \sum_i -\frac{3(\boldsymbol{\mu}_0\cdot\boldsymbol{R}_{0i})(\boldsymbol{\mu}_i\cdot\boldsymbol{R}_{0i}) - R_{0i}^2(\boldsymbol{\mu}_0\cdot\boldsymbol{\mu}_i)}{R_{0i}^5} \tag{13.4}$$

where $\boldsymbol{R}_{0i}$ is the radius-vector drawn from dipole $\boldsymbol{\mu}_0$ to dipole $\boldsymbol{\mu}_i$. The computation of this sum is complicated, but it is not required. Indeed, the answer will depend on the form of the crystal since a homogeneous dipole distribution is created by a macroscopic depolarizing field, which must not be taken into account in the microscopic theory and which can be eliminated if we deal with a crystal placed between the plates of a short-circuited capacitor.

Thus, the field acting on a given dipole may be represented as the sum of the macroscopic field E_{mac} (equal to zero in the case in question) and E_{mic}, which is no longer dependent on the shape of the crystal. As shown by Lorentz, for a homogeneous distribution of dipoles

$$E_{\mathrm{mic}} = -\beta P \tag{13.5}$$

where the coefficient β depends on the crystal structure and the position of the point inside the unit cell at which the field is calculated. For a simple cubic lattice in the center of the unit cell $\beta = 4\pi/3$, and for other cases we have to perform special calculations, which is what has been done in a large number of publications.

Suppose that the macroscopic field is equal to zero (to do so, it is sufficient to short-circuit the plates of the capacitor in which the crystal is placed). The energy of the electrostatic interaction of the dipoles considered will then be equal to (with respect to a single dipole)

$$U_2 = -\frac{1}{2} E_{\text{mic}} P = -\beta e^2 \eta^2 / 2. \tag{13.6}$$

Thus, the dipole–dipole interaction makes a negative, destabilizing contribution. The total change of the energy upon displacement of the central ion is given by

$$U = U_1 + U_2 = U_0 + \frac{1}{2}(a_1 - \beta e^2)\eta^2 + \dots . \tag{13.7}$$

We see that if $a_1 < \beta e^2$, the central position of the sublattice of the ions considered is energetically unfavorable and that at $T = 0$ the crystal is known in advance to have a less symmetric configuration than the original one.

Note that in our treatment use is made of the basic ideas of the derivation of the formula for the dielectric constant of a condensed medium known as the *Clausius–Mosotti equation*. The negative nature of the coefficient at a in (7.1) is equivalent to the condition of the so-called polarization catastrophe.

It should also be noted that the stabilizing role of short-range forces and the destabilizing role of long-range forces in the example given above do not mean at all that these forces will play the same role in other situations as well. If other possible distortions of the original configuration of ions are treated within the framework of the model considered above and if, for example, we assume that the central ions in neighboring unit cells are displaced in different directions, the energy change must be calculated anew and the roles of long-range and short-range forces are not known a priori. Calculations show that in the case of concrete crystals there is possible a situation when the long-range forces play a stabilizing and short-range forces a destabilizing role.

Let us also recall certain qualitative inferences that can be made on the basis of a treatment similar to that given above. Suppose that we are dealing with a phase transition of the displacive type. As was shown in Chap. 8, the temperature of such a phase transition is proportional to the modulus of the coefficient of $u_{\boldsymbol{R}}^2$ in expression (7.1). Let the crystal be subjected to the action of hydrostatic pressure. How will the modulus of the coefficient a change? It is evident that with the ions being arranged more closely the role of short-range forces is enhanced, but the role of long-range forces practically remains unchanged. This means that if the short-range forces make a positive contribution to the coefficient a in (7.1), the transition temperature will fall off and if the contribution is negative, the temperature will increase. Both these cases are observed in perovskites, in which, as has been shown by calculations, for transitions involving a change of translational symmetry the long-range forces play a stabilizing and the short-range forces a destabilizing role. This explains why in perovskites the temperature of transitions involving no change of translational symmetry falls off with increasing pressure and the temperature of transitions with a change of translational symmetry increases.

To sum up, we emphasize that there exist a number of approaches to the explanation of the properties of crystals experiencing structural phase transitions. Each of the approaches is efficient in its own way and has its own sphere of application. The Landau phenomenological theory, which has often been used throughout the book, enables one to reveal the most general features of various phenomena and the relationships between them, but it does not make it possible even to estimate the corresponding material constants and to determine the scope of applicability of the theory itself. On the other hand, calculations from first principles ideally provide exactly the characteristics of a concrete crystal; but it proves to be difficult to distinguish the features common to all ferroelectric crystals. Between these "extremes" there exist numerous different approaches which combine phenomenological and microscopic considerations. Mention may be made here of the so-called phenomenological models, which are successfully used to account for certain properties of ferroelectrics and ferroelastics and ferroelectric liquid crystals.

All the existing approaches lay down a foundation for a detailed explanation of the properties of known ferroelectric substances and for the realization of the final, most important goal — the development of materials with desired properties.

14. Recommended Literature

Ashcroft, N.W. and N.D. Mermin, *Solid State Physics*, Holt, Rinehart and Winston, New York, 1976

Blinc, R. and A. Levanyuk (eds), *Incommensurate Phases in Dielectrics, Parts 1 and 2*, North-Holland, Amsterdam, 1986.

Blinc, R. and B. Zeks, *Soft Modes in Ferroelectrics and Antiferroelectrics*, North-Holland, Amsterdam, 1974.

Bruce, A. and R. Cowley, *Structural Phase Transitions*, Taylor and Francis, Ltd., London, 1981.

Jona, F. and G. Shirane, *Ferroelectric Crystals*, Pergamon Press, New York, 1962.

Landau, L.D. and E.M. Lifshitz, *Statistical Physics*, Part 1, 3rd ed., Pergamon Press, New York, 1980.

Lines, M. and A. Glass, *Principles and Applications of Ferroelectrics and Related Materials*, Oxford University Press, Oxford, 1977.

Mitsui, T., I. Tatsuzaki and E. Nakamura, *An Introduction to the Physics of Ferroelectrics*, Gordon and Breach, New York, 1976.

Pikin, S.A., *Structural Transformations in Liquid Crystals* (in Russian), Nauka (Science), Moscow, 1981.

Smolenskii, G.A., V.A. Bokov, V.A. Isupov et al., *Physics of Ferroelectric Phenomena*, Gordon and Breach, New York, 1984.

Toledano, J.-C. and P. Toledano, *The Landau Theory of Phase Transitions*, World Scientific, Singapore, 1987.

Vaks, V.G., *An Introduction to Microscopic Theory of Ferroelectrics* (in Russian), Nauka (Science), Moscow, 1973.

References

1. Benepe, J.W., Reese, W. (1977): Phys. Rev. **3**, 3032.
2. Bush, G. (1938): Helv. Phys. Acta. **11**, 269.
3. Gonsalo, J.A. (1968): Phys. Rev. **144**, 662.
4. Strukov, B.A., Taraskin, S.A., Fedorikhin, V.A., Minaeva, K.A. (1980): J. Phys. Soc. Japan. **49**, Suppl., 7.
5. Imai, K. (1977): J. Phys. Soc. Japan. **43**, 1320.
6. Strukov, B.A., Amin, M., Koptsik, V.A. (1968): Phys. Stat. Sol. **27**, 741.
7. Zisman, A.N., Kachinskii, V.N., Stishov, S.M. (1980): Zhurn. Eksp. Teor. Fiz., Pis'ma. **31**, 172 (Russian).
8. Fousek, J., Konak, C. (1972): J. Phys. B. **22**, 995.
9. Shirane, G., Sawaguchi, E., Takagi, Y. (1951): Phys. Rev. **84**, 476.
10. Janovec, V. (1966): J. Chem. Phys. **45**, 1874.
11. Ema, K., Ikeda, Y., Katayama, N., Hamano, K. (1980): J. Phys. Soc. Japan. **49**, 181.
12. Ozaki, T. (1980): J. Phys. Soc. Japan. **49**, 234.
13. Dontzova, L.I., Tikhomirova, N.A., Shuvalov, L.A. (1989): Ferroelectrics. **97**, 87.
14. Toshev, S.D. (1963): Kristallografia. **8**, 680 (Russian).
15. Hooton, J.A., Merz, W.J. (1955): Phys. Rev. **98**, 409.
16. Strukov, B.A., Uesu, Y., Arutyunova, V.M. (1982): Zhurn. Eksp. Teor. Fiz., Pis'ma. **35**, 428.
17. Ostrovskii, B.I., Rabinovich, A.Z., Strukov, B.A., et al. (1977): Zhurn. Eksp. Teor. Fiz., Pis'ma. **25**, 80 (Russian).
18. Ostrovskii, B.I., Rabinovich, A.Z., Sonin, A.S., Strukov, B.A. (1980): Ferroelectrics. **24**, 309.
19. Ostrovskii, B.I., Rabinovich, A.Z., Sonin, A.S., Strukov, B.A. (1978): Zhurn. Eksp. Teor. Fiz., **74**, 1748 (Russian).
20. Levstic, A., Zeks, B., Levstic, I. et. al. (1979): J. Phys., Suppl. 4, **40**, C3.
21. Ioshino, K., Nakao, K., Tanigushi, H., Ozaki, M. (1987): J. Phys. Soc. Japan, **56**, 4150.
22. Ostrovskii, B.I., Rabinovich, A.Z., Sonin, A.S., Strukov, B.A., (1983): Ferroelectrics, Izd. Rostovsk. Univ., Rostov-on-Don, (Russian)
23. Ostrovskii, B.I., Pikin, S.A., Chigrinov, V.G. (1979): Zhurn. Eksp. Teor. Fiz., **77**, 1615 (Russian).
24. Sonin, A.S. (1983): An Introduction to Physics of Liquid Crystals, Nauka, Moscow, (Russian).
25. Etxebarria, J., Jubindo, M.A.P., Ezcurra, A., Tello, M.J. (1988): J. Appl. Phys., **63**, 4921.

Index

MIX
Papier aus verantwortungsvollen Quellen
Paper from responsible sources
FSC® C105338

If you have any concerns about our products,
you can contact us on
ProductSafety@springernature.com

In case Publisher is established outside the EU,
the EU authorized representative is:
Springer Nature Customer Service Center GmbH
Europaplatz 3, 69115 Heidelberg, Germany

Printed by Libri Plureos GmbH
in Hamburg, Germany